Dependable Computing and Fault-Tolerant Systems

Edited by
A. Avižienis, H. Kopetz, J. C. Laprie

Volume 9

Springer-Verlag Wien New York

F. Cristian, G. Le Lann,
T. Lunt (eds.)

Dependable Computing for Critical Applications 4

Springer-Verlag Wien New York

Prof. Flaviu Cristian
University of California, La Jolla, CA 92093-0114, USA

Gerard Le Lann
Research Director, INRIA, F-78150 Le Chesnay, France

Teresa Lunt
Program Manager, ARPA/CSTO, Arlington, VA 22203, USA

ISBN-13: 978-3-7091-9398-3   e-ISBN-13: 978-3-7091-9396-9
DOI: 10.1007/978-3-7091-9396-9

© 1995 Springer-Verlag/Wien
Softcover reprint of the hardcover 1st edition 1995

Typesetting: Camera ready by the editors

With 62 Figures

Library of Congress Cataloging-in-Publication Data

Dependable computing for critical applications 4 / F. Cristian, G. Le
    Lann, T. Lunt, eds.
        p.   cm. – (Dependable computing and fault-tolerant systems ;
    v . 9)
      ISBN-13: 978-3-7091-9398-3
      1. Fault-tolerant computing.   2. Application software.
I. Cristian, F. (Flaviu), 1951–   .   II. Le Lann, G. (Gerard), 1943–
. III. Lunt, Teresa F.   IV. Title: Dependable computing for
critical applications four.   V. Series.
QA76.9.F38D463   1995
004.2–dc20

ISSN 0932-5581

FOREWORD

This volume contains the articles presented at the Fourth International IFIP Working Conference on Dependable Computing for Critical Applications held in San Diego, California, on January 4-6, 1994. In keeping with the previous three conferences held in August 1989 at Santa Barbara (USA), in February 1991 at Tucson (USA), and in September 1992 at Mondello (Italy), the conference was concerned with an important basic question: can we rely on computer systems for critical applications?.

This conference, like its predecessors, addressed various aspects of dependability, a broad term defined as the degree of trust that may justifiably be placed in a system's reliability, availability, safety, security and performance. Because of its broad scope, a main goal was to contribute to a unified understanding and integration of these concepts.

The Program Committee selected 21 papers for presentation from a total of 95 submissions at a September meeting in Menlo Park, California. The resulting program represents a broad spectrum of interests, with papers from universities, corporations and government agencies in eight countries. The selection process was greatly facilitated by the diligent work of the program committee members, for which we are most grateful.

As a Working Conference, the program was designed to promote the exchange of ideas by extensive discussions. All paper sessions ended with a 30 minute discussion period on the topics covered by the session. In addition, three panel sessions have been organized. The first, entitled "Formal Methods for Safety in Critical Systems" explored the role of formal methods in specifying and assessing system safety. The second, entitled "Qualitative versus Quantitative Assessment of Security?" debated the role that methods based on mathematical logic and stochastic techniques ought to play in assessing system security. The third panel "Common Techniques for Fault-tolerance and Security" explored techniques that are useful for attaining both fault-tolerance and security.

We are indebted to WG 10.4 chairman Jean-Claude Laprie from LAAS-CNRS, France, for his generous assistance in organizing the conference, and to Martha Chaves for her excellent work in reformatting the papers for this volume.

Flaviu Cristian
General Chair
University of California
San Diego, California

Gerard Le Lann
Program Co-chair
INRIA
Paris, France

Teresa Lunt
Program Co-chair
SRI International
Menlo Park, California

Sponsors

Organized by the IFIP Working Group 10.4 on Dependable Computing and Fault-tolerance, in cooperation with:

IFIP Technical Committee 11 on Security and Protection in Information Processing Systems

IEEE Computer Society Technical Committee on Fault-tolerant Computing Secuirty

EWICS Technical Committee 11 on Systems Reliablity, Safety and Security

University of California at San Diego

Conference Organization

General Chair:
F. Cristian
University of California
San Diego, California
USA

Local Arrangements/Publicity:
K. Marzullo
University of California
San Diego, California
USA

Program Co-Chairs:
G. Le Lann
INRIA
France

T. Lunt
SRI International
USA

Ex-Officio:
J-C. Laprie
LAAS-CNRS
France

Program Committee

J. Abraham
University of Texas at Austin
USA

A. Avizienis
University of California, Los Angeles
USA

D. Bjoerner
UNUIIST
Macau

R. Butler
NASA Langley Research Center
USA

A. Costes
LAAS-CNRS
Toulouse, France

M-C. Gaudel
LRI
France

V. Gligor
University of Maryland
USA

L. Gong
SRI International
USA

H. Ihara
Hitachi
Japan

J. Jacob
Oxford University
United Kingdom

S. Jajodia
George Mason University
USA

J. Lala
CS Draper Laboratory
USA

C. Landwehr
Naval Research Laboratory
USA

K. Levitt
University of California, Davis
USA

C. Meadows
Naval Research Laboratory
USA

J. McLean
Naval Research Laboratory
USA

M. Melliar-Smith
University of California, Santa Barbara
USA

J. Meyer
University of Michigan, Ann Arbor
USA

J. Millen
MITRE
USA

D. Parnas
McMaster University
Canada

B. Randell
University of Newcastle upon Tyne
United Kingdom

G. Rubino
IRISA
France

R. Schlichting
University of Arizona
USA

J. Stankovic
University of Massachusetts, Amherst
USA

P. Thevenod
LAAS-CNRS
Toulouse, France

Y. Tohma
Tokyo Institute of Technology
Japan

Contents

Formal Methods for Critical Systems   1

On Doubly Guarded Multiprocessor Control System Design   3
W. M. Turski (Warsaw University, Poland)

Using Data Consistency Assumptions to Show System Safety   15
G. Bruns, S. Anderson (University of Edinburgh, United Kingdom)

PANEL SESSION: Formal Methods for Safety in Critical Systems   29

Are Formal Methods Ready for Dependable Systems?   31
R. W. Butler (NASA Langley Research Center, USA)

Industrial Use of Formal Methods   33
S. P. Miller (Rockwell International, USA)

Formal Methods for Safety in Critical Systems   37
M. J. Morley (National Research Center for Computer Science, Germany)

Can We Rely on Formal Methods?   41
N. Shankar (SRI International, USA)

A Role for Formal Methodists   43
F. B. Schneider (Cornell University, USA)

Combining the Fault-Tolerance, Security and Real-Time Aspects of Computing   47

Toward a Multilevel-Secure, Best-Effort Real-Time Scheduler   49
P. K. Boucher, I. B. Greenberg (SRI International, USA)
R. K. Clark (Concurrent Computer Corporation, USA)
E. D. Jensen (Digital Equipment Corporation, USA)
D. M. Wells (Connection Technologies, USA)

Fault-Detecting Network Membership Protocols for Unknown Topologies   69
K. Echtle, M. Leu (Universität Dortmund, Germany)

Secure Systems   91

Denial of Service: A Perspective   93
J. K. Millen (The Mitre Corporation, USA)

Reasoning About Message Integrity   109
R. Kailar, V. D. Gligor (University of Maryland, USA)
S. G. Stubblebine (University of Southern California, USA)

On the Security Effectiveness of Cryptographic Protocols   139
R. Kailar, V. D. Gligor (University of Maryland, USA)
L. Gong (SRI International, USA)

Assessment of Dependability   159

Assessing the Dependability of Embedded Software Sytems Using
the Dynamic Flowgraph Methodology   161
C. Garrett, M. Yau, S. Guarro, G. Apostolakis (University of California at Los Angeles, USA)

On Managing Fault-Tolerant Design Risks   185
D. Ball (The Mitre Corporation, USA)
A. Abouelnaga (TRW, Inc., USA)

PANEL SESSION: Quanlitative versus Quantitative Aspects of Security   209

Qualitative vs. Quantitative Assessment of Security: A Panel Discussion   211
T. Lunt (SRI International, USA)

A Fault Forecasting Approach for Operational Security Monitoring   215
M. Dacier (LAAS-CNRS, France)

Measurement of Operational Security   219
B. Littlewood (City University, London, United Kingdom)

Quantitative Measures of Security   223
J. McLean (Naval Research Laboratory, USA)

The Feasibility of Quantitative Assessment of Security   227
C. Meadows (Naval Research Laboratory, USA)

Quantitative Measures vs. Countermeasures   229
J. K. Millen (The Mitre Corporation, USA)

Basic Problems in Distributed Fault-Tolerant Systems   231

Continual On-Line Diagnosis of Hybrid Faults   233
C. J. Walter, N. Suri, M. M. Hugue (AlliedSignal MTC, USA)

The General Convergence Problem: A Unification of Synchronous Systems   251
M. H. Azadmanesh (Standard Manufacturing Company, USA)
R. M. Kieckhafer (University of Nebraska, USA)

Specification and Verification of Distributed Protocols   269

Specification and Verification of Behavioral Patterns in Distributed Computations   271
Ö. Babaoğlu (University of Bologna, Italy)
M. Raynal (IRISA, France)

Specification and Verification of an Atomic Broadcast Protocol   291
P. Zhou, J. Hooman (Eindhoven University of Technology, The Netherlands)

Trace-Based Compositional Refinement of Fault-Tolerant Distributed Systems   309
H. Schepers (Philips Research Laboratories, The Netherlands)
J. Coenen (Eindhoven University of Technology, The Netherlands)

Design Techniques for Robustness   325

A Modular Robust Binary Tree   327
N. A. Kanawati, G. A. Kanawati, J. A. Abraham (University of Texas at Austin, USA)

Secondary Storage Error Correction Utilizing the Inherent Redundancy
of the Stored Data   349
R. Rowell (BNR, Inc., USA)
V. S. S. Nair (Southern Methodist University, USA)

PANEL SESSION: Common Techniques in Fault-Tolerance and Security   371

Common Techniques in Fault-Tolerance and Security   373
K. N. Levitt, S. Cheung (University of California at Davis, USA)

Improving Security by Fault-Tolerance   379
Y. Deswarte (LAAS-CNRS and INRIA, France)

The Need for A Failure Model for Security   383
C. Meadows (Naval Research Laboratory, USA)

Reliability and Security   387
P. G. Neumann (SRI International, USA)

Fault Tolerance and Security   389
B. Randell (University of Newcastle upon Tyne, United Kingdom)

Common Techniques in Fault Tolerance and Security (and Performance!)   393
K. D. Wilken (University of California at Davis, USA)

Real-Time Systems   397

Upper and Lower Bounds on the Number of Faults a System Can Withstand
Without Repairs   399
M. Goemans, N. Lynch, I. Saias (Massachusetts Institute of Technology, USA)

Scheduling Fault Recovery Operations for Time-Critical Applications   411
S. Ramos-Thuel, J. K. Strosnider (Carnegie Mellon University, USA)

Evaluation of Dependability Aspects   433

Effects of Physical Injection of Transient Faults on Control Flow and Evaluation
of Some Software-Implemented Error Detection Techniques   435
G. Miremadi, J. Torin (Chalmers University of Technology, Sweden)

System-Level Reliability and Sensitivity Analyses for Three Fault-Tolerant
System Architectures   459
J. B. Dugan (University of Virginia, USA)
M. R. Lyu (Bellcore, USA)

Improving Availability Bounds Using the Failure Distance Concept   479
J. A. Carrasco (Universitat Politecnica de Catalunya, Spain)

Formal Methods
for Critical Systems

On Doubly Guarded Multiprocessor Control System Design

Władysław M. Turski
Institute of Informatics
Warsaw University
Banacha 2, 02-097 Warsaw, Poland
wmt@mimuw.edu.pl

Abstract

In this paper we consider a behavioural specification for a multiprocessor control system. The specified behaviour exhibits two desirable properties: it reacts to spontaneous changes of the controlled process and its environment with minimal delay and it is insensitive to a large class of random processor malfunctions. Both properties improve with increased number of processors involved.

1   Preliminaries

For historic reasons, programming (and its theory, as well as methodology) evolved from the *computing paradigm*. Many computer applications in common use do not fit this paradigm well. Neither an operating system, nor a word processor *compute* anything, even if their operation involves many computations.

Here we follow another paradigm, *viz.* that of *system behaviour*, in which finite *actions* are undertaken and accepted in specific circumstances. In this approach there is no notion of action sequencing, therefore we avoid all problems of synchronization and global time. Instead, we recognize that action executions are not instantaneous and—therefore—the circumstances under which a particular action was deemed desirable may have changed during an execution to ones under which its outcome is no more needed. (Establishing a telephone connection in a modern network is an action which certainly takes some time during which the caller may decide to hang up; should this be the case, the laboriously established connection may safely be dissolved.)

The chosen paradigm accommodates an arbitrary degree of parallelism in the sense that an unspecified number of actions may be executed concurrently and, as far as the execution processes are concerned, independently. All actions that *can* be initiated in a given state, *are* initiated, regardless of any possible future conflict; conflicts are resolved in the accepting state. (When a client asks the bank to cash a cheque, the teller starts counting the money and checks the balance only when handing the cash out.)

Observable system behaviour obtains from successful executions of actions. For each system the collection of actions is determined by its specification which associates each action with *two* guards. A *preguard* characterizes the set of system *states* in which an action may be undertaken, a *postguard* characterizes the set of states in which the outcome of the executed action is acceptable i.e. may be reflected in the system state. Each action is assumed finite if undertaken in a state satisfying the corresponding preguard.

Actions are performed by *processors (agents)* of which there is an unspecified number (often it is essential that the number of processors exceeds one). Each processor is capable of executing any action, but different agents may need different lengths of time to complete the same action. No processor is allowed to idle in a system state in which a preguard is satisfied. Actions are assigned to processors in a fair way: when more than one can be assigned, the actual choice is unbiased.

If, at the instant of action completion the system state satisfies the corresponding postguard, the action outcome is instantaneously accepted.

In other words, the paradigm[1] aims to capture behaviours exhibited by systems in which actions take time (and may be futile) but all controls and state-updates are instantaneous.

In this paper we apply the new specification paradigm to some simple control problems and probe its relevance for a class of reliability concerns.

The basic building block for behavioural specification is a *doubly guarded* action (dga). Let P and Q be predicates on a suitably chosen state space, $TRUE$ be a predicate satisfied everywhere in this space, and p be a program satisfying $P \Rightarrow wp(p, TRUE)$ with wp denoting the usual Dijkstra's weakest precondition. Then a dga

$$(P, Q) \rightarrow [p]$$

specifies an action which may be initiated in a state satisfying P and executed on a private copy of the state space instantaneously and atomically obtained at the initiation. The execution of this action consists in performing program p, which consumes a finite amount of time. It is not assumed that each execution of p uses the same amount of time, but all executions are finite (by virtue of the assumption about p and P). When the execution terminates, the predicate Q

---

[1]The programming paradigm used here, its background and motivation have been described in several articles published in various journals. Reference [2] contains both a fairly thorough exposition of motivation and a bibliographical listing, reference [1] provides a number of examples from classical applications.

is evaluated on the public state of the state space. Should this state satisfy Q, it is instantly and atomically updated by the results of just executed program, otherwise the executed program is ignored.

Whenever it is not entirely obvious what the results of an action are, we list the state space components which are to be updated

$$(P, Q) \rightarrow [p \mid y] \ .$$

Sometimes, for symmetry (and to indicate possible savings in implementation) we shall write

$$(P, Q) \rightarrow [x \mid p \mid y]$$

in which case the x, y pair constitutes the frame for program p.

A dga

$$(P, TRUE) \rightarrow [p]$$

specifies an action which once initiated is never ignored; it is expected that actions which modify real environment (e.g. mechanical outputs and manipulations of actuators) are specified in this fashion.

A system behaviour will be usually specified by a set of dgas

$$(P_i, Q_i) \rightarrow [p_i] \ , 0 \leq i \leq N$$

Normally it is assumed that there are more than one processors available for execution of actions; the actual number of agents is not related in any way to the number of dgas constituting a behavioural specification. Processors are assumed identical with respect to their capability, i.e. each one can execute any program p_i, $0 \leq i \leq N$; it is not assumed that the processors are equally fast.

Note 1. In the proposed style it is relatively easy to write a behavioural specification that cannot be executed by a single processor, for instance $(P, \neg P) \rightarrow [p]$, with any terminating p, is a specification meaningful only when there are at least two available processors: while one is executing p, at least one more is needed to change the global state from one satisfying P to one satisfying $\neg P$. *End of Note 1.*

Note 2. If there is only one processor available, the dga $(P, Q) \rightarrow [p]$ is equivalent to Dijkstra's guarded statement $P \wedge Q \rightarrow p$. Thus, at least for a single-processor environments, we possess a well-defined and fully investigated semantics for dgas. In this paper, however, we pursue a more interesting case when there are more than one processor. *End of Note 2.*

It is assumed that processors do not idle when there is work to do. As soon as a processor becomes free it seizes a dga with satisfied preguard P_i and executes p_i. When the execution is completed, the processor becomes free (regardless of whether the results were accepted or ignored). Only when none of the P_i, $0 \leq i \leq N$, are satisfied the processors are allowed to idle. Note that we assume a random choice among actions with satisfied preguards. This does not exclude a long run of inimical choices with respect to a particular action; however, the longer is such a run, the smaller is the probability of it occurring. No correlation between choices made by individual processors is admitted. (These are powerful constraints on implementation, but here we consider an ideal case.)

2   A simple observer

$$\begin{Bmatrix} x = 0, x = 1 \\ x = 1, x = 0 \end{Bmatrix} \;\; \rightarrow \;\; \begin{bmatrix} n \\ n \end{bmatrix} \; n := n + 1 \; \begin{bmatrix} n \\ n \end{bmatrix} \quad \text{(SO)}$$

specifies a counter n (presumably initiated to zero) in which the number of state-changes of a flip-flop variable x (from 0 to 1 and vice versa) are recorded. The flip-flop itself is not specified: it is merely observed.

How good is the specification (SO)? Assume we have just one processor, then during the execution of $n := n + 1$ either the flip-flop changes its state and the increased value of n is accepted, or the flip-flop stays in its prior state, in which case the increased value of n is ignored. So far, so good.

Our counter fails, however, when the flip-flop is too fast: two (or any even number of) changes of the flip-flop's state during an execution of $n := n + 1$ are unrecorded, any odd number of changes is registered as 1. This is — in general — unavoidable. No discrete observer can count arbitrarily frequent phenomena unless it controls them, i.e. effectively prevents their fresh occurrence before the "old" one has been dealt with. (This is also know as synchronization, monitoring etc., cf. semaphores, critical regions etc.)

Indeed, if the observer's internal-state switching time is limited from below by a (duration) constant Δt, then any process with characteristic frequency higher than $1/\Delta t$ will be eventually misrepresented by this observer.

Note. This observation is quite general, certainly not restricted to our chosen specification paradigm. That it is usually ignored in other work on digital control systems is, perhaps, indicative of a confusion between notions of "digital observation" and "digital control" (going all the way back to the TEST-AND-SET instruction). However, when we are increasingly concerned with control systems applied to objects implemented in the same technology as controls, we should not be oblivious of this version of Heisenberg's principle. *End of Note.*

Another limitation of counter (SO) relates to the delay between an occurrence of an x-switch and corresponding increase in the value of n, which may be as large as the execution time of the assignment in the right-hand side of (SO) augmented by the time needed to evaluate postguard and update global state.

Both shortcomings of the design (SO) become irrelevant as the frequency of the flip-flop decreases with respect to the internal time-constant of the observer. No logical analysis of an actual design can warn us about how serious it is in any particular case; the best one can hope for is a well-designed experimental verification, with usual caveats on validity of experimental verification.

3 A multiagent observer

Assume now that the specification (SO) is given to a large number of processors. According to the rules, only one processor at a time is launched in a state s satisfying a preguard. Therefore, if the state s lasts for some time, a number of processors will be started with identical private copies of the state-space (consisting in the considered case of a single variable, n) but at slightly different times (physical instants). The exact distribution of the starting instants depends, of course, on the details of implementation, as does the distribution of execution times. Nevertheless, it is fair to expect that the instants of termination will cover (not necessarily uniformly) certain interval. If the flip-flop switch happens to fall in this interval, it will be detected with a delay rather smaller than that of a single observer of comparable execution speed.

In fact, if the flip-flop's average time to switch is significantly larger than the average execution time of the right-hand side program in (SO), one can expect the detection delay to be inversely proportional to the number of processors executing (SO), see Appendix A.

Unfortunately, with multiple processors, our design exhibits a weakness absent in the single-processor implementation. If among the processors there is (at least) one significantly slower than the others, it may upset the main property of the event counter, viz. its monotonicity. Indeed, such a processor may carry in its private state an obsolete value of n, and, having increased it by 1, under propitious circumstances[2] it may update the public value of n to this ancient value $+1$, thus effectively decreasing the observed count. To prevent errors of this (time-warp) kind we use another design

$$
\begin{aligned}
(n = m \wedge x = 0, n = m \wedge x = 1) &\rightarrow [n|\ n := n+1\ |n] \\
(n = m \wedge x = 1, n = m \wedge x = 0) &\rightarrow [n|\ n := n+1\ |n] \\
(n = m+1, n = m+1) &\rightarrow [n|\ m := n\ |m] \\
(n < m, n < m) &\rightarrow [m|\ n := m\ |n]
\end{aligned}
\qquad \text{(MO)}
$$

in which the observed count is now represented by value of m. The conjuncts $n = m$ in preguards of the first two dgas of (MO) disable the n-increasing actions in abnormal ($n \neq m$) situations, while in postguards they prevent (at least some) unnecessary updates of the public value of n already increased. The third dga of (MO) ensures that the desirable invariant of the design, $m \leq n \leq m+1$, is preserved, while the fourth dga defines the action to be executed if

---

[2]Circumstances propitious for the slow processor are calamitous for the system performance!

a time-warp-causing "observer" processor manages to decrease the value of n. Postguards in the last two dgas of (MO) prevent (some) unnecessary (vacuous) updates of public variables.

Regrettably, the design (MO) is not perfect yet. A slow processor executing the third dga may set m to an obsolete value (although the acceptance guard, $n = m + 1$, makes it a very unlikely event). Observing that the fourth dga is immune against a similar threat (an obsolete value of n will necessarily be too small, thus setting public n to this value will enable the very same dga again) convinces us to use a similar form of guards for the third dga:

$$\begin{array}{rcl}
(n = m \wedge x = 0, n = m \wedge x = 1) & \rightarrow & [n|\ n := n + 1\ |n] \\
(n = m \wedge x = 1, n = m \wedge x = 0) & \rightarrow & [n|\ n := n + 1\ |n] \\
(n > m, n > m) & \rightarrow & [n|\ m := n\ |m] \\
(n < m, n < m) & \rightarrow & [m|\ n := m\ |n]
\end{array} \qquad \text{(MO1)}$$

The design (MO1) is now immune against time-warp errors resulting from some processors being too slow. The first ("fastest") processor completing a logically correct action updates the public state to a configuration consistent with "physics" of the system, any delays in execution are corrected. By the same argument, no processor failing to complete its action can spoil the overall behaviour of the counter m, as long as there are enough (at least one!) other active processors.

As a bonus, the design (MO1) is also immune against some spontaneous errors: if values of m or n "by itself" become too small, they are corrected. If, however, any of these values becomes spontaneously too large, the counter will represent it as the actual count of flip-flop events. The design (MO1) is also vulnerable to spontaneous simultaneous decrease of n and m values.

Note. There are two kinds of threats to the dependability of our counter. One is related to the design: what happens if a processor is too slow? (It cannot be too fast, the guards dependent on flip-flop values take care of this.) The improvements introduced in the design eliminate _this_ threat. Another kind of threat is related to processors producing spurious values all by themselves (e. g. 125 instead of 76). We did _not_ aim at the elimination of consequences of such errors, hence the term "bonus" applied to the obtained elimination of spurious undercounts, even if spurious overcounts could remain uncorrected. _End of Note._

4 Multiagent controller

The design method illustrated in sections 2 and 3 can be easily extended to on-line control problems.

Consider a simple version of such a problem. An aircraft flies in a plane on a course determined by its current position (x, y) and coordinates of the goal (X, Y). We assume that X and Y are fixed, and the instantaneous values of x and y are constantly available as values of the public variables x and y, respectively. Function $f(x, y, X, Y)$ yields the setting of aircraft's controls compatible with the optimal trajectory between (x, y) and (X, Y). On board of the aircraft there is a radar which sweeps the plane in a more or less uniform circular scan. At each instant the value of public variable ϕ indicates the current measure of the angle between aircraft's trajectory and the direction of the radar's beam. If an obstacle is detected two public variables d and v_r provide values of measured distance to the obstacle and its radial (Doppler) velocity towards the aircraft.

Predicate $A(\phi, d, v_r)$ is satisfied iff the obstacle presents a danger to the aircraft. Function $g(\phi, d, v_r)$ yields the setting of aircraft controls compatible with the optimal avoidance manoeuvre. It is assumed that A is a total predicate, i.e. it is defined for all values of variables ϕ, d, v_r, including those which correspond to no obstacle detected (e.g. $d = \infty$) or an obstacle moving away from the aircraft ($v_r < 0$); similarly, g is assumed to be a total function, yielding some reasonable value for setting of controls when evaluated in a state where $\neg A$ holds. A simple controller may be specified as follows:

(SC):   (*normal, normal*) $\rightarrow [x, y, \phi, d, v_r|$
                     read(x, y, ϕ, d, v_r);
                     if $A(\phi, d, v_r)$ **then** *normal* := **false**
                                     else *controls* := $f(x, y, X, Y)$
                     | *normal, controls*]
   (¬*normal,* ¬*normal*)$\rightarrow [x, y, \phi, d, v_r|$
                     read(x, y, ϕ, d, v_r);
                     if $A(\phi, d, v_r)$ **then** *controls* := $g(\phi, d, v_r)$
                                     else *controls* := $f(x, y, X, Y)$;
                     normal := **true**
                     | *normal, controls*]
   (¬*normal, normal*) $\rightarrow [x, y, \phi, d, v_r|$
                     read(x, y, ϕ, d, v_r);
                     if $A(\phi, d, v_r)$ **then** *normal* := **false**
                                     else *controls* := $f(x, y, X, Y)$
                     | *normal, controls*]

The read operations in the right-hand side actions have been included in order to underscore the fact that, even though each processor assigned to an action gets a copy of the listed part of the state, corresponding variables may have to be polled (in the private copy of the state) and their values may have to be converted; in other words, it may take some time before the values needed for computations are available.

The first dga of (SC) specifies an action that may be started in a normal situation (roughly: when no danger is perceived) and whose outcome is intended to be accepted also in a normal situation. This action, apart from reading the current position and radar data, consists of two mutually exclusive options: either a threat is detected and the public situation must be reclassified to abnormal, or a fresh setting of controls is to be worked out by evaluating f.

The second dga of (SC) specifies an action that may be started in an abnormal situation and whose outcome is intended to be accepted in an abnormal situation, too. This action — when accepted — restores the normal classification of the public state either because the test indicates that the threat is there no more and controls are set according to f algorithm, or because controls are set according to g algorithm.

Finally, the third dga of (SC) specifies an action that starts in an abnormal state and whose outcome is accepted in a normal one. If the test confirms the presence of a threat the public classification will be changed to abnormal. If the test fails (there is no imminent danger), controls are set according to f algorithm.

Note that the initialization of actions and acceptance of their results is governed by the value of *normal*, not by the results of evaluation of A; this illustrates an important principle of behaviour specifications: actions are selected according to the state classification, the classification procedure proper is a part of the actions.

Note also that the first dga of (SC) initiated in a state classified as normal terminates as soon as this classification is negated by observations, without evaluating new settings for controls (it is assumed that if no value is assigned to an output variable no update happens and public variables retain their values). This reflects another design principle: abnormal (dangerous) situations need to be dealt with by actions designed for such situations; in this sense the similarity of actions in (SC) is somewhat misleading.

In a more realistic treatment, the algorithms for control settings in the first and third dgas may differ, e.g. the algorithm for f in the the third dga may take the advantage of the fact that it is meant to set optimal controls for ending the avoidance manoeuvre and getting back onto normal course (thus it may be different from the algorithm for f in the first dga). If, however, as in the design (SC), the right-hand sides of both dgas are identical, we may use one of the transformation laws for dgas (cf.[2]) and combine these two dgas into one, with common right-hand side and postguard, and with preguard given by *normal* $\lor \neg normal$ = **true**, thus obtaining design

(SC1):   (**true**, *normal*) $\rightarrow [x, y, \phi, d, v_r|$
                              $\textbf{read}(x, y, \phi, d, v_r)$;
                              if $A(\phi, d, v_r)$ **then** *normal* := **false**
                              else *controls* := $f(x, y, X, Y)$
                              $| \ normal, controls]$
        $(\neg normal, \neg normal) \rightarrow [x, y, \phi, d, v_r|$
                              $\textbf{read}(x, y, \phi, d, v_r)$;
                              if $A(\phi, d, v_r)$ **then** *controls* := $g(\phi, d, v_r)$
                                                else *controls* := $f(x, y, X, Y)$;

$$normal := \textbf{true}$$
$$| \; normal, controls]$$

which neatly demonstrates two kinds of behaviour: one, permanently enabled (preguard **true**), setting controls for "normal" flight and signalling whenever a threat is detected, and another, engaged in while the situation is abnormal, setting the controls for an evasive manoeuvre.

As with the flip-flop-counter design, the multiagent implementation increases the chances for an early detection of a switch from dangerous to normal (and vice versa). Now, however, the prevention of the time-warp effects is a little more complex. We start again by considering the *controls* set in actions of (SC1) to be the "proposed" settings and by introducing *act_controls* for actual settings. Of course, we shall need dgas to execute the update of actual controls, (...,**true**) → [... *act_controls* := *controls* ...], but we lack a simple means of detecting obsolete values, such as was provided by the ratchet-like nature of the counter.

One way out of the difficulty would be to introduce a time-stamping mechanism and rely on its ratchet-like properties (control settings will be accepted as *act_controls* value only if they are fresher than the current value of *act_controls*).

There is, however, a different way, much more general, and directly related to the "physics" of the problem.

Denote a consistent[3] tuple (x, y, ϕ, d, v_r) by *observ*, hence $observ := \textbf{read}(x, y, \phi, d, v_r)$ is the action-internal operation representing the observation gathering. Let *hist* be a sequence of pairs $(observ, c)$, where c is of type **control_settings**. The maximal size of *hist*, its structure (list, queue, array, etc.) is left undecided in this paper. It is, however, assumed that:

(i) there is available an operation *update(hist,observ,c)* which updates the *hist* variable by making the pair *observ,c* its "latest" element. (The undefinedness of *hist* size and structure allows us to ignore questions about deletions etc. contingent on execution of an update.)

(ii) there is available a test *compatible(hist,observ)* which yields **true** iff *observ* is a physically possible extension of *hist*. Some elements (conjuncts) of this test are quite obvious, e.g. $(\phi_{observ} - last \cdot \phi_{hist} \geq 0) \bmod 2\pi$, others may depend on a sophisticated extrapolation, sensitive to admitted flight manoeuvres.

(iii) both control setting functions f and g may advantageously use information contained in *hist*.

The controller specification becomes now:

(BC): $(\textbf{true}, normal) \rightarrow [x, y, \phi, d, v_r, hist|$

---

[3]"Consistent" here means "relating to the same instant of observation".

$$observ := \textbf{read}(x, y, \phi, d, v_r);$$
$$\textbf{if } A(\phi, d, v_r) \textbf{ then } normal := \textbf{false}$$
$$\quad \textbf{else } controls := f(hist, x, y, X, Y)$$
$$\mid normal, observ, controls]$$

$(\neg normal, \neg normal) \rightarrow [x, y, \phi, d, v_r, hist\mid$
$$observ := \textbf{read}(x, y, \phi, d, v_r);$$
$$\textbf{if } A(\phi, d, v_r) \textbf{ then } controls := g(hist, \phi, d, v_r)$$
$$\quad \textbf{else } controls := f(hist, x, y, X, Y);$$
$$normal := \textbf{true}$$
$$\mid normal, observ, controls]$$

$(controls \neq act\_controls, \textbf{true}) \rightarrow$
$$[observ, control, hist\mid$$
$$\textbf{if } compatible(hist, observ) \textbf{ then}$$
$$\quad \textbf{begin}$$
$$\quad\quad act\_controls := controls;$$
$$\quad\quad update(hist, observ, controls)$$
$$\quad \textbf{end}$$
$$\mid act\_controls, hist]$$

Further improvements to the design (BC) are possible. The test upon which the setting of actual controls depends can be made sensitive not only to *hist* and *observ*, but also to the proposed controls, thus eliminating not only time-warp effects due to delayed processors but also a class of spuriously erroneous settings. Thus the design (BC) has similar properties to those of (MO1).

In conclusion, it is perhaps worth observing that the described method of specifying control systems by doubly guarded actions of multiple processors totally avoids difficulties usually associated with integration of continuous and discrete components. Indeed, the continuous operation of the aircraft's radar (i.e. continuous changes of ϕ) and even the obviously continuous nature of its flight parameters (x, y) present no difficulty: corresponding variables are read only, i.e. instantaneous "snapshots" are made of their readings. This is quite apart from any mathematical constraints the continuity of the physical system imposes on the form of algorithms for functions f and g and on the form of condition *compatible*.

A   Simulation

In order to get a feel for the actual behaviour of multiprocessor controls, a simulation of (MO1) was run for up to a hundred observers ($N = 100$). All observations were made over 10 units of simulated time.

In the following table FFMTS stands for the flip-flop mean time to switch, i. e. the average time it takes to switch x from 0 to 1 or vice versa. When FFMTS is 0.3, the actual switches occur with uniform distribution within interval (0.2, 0.4), when FFMTS is 0.45, the corresponding interval is (0.30, 0.60).

OMC represents the observer mean cycle, i. e. the average time for execution of the guarded statement $n := n + 1$. When OMC is 0.15, the execution times are drawn uniformly from the interval (0.10, 0.20), when it is 0.075, the execution times cover the interval (0.050, 0.100).

Finally, AD stands for the average delay in the detection of a flip-flop switch, i. e. for the observed average delay between the actual instant of the flip-flop switch and the instant at which the increased value of n was posted.

FFMTS	0.3		0.3	0.45
OMC	0.15		0.075	0.15
N	AD	OMC/N	AD	AD
2	0.0500	0.0750	0.0249	0.0456
5	0.0303	0.0300	0.0167	0.0254
10	0.0145	0.0150	0.0064	0.0137
20	0.0074	0.0075	0.0034	0.0067
50	0.0029	0.0030	0.0014	0.0028
100	0.0016	0.0015	0.0008	0.0015

As expected, the average delays are inversely proportional to the number of observer-processes and to their average speed. Thus, to obtain a more reliable (finer) observations one can simply increase the number of observer-processes without necessarily making them any faster.

Acknowledgements

The research reported herein was partially supported by grants: KBN 211999101 (Polish Research Committee), CRIT-1 (EU) and by US (DoD) Contract F61708-94-C0001.

References

[1] Turski, W.M.: Specification of Multiprocessor Computing. *Acta Informatica* **27**(1990) 685-696.

[2] Turski, W.M.: Programming for Behaviour. In: R.S.Bird, C.C.Morgan and J.C.P.Woodcock (Eds.) *Mathematics of Program Construction*. LNCS 669, Springer-Verlag (1993) 18–31.

Using
Data Consistency Assumptions
to Show System Safety

Glenn Bruns and Stuart Anderson
Laboratory for Foundations of Computer Science
University of Edinburgh
Edinburgh EH9 3JZ, UK
bruns@lfcs.ed.ac.uk, soa@lfcs.ed.ac.uk

Abstract

Systems cannot usually be proved safe unless some failure assumptions are made. Here we prove that the water level in a generic boiler system is always within its safe range by assuming that device failures result in inconsistent readings. Key parts of our approach are a *failure-reporting* strategy that determines failures from consistency conditions, and a *level-calculation* strategy that gives a best estimate of boiler level in light of the reported failures. These strategies are generic and could be used in other safety-critical applications.

1   Introduction

It is better to prove that accidents are impossible than to estimate that they are unlikely. Unfortunately, system safety cannot usually be proved, because it is impractical to formalise every component failure and environmental condition. However, we need not give up deterministic reasoning altogether. Systems *can* be proved safe relative to some assumptions. By estimating the likelihood that these assumptions hold, probabilistic and logical methods can be combined.

Since safety systems are often built on the principles of diversity and redundancy, assumptions about the consistency of data from diverse or redundant sources are of particular interest. In this paper we show, using a generic boiler system example [3], how to prove system safety

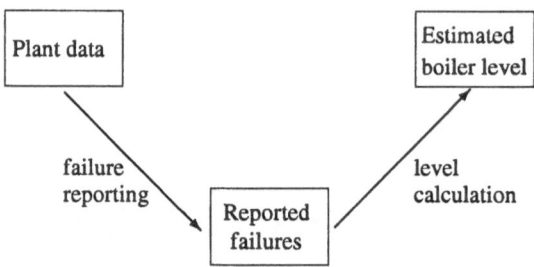

Figure 1: A Strategy for Proving Boiler Safety

using data consistency assumptions. These assumptions state that the data from a set of devices is consistent just when no device in the set has failed. The two main parts of our approach are a *failure-reporting* strategy, which uses data consistency to detect failures, and a *level-calculation* strategy, which makes the most accurate estimate of the water level in the boiler in light of the reported failures (see Figure 1). These strategies are generic and could be used in other safety-critical applications.

In the rest of the paper we present a model of the boiler system in the notation TLA, and prove that certain important safety properties hold of the model. We do not quantify the safety of the system in probabilistic terms, but we do discuss the problems involved in estimating the likelihood that our specific assumptions hold.

2 Notation

We use Lamport's Temporal Logic of Actions (TLA) to describe both the boiler system and its properties. We briefly describe TLA here; for more details see [4].

The atomic formulas of TLA are *predicates* and *actions*. A predicate is a boolean expression built from variables and values, such as $x > 1$. A predicate can be viewed as a function from states to booleans, where a *state* is a mapping from variables to values. An action is a boolean expression built up from variables, primed variables, and values, such as $x' = x + 1$. An action can be viewed as a relation on states, in which primed variables refer to a "new" state and unprimed variables to an "old" state. Thus, $x' = x + 1$ holds between two states if the value of x in the new state is one greater than the value of x in the old state.

The syntax of TLA formulas is as follows, where P ranges over predicates and $\mathcal{A}$ ranges over actions:

$$F ::= P \mid \Box \mathcal{A} \mid \neg F \mid F_1 \wedge F_2 \mid \Box F$$

Formulas are interpreted relative to an infinite sequence of states. Predicate P holds of sequence σ if the first state of σ satisfies P. Formula $\Box\mathcal{A}$ holds of σ if $\mathcal{A}(s_1, s_2)$ holds of every pair (s_1, s_2) of states such that s_1 is immediately followed by s_2 in σ. Formula $\neg F$ holds of σ if F does not hold of σ. Formula $F_1 \wedge F_2$ holds of σ if both F_1 and F_2 hold of σ. Finally, $\Box F$ holds of σ if F holds of every suffix of σ.

Systems are always represented in TLA by formulas of the form $P \wedge \Box\mathcal{A}$, where P represents an initial condition. For example, $(x = 0) \wedge \Box(x' = x + 1)$ represents a system in which x is initially 0 and is incremented in every successive state. To express the correctness condition that a property F holds of a system *Sys*, we write *Sys* $\Rightarrow F$. For example, letting *Sys* be the formula $x = 0 \wedge \Box(x' = x + 1)$, we write *Sys* $\Rightarrow \Box(x' > x)$ to express that x increases in every successive state of *Sys*.

Some basic TLA proof rules, taken from [4], are listed in Appendix 7.

We use the notation $\bigwedge_{i \in I} F_i$ rather than $\forall i \in I.F_i$ because we always quantify over finite sets, and are therefore using only simple propositional logic, not predicate logic. Similarly we use $\bigvee_{i \in I} F_i$ rather than $\exists i \in I.F_i$. Finally, $\sharp i.F_i$ stands for the number of i's for which the formula F_i holds.

Some parameters of the boiler, such as the boiler level, are modelled as a range of values to take possible measurement error into account. All model variables of type range have names beginning with an upper-case letter. Formally, a range is a pair (x, y) in which x and y are real and $x \leq y$. We define some operations on ranges:

$$(x, y)_1 \overset{\text{def}}{=} x$$
$$(x, y)_2 \overset{\text{def}}{=} y$$
$$A \subseteq B \overset{\text{def}}{=} A_1 \geq B_1 \text{ and } A_2 \leq B_2$$
$$A + B \overset{\text{def}}{=} (A_1 + B_1, A_2 + B_2)$$
$$A - B \overset{\text{def}}{=} (A_1 - B_2, A_2 - B_1)$$

Notice that ranges are closed under the $-$ operator. We will later use the fact that $+$ and $-$ are monotonic with respect to $\subseteq$, so that $A \subseteq B \Rightarrow (A + C) \subseteq (B + C)$, and similarly for the $-$ operator.

For convenience, certain plant variables that really represent scalar values are also represented as ranges of the form (x, x).

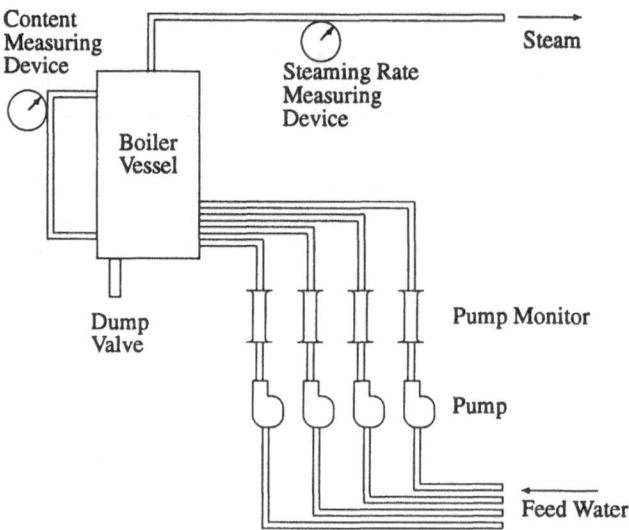

Figure 2: A Generic Boiler

3   Modelling the Boiler System

Our model of the boiler system is based on the generic boiler specification given in [3]. The model captures the physical plant, shown in Figure 2, and a monitoring system. The monitoring system periodically reads the measurement devices, pump monitors, and pump status indicators, and determines whether to shutdown the system. The pumps are turned on and off by a control system, which we do not model. Neither do we model the data-communications system, which sends plant measurements to the monitoring and control systems.

We begin by describing the model variables. Our specification contains the following plant variables:

L boiler content level

S steam rate

P pump rate

np number of operating pumps

L_m metered content level

S_m metered steam flow rate

pi_i motor on/off indicator for pump i

pm_i monitor for pump i

the following control variables:

L_c	calculated boiler content level
S_c	calculated steam rate
P_c	calculated net pump rate
f_d	actual failure of device d
r_d	reported failure of device d
c_D	consistency of readings from devices in set D
up	shutdown variable

and the following constants:

L_p	physical limits of boiler level
S_p	physical limits of steam flow
P_p	physical limits of net pump flow
Safe	safe boiler level range
K	pump flow per pump

The subscript d of variables f_d and r_d range over elements of *Dev* $\stackrel{\text{def}}{=} \{l, s, p\}$, where l stands for the content level meter, s stands for the steaming valve meter, and p stands for the pumps. In writing a consistency variable c_D, we abbreviate device sets as strings of symbols from *Dev*, for example, we write c_{sp} instead of $c_{\{s,p\}}$.

The boiler system model has five components, which model the boiler behaviour, shutdown behaviour, data fusion, data consistency, and level determination. The boiler model states the relationships between physical plant variables, and also contains failure assumptions:

$$
\begin{aligned}
\textit{Boiler} \ \stackrel{\text{def}}{=} \ & L' = L + (P' - S') \\
\wedge \ & P = K \cdot np \\
\wedge \ & S \subseteq S_p \\
\wedge \ & P \subseteq P_p \\
\wedge \ & \neg f_l \Rightarrow L \subseteq L_m \\
\wedge \ & \neg f_s \Rightarrow S \subseteq S_m \\
\wedge \ & \neg f_p \Rightarrow np = \#i.pm_i \\
\wedge \ & \bigwedge_{d \in \textit{Dev}} \left(f_d \Rightarrow \bigvee_{D \supseteq \{d\}} \neg c_D \right) \\
\wedge \ & \bigwedge_{D \subseteq \textit{Dev}} \left(\neg c_D \Rightarrow \bigvee_{d \in D} f_d \right)
\end{aligned}
$$

The model is a discrete approximation of the physical plant behaviour. The first three failure assumptions state that the meters are accurate in the absence of failure. The fourth states that failure of a device d produces some inconsistency involving d. The final assumption states that every inconsistency is produced by some failed device.

The shutdown model determines the value of the variable up, which holds whenever the calculated boiler level is within the safe bounds:

$$Shutdown \stackrel{\text{def}}{=} up = L_c \subseteq Safe$$

The data fusion model determines the value of failure reporting variables. The model states that if an inconsistency exists for some device set D, then all devices in D are reported as failed. This pessimistic failure reporting strategy is necessary when calculating a level range that is guaranteed to contain the actual boiler level.

$$Fusion \stackrel{\text{def}}{=} \bigwedge_{d \in Dev} \left(r_d = \bigvee_{D \supseteq \{d\}} \neg c_D \right)$$

Notice that the failure reporting strategy is given as a function of consistency conditions only.

The consistency model states consistency conditions between device readings. Informally, c_D holds if the devices named in D give consistent values. Thus, c_l holds if the level meter reading taken alone is consistent, i.e., if it is not outside the possible physical range. Similarly, c_{sp} holds if the steam and pump readings are consistent. Note that $\neg c_{d_1 d_2}$ does not imply $\neg c_{d_1}$ or $\neg c_{d_2}$.

$$
\begin{aligned}
Cons \stackrel{\text{def}}{=} \quad & c_s = S_m \subseteq S_p \\
\wedge \quad & c_p = \bigwedge_i pm_i = pi_i \\
\wedge \quad & c_l = L_m \subseteq L_p \\
\wedge \quad & c_{sp} = \textit{true} \\
\wedge \quad & c_{sl} = \textit{true} \\
\wedge \quad & c_{pl} = \textit{true} \\
\wedge \quad & c'_{spl} = L'_m \subseteq L_m + (\#i.pm'_i - S'_m)
\end{aligned}
$$

The final component determines the estimated level variables from reported failures. The calculated level is the best estimate of the actual level given the current reported failure conditions. For example, if the level meter is not reported as failed, then the best estimate of the actual level is by the level meter.

$$
\begin{aligned}
Level \stackrel{\text{def}}{=} \quad & \neg r'_l \Rightarrow L'_c = L'_m \\
\wedge \quad & r'_l \Rightarrow L'_c = L_c + (P'_c - S'_c)
\end{aligned}
$$

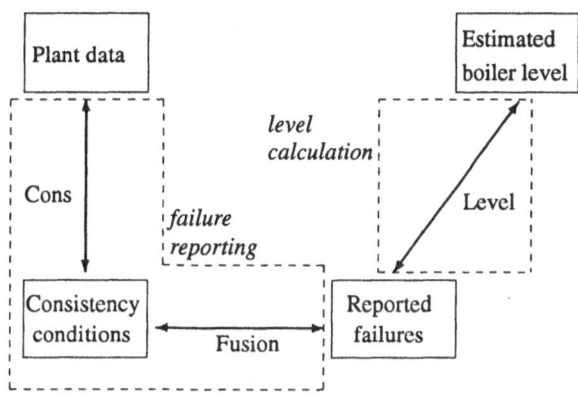

Figure 3: Relationships defined in the Boiler Model

$$\begin{aligned} &\wedge \quad \neg r_s \Rightarrow S_c = S_m \\ &\wedge \quad r_s \Rightarrow S_c = S_p \\ &\wedge \quad \neg r_p \Rightarrow P_c = K \cdot (\sharp i.pm_i) \\ &\wedge \quad r_p \Rightarrow P_c = P_p \end{aligned}$$

The action *Step* represents the combined behaviour of the preceding components:

$$Step \overset{\text{def}}{=} Boiler \wedge Shutdown \wedge Fusion \wedge Cons \wedge Level$$

Figure 3 shows the relationships defined by the model, and how the model embodies the general strategy described in the introduction.

The initial conditions include the conjunct $L_c = L$, which states that we must initially know the actual level of the boiler.

$$Init \overset{\text{def}}{=} L_c = L \wedge up \wedge L \subseteq Safe$$

The top-level boiler system description has the standard form of a TLA specification:

$$BSys \overset{\text{def}}{=} Init \wedge \Box Step$$

4   Failure Properties

Reports of device failure in the boiler system model are based on the consistency of sensor data. There are two important properties to show of the failure reports. First, if a failure occurs it should be reported. This is critical to the proof of safety. Letting D_r be the the set of devices reported as failed, and D_f be the set of actually failed devices, the property can be formalised as follows:

$$Pessimism \stackrel{\text{def}}{=} \Box(D_f \subseteq D_r)$$

Theorem 1 $BSys \Rightarrow Pessimism$

Proof.

$$
\begin{array}{ll}
D_f = \{d \in Dev \mid f_d\} & \text{by definition of } D_f \\
f_d \Rightarrow \bigvee_{D \supseteq \{d\}} \neg c_D & \text{in def. of } Boiler \\
D_f \subseteq \{d \in Dev \mid \bigvee_{D \supseteq \{d\}} \neg c_D\} & \text{propsitional logic, set theory} \\
D_f \subseteq \{d \in Dev \mid r_d\} & \text{by definition of } r_d \\
D_f \subseteq D_r & \text{by definition of } D_r
\end{array}
$$

Clauses of *Step* were used to prove $D_f \subseteq D_r$, so by the deduction principle we have that $Step \Rightarrow D_f \subseteq D_r$. TLA rule STL4 then gives us that $\Box Step \Rightarrow \Box(D_f \subseteq D_r)$, and from this $BSys \Rightarrow Pessimism$ easily follows.   □

The second property is that if one or more devices are reported as failed, then at least one of the reported devices must have actually failed. This property can be formalised as follows:

$$No\ False\ Alarms \stackrel{\text{def}}{=} \Box(D_r \neq \emptyset \Rightarrow \bigvee_{d \in D_r} f_d)$$

Theorem 2 $BSys \Rightarrow No\ False\ Alarms$

Proof. By definition, $D_r = \{d \in Dev \mid r_d\}$. Expanding the definition of r_d and simplifying, we equivalently have that $D_r = \bigcup\{D \subseteq Dev \mid \neg c_D\}$. Writing the finitely many elements of the set $\{D \subseteq Dev \mid \neg c_D\}$ as $\{D_1, \ldots, D_n\}$, we have that $D_i \subseteq D_r$ and that $\bigvee_{d \in D,} f_d$ for $1 \leq i \leq n$. Therefore, since $D_r \neq= \emptyset, \bigvee_{d \in D_r} f_d$.   □

5   Safety Properties

The most important property to show is that if the shutdown variable up is true, then the boiler level is within its safe bounds. This notion of safety can be formalised in TLA as the formula:

$$Safety \stackrel{\text{def}}{=} \Box(up \Rightarrow L \subseteq Safe)$$

Before proving the safety property we present a few lemmas. Of these, the third is the most important, stating that the actual boiler level is always within the calculated level range.

Lemma 1 *Step* $\Rightarrow S \subseteq S_c$

Proof. Assume that *Step* holds. If $\neg r_s$ then $S_c = S_m$. By the *Pessimism* property we know that $\neg r_s \Rightarrow \neg f_s$, and since $\neg f_s \Rightarrow S \subseteq S_m$, we have $S \subseteq S_c$. If r_s, then $S_c \subseteq S_p$, and since $S \subseteq S_p$, we have $S \subseteq S_c$. Thus *Step* $\Rightarrow S \subseteq S_c$. $\qquad\square$

Lemma 2 *Step* $\Rightarrow P \subseteq P_c$

Proof. We use a similar argument as in the proof to the previous theorem. If $\neg r_p$, then $P_c = K \cdot (\#i.pm_i)$ and $np = \#i.pm_i$, so $P_c = P$. If r_p, then $P \subseteq P_c$ just as for r_s. $\qquad\square$

Lemma 3 *BSys* $\Rightarrow \Box(L \subseteq L_c)$

Proof. To simplify the proof, we use the following derived TLA proof rule (where I is a predicate):

$$\frac{\mathcal{A} \Rightarrow Q \quad I \wedge (\mathcal{A} \wedge Q') \Rightarrow I'}{I \wedge \Box \mathcal{A} \Rightarrow \Box I}$$

To see that the rule is sound, observe that we get $\Box \mathcal{A} \Rightarrow \Box(\mathcal{A} \wedge Q')$ from the first premise by TLA rule STL4 and other rules of simple temporal logic, and that we get $I \wedge \Box(\mathcal{A} \wedge Q') \Rightarrow \Box I$ from the second premise and TLA rule INV1. Putting these together gives the conclusion.

We define R to be $(\neg r_l \Rightarrow \neg f_l) \wedge (S \subseteq S_c) \wedge (P \subseteq P_c)$. The first conjunct is a clause of *Step*, and the others were shown to be implied by *Step* in the previous lemmas, so *Step* $\Rightarrow R$.

We now show that $(L \subseteq L_c) \wedge (Step \wedge R') \Rightarrow (L \subseteq L_c)'$. If $\neg r_l'$, then $L_c' = L_m'$ by a clause of *Level*. Knowing $\neg r_l'$ also gives us $\neg f_l'$, by a conjunct of R', and therefore $L' \subseteq L_m'$, giving $L' \subseteq L_c'$. If r_l', then $L_c' = L_c + (P_c' - S_c')$. By $L \subseteq L_c$, R', and since $+$ and $-$ are monotonic on ranges, we have that $L + (P' - S') \subseteq L_c + (P_c' - S_c')$, so $L' \subseteq L_c'$.

Applying the derived proof rule we get $(L \subseteq L_c) \wedge \Box Step \Rightarrow \Box(L \subseteq L_c)$. Furthermore, $L \subseteq L_c$ holds initially, giving *Init* $\wedge \Box Step \Rightarrow \Box(L \subseteq L_c)$. $\qquad\square$

Theorem 3 *BSys ⇒ Safety*

Proof.

$$
\begin{array}{lll}
BSys & \Rightarrow & \Box Step & \text{by def. of } BSys \\
     & \Rightarrow & \Box(up \Rightarrow L_c \subseteq Safe) & \text{by def. of } Shutdown \\
BSys & \Rightarrow & \Box(L \subseteq L_c) & \text{Lemma 3} \\
BSys & \Rightarrow & \Box(up \Rightarrow L \subseteq Safe) & \text{by TLA rule STL5 and trans. of } \subseteq
\end{array}
$$

□

6 Assessing the Failure Assumptions

We have proved some important safety properties of the boiler system relative to some failure assumptions. As mentioned in the Introduction, the probability that the assumptions hold could be estimated, allowing an overall estimate of system safety to be made. We will not perform a probabilistic analysis here, but will review the failure assumptions and the related consistency conditions.

First, we have assumed that devices report values within their specified accuracy when they have not failed. Estimating the likelihood that this condition holds could probably be done.

Second, we have assumed that failed devices report inconsistent values. Calculating the likelihood of this assumption holding may be difficult, as a detailed knowledge of the likelihood and behaviour of failure modes is needed. Furthermore, the system context is relevant. For example, a meter that fails to a 0 reading will produce a consistent failure in contexts where a 0 reading is expected.

Third, we have assumed that inconsistencies arise only in the presence of failures. The difficulty of assessing this assumption depends on the particular consistency conditions adopted.

The consistency conditions play no part in the deterministic analysis, but do affect the probabilistic analysis. The conditions chosen in our model could certainly be augmented. For example, conditions c_p and c_l could be strengthened by additionally requiring that the change in pump and level values in the last step are within a certain range.

The condition c_{spl} holds when the level meter reading is consistent with the old level reading and the calculated net flow. Note that this condition depends on values in both the current and previous states. The use of values from the current and previous state in determining consistency could be extended so that a sequence of past values is used. For example, a sequence of flow values could be recorded and tested. An important part of the design of the boiler system is the selection of consistency conditions that are easy to compute and likely to uncover device failures. Furthermore, a good consistency condition is one for which it is possible to estimate the probability that the condition fails just when some device fails.

7 Conclusions

We have proved some important safety properties of a generic boiler system from assumptions about the consistency of data from diverse and redundant sources. An upper bound on the likelihood of the safety properties failing to hold can be obtained if the likelihood of the consistency assumptions failing to hold is known. In this way, it is possible to combine logical and probabilistic reasoning about the system. The analysis of a storage management system in [2] also combines logical and statistical reasoning, but there failures make data unavailable but not corrupted. For example, if a processor crashes during a disk write then the half-written block is assumed to be unreadable. The correctness of the system is proved from the assumption that at least one disk is always error-free.

As often happens when specifying systems, the process of formalising the system properties helped to clarify some key issues. For example, should all possibly failed devices be reported, or just the the smallest set of devices having at least one failed device? The latter reduces unnecessary diagnosis and repair, but was found to be inadequate to ensure the main safety property.

The process of proving the properties was also beneficial. Attempting the proofs often uncovered missing details. On the other hand, superfluous parts of the specification were found by examining the finished proofs, and noting the dependencies on the specification. For example, we had initially included a single-failure assumption in the model, and realised later, after looking at our proofs, that this assumption was not needed.

The data fusion model embodies a general strategy for detecting failure from inconsistencies between measured values. We chose a pessimistic strategy that reports a device failed if a measurement from the device is inconsistent with other values. Less pessimistic strategies may be better in other contexts. For example, another strategy is to select the smallest set of devices such that some device in the set is guaranteed to have failed. Thus if the $\neg c_{spl}$ and $\neg c_{pl}$, the devices p and l might be reported as failed, but not s. Failure assumptions could also be incorporated into data fusion model, allowing a smaller set of devices to be reported as failed. For example, if $\neg c_{sp}$ and $\neg c_{sl}$, and if f_l implies $\neg f_s$ and $\neg f_p$, then one can conclude that s has definitely failed.

The level calculation strategy is also quite generic. This strategy is related to various software fault-tolerance schemes, such as recovery blocks [5, 1]. The resemblance may be easier to see if the strategy is given in pseudo-code (see Figure 4) rather than in logic. In a recovery block, alternative computations are tried until an acceptable result is delivered. Here, alternative

if level meter ok then L_c := level meter reading
else if steam meter ok then S_c := steam meter reading
else S_c := physical steam limit
if pump mon's ok then P_c := sum of pump mon. readings
else P_c := sum of physical pump limits
$L_c := L_c + (P_c - S_c)$

Figure 4: The Level Calculation in Pseudo-code

computations are ordered according to the accuracy of the result. A specific computation is chosen according to which devices have failed.

Acknowledgements

We would like to thank Peter Bishop of Adelard and the anonymous referees for their helpful comments. This work was supported by SERC/IED project 1224, "Mathematically Proven Safety Systems".

Appendix: Some TLA Proof Rules

STL1 $\dfrac{F \text{ provable by propositional logic}}{F}$

STL2 $\vdash \Box F \Rightarrow F$

STL3 $\vdash \Box\Box F \equiv \Box F$

STL4 $\dfrac{F \Rightarrow G}{\Box F \Rightarrow \Box G}$

STL5 $\vdash \Box(F \wedge G) \equiv (\Box F) \wedge (\Box G)$

STL6 $\vdash (\Diamond\Box F) \wedge (\Diamond\Box G) \equiv \Diamond\Box(F \wedge G)$

INV1 $\dfrac{P \wedge \mathcal{A} \Rightarrow P'}{P \wedge \Box\mathcal{A} \Rightarrow \Box P}$

INV2 $\vdash \Box P \Rightarrow (\Box\mathcal{A} \equiv \Box(\mathcal{A} \wedge P \wedge P'))$

References

[1] T. Anderson and P.A. Lee, editors. *Fault Tolerance: Principles and Practice.* Prentice Hall, 1981.

[2] Flaviu Cristian. A rigorous approach to fault-tolerant programming. *IEEE Transactions on Software Engineering*, SE-11(1), January 1985.

[3] Specification for a software program for a boiler water content monitor and control system. Institute for Risk Research, 1992.

[4] Leslie Lamport. The temporal logic of actions. Technical Report 79, Digital Systems Research Center, 1991.

[5] B. Randall. System structure for software fault tolerance. *IEEE Transactions on Software Engineering*, SE1(2), 1975.

Panel Session: Formal Methods for Safety in Critical Systems

Formal Methods Panel: Are Formal Methods Ready for Dependable Systems?

Ricky W. Butler (moderator)
NASA Langley Research Center
130/Assessment Technology Branch
Hampton, Virginia 23681-0001
U.S.A.
R.W.Butler@Larc.nasa.gov

Formal specification and verification techniques have been advocated for life-critical applications for over two decades but have only been applied to a few deployed systems. Nevertheless, formal methods have been successfully used on increasingly difficult research problems over the last 5-10 years. A key question arises: "Have we finally reached the place where the transfer of this technology to industry is practical?"

Many different kinds of formal methods have been advocated, including the following:

1. code verification

2. hierarchical specification and design proof

3. gate-level hardware verification

4. abstract modeling and requirements analysis

5. model-checking (e.g. Binary Decision Diagrams)

6. process algebras

7. temporal logic

8. pencil-and-paper specification methods

9. general purpose theorem proving

10. VHDL analyzers

Which of these methods are relevant and practical and which ones should be discarded?

The panel on formal methods will examine the state-of-research in formal methods and debate where the greatest oportunities lie in formal methods. The panelists will discuss where the major obstables are and what should be done to overcome them. Towards this end, the panelists have been asked to address the following specific questions:

1. Is formal specification without mechanical proof the only practical approach to system design?

2. Is code verification dead? Will it forever remain impractical? The ideas have been around since the 60's; why then has it never been employed (with the possible exception of SACEM and a few others)?

3. Are pencil-and-paper techniques such as Z so error-prone that they merely give an illusion of clarifying the specification?

4. Is the availability of so many theorem provers a serious hindrance to the advancement of formal methods? Does the abundance of formal methods hinder or aid technology transfer?

Industrial Use of Formal Methods

Steven P. Miller
Rockwell International
Collins Commercial Avionics
Cedar Rapids, Iowa 52498
U.S.A.
spmiller@pobox.cca.cr.rockwell.com

The views expressed in this paper are those of the author and do not necessarily reflect the position of Collins Commercials Avionics or Rockwell International.

Opinions abound on the industrial use of formal methods, ranging from claims they will never prove practical to demands for nothing less than complete proofs of correctness now. The first position overlooks the widespread success of mathematics and logic in engineering today, while the latter unrealistically minimizes the problems of placing a complex, new technology into practice.

Much of the confusion I see regarding the use of formal methods results from failing to distinguish between formal specification and formal verification. Formal specification, i.e., the use of precise mathematical models and languages to define the system of interest, can provide important advantages today when used properly and in conjunction with good design practices. Formal verification, i.e., mechanical proofs of correctness, offers great promise, but its not clear if its cost can be justified for other than key properties or critical components. However, there is an entire spectrum of use between these extremes.

The most cost effective use of formal methods appears to be in the analysis of requirements. The process of constructing an abstract model helps to flush out requirements and constraints that have been overlooked. In particular, it encourages the designers to look in the "corners" at boundary cases that are a common cause of error. The presence of a formal model also provides a ready framework for checking if informally specified functional requirements can be met. Because such models abstract away from implementation detail, they are usually small and formal verification can often be applied to establish key properties. We have had good success formally specifying the requirements for a microprocessor and a small real-time executive.

Another important benefit I've seen from the use of formal specification is that it provides a precise interface between the user of a system and the implementor. A specification written in English and diagrams can leave this interface blurred and open to interpretation. The user of the system is likely to construct a mental model at a fairly high level of abstraction and to assume the system exhibits behavior consistent with this model. The implementor, with full access to the details of implementation, is likely to construct the simplest interface that falls within this gray area of interpretation. Since they are discovered during system integration, the resulting "glitches" are usually resolved in favor of the implementor. By explicitly creating a mathematical model of the system at the user's level of abstraction, both the user and the implementor are provided with a precise definition of what the system is to do. While this may be more difficult to implement than would be the case with an informal specification, the rework avoided by having a well understood interface should more than justify the expense.

Contrary to popular belief, well written formal specifications can be read and understood by practicing engineers. We have conducted very successful inspections of formal specifications after providing our engineers with only a few hours of training. Once the syntax and a few new concepts, such as functions of functions, have been mastered, many formal specification languages are simpler to read than programming languages.

Our experiences with inspections of formal specifications has been very illuminating. On one project, we produced several pen-and-pencil formal specifications. While these were of some value in the very early stages of design, they weren't appropriate once we moved from informal reviews to inspections, simply because so much effort was diverted to relatively minor issues of syntax. In a later project, we used a system that provided mechanical syntax and type checking and found the inspections went very well. The precision of the specification language improved the inspections, and the inspections served as an important vehicle for education and arriving at a consensus on issues of style and form. Particularly interesting was the use of automatically generated proof obligations for simple type checking conditions. Review of these during the inspections routinely exposed errors in the specifications.

There are three issues I feel need to be addressed for formal methods to achieve widespread use in industry. First, the industrial community must master the basic skills involved. Formal specifications are not inherently well written. Effort has to be invested to ensure a specification is well organized and readable. Many issues, such as coupling, cohesion, and information hiding are already well understood and appreciated. Others are not. Abstraction, in particular seems to be a difficult, even counter-intuitive, skill to master, yet I would argue that it is the single most important method of dealing with the increasing complexity of today's systems. Unfortunately, there is remarkably little guidance available for the average engineer on the methods to be used in applying formal methods.

Second, the formal methods community needs to develop specification styles and languages that are more natural and intuitive. For example, tables are a simple, familiar format for most engineers, yet few specification languages have an actual construct for tables. Graphical styles, such as statecharts, are readily adopted by engineers when they fit the problem at hand and should be available in the specification language of choice. Selecting the right mix of graphical and tabular notations and integrating those within the more general framework already provided by existing languages will determine which languages achieve widespread use. I also expect to see many specification methods tailored for particular problem domains.

Third, reliable tools need to be built to support the methods and languages that evolve in response to the first two issues. We have found simple syntax and type checkers to be essential. Automated tools are also needed to support the organization of large specifications, graphical specification styles, the production of readable documents, and reuse of specifications and proofs. I would especially like to see more tools to support inspections and reviews, perhaps by generating proof obligations to be considered during inspections.

There are several ways to address these issues. Extending, and perhaps simplifying, an existing methodology with a formal syntax and methods and tools for verification may be the simplest route. A good candidate is object oriented design, which is winning widespread support in industry and shares many of the same roots as formal methods. Another approach would be to focus on support for inspections. Inspections provide a natural vehicle for education, training, and enforcing the quality of specifications. I can think of no better vehicle for selecting among specification and verification methods than forcing practicing engineers to work with them.

In the preceding, I have said relatively little about the role of formal verification in industry. This is not because I believe it infeasible. Formal verification is often the method of choice in establishing key properties of a formal specification. At high levels of abstraction, it can help to increase one's confidence in the basic soundness of a design or algorithm. The simple act of formulating a postulate to be proven can provide (and capture) new insights. For the most critical systems, even complete proofs of correctness may be justified. However, these techniques currently seem to require skilled formal methods experts and time, both of which are in short supply. Just as work needs to be done to make specification languages more natural, work also needs to be done to make verification simpler. For all of these reasons, work should continue on formal verification, but the industrial focus should be on determining what specification styles and methods work, keeping in mind that the ultimate goal is a unified process that leads from validated requirements to appropriately verified implementations.

Formal Methods for Safety in Critical Systems

M. J. Morley
GMD SET-SKS, *Schloss Birlinghoven*
53757 St. Augustin, Bonn
Germany
morley@gmd.de

My experience in the application of formal methods covers process algebra, higher-order logic and temporal logic. I have had some success in applying these methods in the analysis of certain aspects of a computer controlled railway signalling system. In seeking to improve confidence in the design and overall safety of these systems, the challenge was to implement a means to mechanically verify the correctness of the data driving the generic control program. The main difficulty is perhaps unsurprising: it is very hard to bridge the divide between theoretical computer science (the needs of formal methods) and engineering practice (in this case, the needs of signal engineers).

Since a useful—perhaps even the best—approach to using formal methods is to tailor general techniques to specific application domains, the differences between formal methods expert and practicing engineer should be addressed somehow. I believe this is much more than just a matter of engineers acquiring the right mathematical skills. It has, for example, become fashionable to embed languages like VHDL in higher-order logic. Just one of the intentions here is that designers familiar with the hardware description language can begin to reason formally about their circuit descriptions. However, in order for the tool (method) to be useable at all to the engineering community a great deal of insight, both engineering and mathematical, needs to go into selecting the right semantic embedding. A naive approach may render the tool wholly unusable due to efficiency considerations. More seriously, the tool has to provide the right kind of abstraction facilities, and these come to light not through test cases, but in trying to conduct large scale applications, managing enormous proofs. There is therefore a highly complex relationship between the engineering and mathematical aspects, the theory and practice, of applied formal methods which presents a significant challenge to their wider uptake in industry.

*National Research Centre for Computer Science, System Design Technology Institute

It seems likely that picking the wrong formalism may be costly in the industrial context. When selecting a programming language for a particular application there are normally simple criteria which guide the correct choice, and there exists a wealth of experience against which to make that judgement. So it ought to be with formal methods. The trouble is that there is currently a paucity of guidance available to systems engineers in the form of documented case-studies, teaching texts and of course clarifying industrial and international standards. This situation must improve; in particular the guidance that standards can offer should become clearer once a broad body of industrial practice has been established. This problem need not be cyclic, but industry and academia need to work together in a concerted fashion, perhaps motivated by strategic funding initiatives, to publicise case-study material. While seemingly the proprietary nature of the products for which formal methods are being proposed presents a snag, when safety is the *motivation* for adopting formal methods, property should always give way to openness in making the safety case.

A mixture of methods are needed to analyse complex systems. No single formalism will address all the issues raised from requirements capture through program code to analogue and digital hardware. And different methods are effective at different levels of abstraction. For example, process algebra is a valuable tool for formalising and animating system descriptions at high levels of abstraction; higher-order logic is probably better suited to formal verification at the gate level of digital hardware. I certainly do not cast these distinctions in stone! Good understanding may come from manipulating digital circuits algebraically, but the point is that it would be a mistake to expect of a formal method the best of both these worlds. The reasons are again semantic, though also pragmatic as we seek efficient representations of our formal specifications. This is an argument for plurality when adopting formal methods, although it may entail additional costs as the necessary skills are difficult to acquire.

For serious applications I see little distinction between methods and the tools which mechanise them. One difficulty with the current swatch is that they are provided more with the view of supporting the formal method than of supporting applications and engineers. Commercially, one can not expect to sell a programming language without supplying a compiler—but the analogy is false since formal methods are necessarily much more difficult to use than compilers. This situation will improve as feedback from design case-studies appears (so we should take heart from the growing industrial interest in the Lambda[1] system, for example), but there are also fundamental difficulties. The argument for plurality means that a combination of tools are needed to approach realistic designs and safety assessments. Ensuring that the tools supporting diverse modelling paradigms sit together comfortably within a coherent semantic framework is delicate. It is essential to work towards providing clean, mathematically precise, interfaces so that tools can communicate, so that different specification and verification styles can be brought to bear alongside more traditional methods based on logic analysis, testing and simulation.

So why use formal methods? It is certainly infeasible to expect totally verified systems, yet much insight can be drawn from the difficult exercise of formally specifying systems and

---

[1] Abstract Hardware Ltd.

their properties, of validating the formal models so derived and in conducting proofs to verify properties. Although formal methods deal with abstractions the formal analysis brings about a deep understanding of *why* a system is safe; this is far more valuable than the proof of correctness itself, and is an understanding that remains unobtainable by quantitative methods alone. The lesson is in *attempting* the proof. Formal methods may actively support the design process, when introduced into that process incrementally, and will thereby improve everyone's confidence in the end product.

In the computing business we seem to be forever on the verge of the next technology revolution, so it is worth reflecting on the central role that engineering mathematics plays in the design and analysis of "traditional" engineering structures; I think this is the role we would wish formal methods to serve in the engineering of digital systems.

Can we rely on Formal Methods?

Natarajan Shankar
Computer Science Laboratory
SRI International
Menlo Park, California 94025
U.S.A.
shankar@csl.sri.com

We can all agree that safety-critical systems already contain software and hardware components of immense complexity. The reliability of these discrete systems is obviously a matter of concern to those involved in building or using such systems. We can also agree that there is a major gap between the reliability levels that are required of these systems and those that are obtainable using traditional hardware and software design methods. The basic question then is: *Can formal methods play a useful role in filling this reliability gap?* I will argue that formal methods technology has matured to a point where it ought to be a major component of any process for building reliable systems. Like any other technology, it will have to be used judiciously and in conjunction with other effective design tools and techniques. I will also argue that if it makes sense to apply formal methods, then it also makes sense to employ mechanized tools to support the formal process. I will then discuss some of the prospects and challenges for formal methods. My opinions are mostly based on my personal experience with building and using mechanized tools and methodologies for specification and verification.

What is the point of using formal methods? No formal method can guarantee the absence of errors. The point of using expressive notation and meaningful abstractions in the design process is to gain intellectual mastery over a complex design. The goal of a formal process is not to eliminate errors but to localize them by means of firewalls of assumptions and guarantees between components and between description layers. A good formal method provides a medium for conceptualizing, constructing, analyzing, communicating, and generalizing specifications and designs of digital systems.

When are formal methods most effective? The answer depends on the nature of the project and the method being employed. For safety-critical systems, the biggest bang for the buck is in applying formal methods early in the design lifecycle to the functional abstractions, interfaces, and key algorithms and protocols. The use of formal methods at an early stage can ensure that requirements are captured in precise terms, and that the design is based on a sensible functional decomposition and a sound algorithmic foundation. The formal process is akin to programming with properties (of components and interfaces) where program execution is replaced by proofs. With regard to code verification, the cost-to-benefit ratio here is quite high so that it is likely to be useful only in rare instances.

Others have argued that formal specification is useful in constructing meaningful models, but that formal verification is too hard and not demonstrably useful. I disagree. The verification process uncovers many more errors of far greater subtlety than are revealed merely by formal specification. A specification notation without decent proof rules is of negligible value since the point of a specification is that it can be probed and analyzed. The difficulty of formal verification has been greatly exaggerated. In practice, the effort required by verification is of the same degree as that of a careful walkthrough.

Formal methods that are not mechanized (or at least mechanizable) cannot be regarded as either formal or as methods. Formal methods can be studied and developed independently of any mechanization, but mechanization helps to achieve the scale, precision, and formality that make these methods useful. Mechanized tools for formal methods have been regarded as more of a hindrance than a help, but this situation is bound to improve as the technology comes to be more widely used.

Why are formal methods not used more widely? The basic reason is that formal methods are not taught very widely. A second reason is that only parts of the subject are mature enough to be widely used. Important topics such as the verification of concurrent, fault-tolerant, and real-time systems are still being hotly researched. A third reason is that despite the name, there are few *methods* that can be systematically applied to the modelling and verification of complex systems. A fourth reason is that the tools that are currently available are primitive in terms of functionality and interfaces (to users as well as other tools).

The outlook for formal methods is, however, bright. On the one hand, existing non-formal design and analysis methods can no longer fully cope with the complexity of modern digital systems. The reliability requirements on these systems have also become more stringent. On the other hand, many formal techniques are reaching moderate levels of maturity and are being employed by well-qualified people in nontrivial projects. This in turn has created a demand for better tools that is only slowly being met. The challenges, though significant, have never been clearer or more exciting.

A Role for Formal Methodists*

Fred B. Schneider
Department of Computer Science
Cornell University
Ithaca, New York 14853
U.S.A.
fbs@cs.cornell.edu

Proving "correctness" of entire systems is not now feasible, nor is it likely to become feasible in the foreseeable future. Establishing that a large system satisfies a non-trivial specification requires a large proof. Without mechanical support, building or checking such a proof is not practical. Even with mechanical support, designing a large proof is at least as difficult as designing a large program. We are barely up to the task of building large and complex systems that almost work; we are certainly not up to building such systems twice — once in a programming language and once in a logic — without any flaws at all.

On the other hand, the use of formal methods for determining whether small but intricate protocols satisfy subtle properties is well documented. Many large systems are constructed from such small protocols, so formal methods certainly have a role here. Sadly, this role has been a small one. It requires identifying the protocols and constructing abstractions of the environment. Both activities require time, effort, and taste. Practitioners of formal methods are rarely in a position to build the abstractions; they lack the necessary detailed knowledge about the system. Experts on any given system are rarely in a position to apply formal methods; they lack the knowledge about the use of these methods.

Formal methods can play another role in analyzing large systems. In fact, it is one that I find quite promising. If the property to be proved can be checked automatically, then the size of the system being analyzed is irrelevant. Type checking is a classical example. Establishing that a program is type-correct is a form of theorem proving. It is a weak theorem, but nevertheless, one that is useful to have proved. More recently, we see tools to check whether interfaces

---

*This material is based on work supported in part by the Office of Naval Research under contract N00014-91-J-1219, the National Science Foundation under Grant No. CCR-8701103, and DARPA/NSF Grant No. CCR-9014363. Any opinions, findings, and conclusions or recommendations expressed in this publication are those of the author and do not reflect the views of these agencies.

in a large system are being used properly. And, in circuit design, we see model checkers to determine if a digital circuit satisfies elements of its specification. The properties to be checked are "small," the system being checked might be large, so the leverage can be great.

For large systems, though, it is unlikely that we can anticipate *a priori* all of the properties that might be of interest. Therefore, the chances are slim that the literature will contain the formal method that we seek or that a suitable tool will have been implemented. This means we must be prepared to design and use "custom" formal methods for the problem at hand. For example, in the design of the AAS system, the IBM/FAA next-generation U.S. air traffic control system, we found it necessary to avoid certain types of race conditions. This form of race was specific to the application and the design; it was not anticipated in the system requirements. Checking for these races was simple: it involved writing equations that described data dependencies for the system (they occupied about 2 pages) and implementing a program to check these equations for circularities. In short, benefits accrued from the freedom to identify and check properties on-the-fly as the design proceeded.

Generalizing, I see here a new role for formal methods in large systems. We must be prepared to design formal methods for the task at hand rather than design our systems so they can be checked with the formal method at hand. This new role has implications for how projects are staffed, how formal methods are taught, and what support tools are required.

Staffing. Having a "formal methodist" associated with a project allows targets of opportunity to be identified — properties of interest that can be checked mechanically. The formal methodist must be a full-fledged participant in the system project, for only then can he appreciate what properties are important to check and what assumptions are plausible in building models. Other project participants need not be practitioners of any formal method, but will still benefit from having the formal methodist working with them.

Education. In order to be effective, the formal methodist must be comfortable using a wide collection of formal methods. Polytheism is better than monotheism, since different formalisms are suited for different tasks. In fact, the formal methodist must be comfortable with the prospect of learning or designing a new formal method — almost at the drop of a hat. This suggests a rather different education than is traditional for today's practitioner of formal methods. Rather than producing zealots in one or another method, we must educate formal methodists who are non-partisan.

Support Tools. Not only must the formal methodist have access to tools that support specific formal methods, but access is also needed to tools that enable special-purpose formal methods to be designed and used. We need open systems, rather than closed, monolithic, theorem-proving environments. The formal methodist should be able to quickly cobble together software to support a new formalism: editors, checkers, printers, etc.

The time has come to admit that our vision for formal methods has been flawed. For whatever reason, programmers do not regularly employ program correctness proofs in building programs. Commercial software vendors do not perceive great benefits in writing formal specifications for their products, so they don't write them. The formal-methods "solution" to the problem of constructing high-integrity software is too often ignored.

Some believe the problem is education — today's programmers are just not comfortable with formalism. Others believe the problem is a lack of tools — there is inadequate automated support for programmers who seek to use formal methods. I believe the problem is our vision. We are too concerned with the grand challenge — provably "correct" systems — to appreciate the real and immediate opportunity: special-purpose and custom formal methods that provide leverage to the system designer or implementor.

Combining the Fault-Tolerance, Security and Real-Time Aspects of Computing

Toward a Multilevel-Secure, Best-Effort Real-Time Scheduler

Peter K. Boucher*, Raymond K. Clark†, Ira B. Greenberg*,
E. Douglas Jensen‡, Douglas M. Wells§
*SRI International, Menlo Park, California 94025
†Concurrent Computer Corp., Westford, Massachusetts 01886
‡Digital Equipment Corp., Maynard, Massachusetts 01754
§Connection Technologies, Harvard, Massachusetts 01451
U.S.A.

Abstract

Some real-time missions that manage classified data are so critical that mission failure might be more damaging to national security than compromising the data. The conflicts between computer security requirements and timeliness requirements are described in the context of large, distributed, supervisory-control systems that are intended for use in such critical missions. The Secure Alpha[3] approach to addressing these conflicts is introduced.[1] A prototype tradeoff mechanism is described, as are the results of testing the mechanism.

Keywords: *multilevel security, real-time, covert channel, tradeoff.*

1 Introduction

Small, low-level, real-time (RT) computer systems are used to manage controlled, predictable physical systems and are designed to guarantee that all tasks are accomplished by their deadlines. Supervisory-control RT operating systems, in contrast, cannot always guarantee that all tasks are accomplished by their deadlines, because these systems are used to manage large, asynchronous

---

[1] This research was supported by Rome Laboratory under contract No. F30602-91-C-0064.

physical systems (which may be composed of a collection of low-level, RT systems). This paper is primarily concerned with the implications of imposing high-assurance multilevel security (MLS) requirements on supervisory-control RT systems such as military systems that contain classified information and control automated weapon, communication, or command and control systems [1, 10, 11]. This type of system is too complex to be controlled in the absolute sense that low-level RT systems are controlled (e.g., static resource allocation, and/or verification of all the application software), yet their ability to respond appropriately in real-time, as well as their security enforcement, may be critical to national security.

An RT system manages its resources to attempt to satisfy the collection of timing constraints imposed by an application. Secure systems tend to separate and encapsulate resources to control access and minimize sharing. Existing security-enhancement approaches cannot handle these combined requirements except when the environment or application has been significantly constrained. For example, if an application can be designed as a collection of periodic tasks with predetermined interactions and if the environment can be carefully controlled so that a processing overload does not occur, then tasks of different security access classes[2] can be prevented from interfering with one another while still satisfying their time constraints. Similarly, if an application is composed exclusively of tasks of a single access class or if the application can be partitioned into a set of independent subsystems, each containing tasks of only a single access class [16], then RT issues can be addressed without compromising security concerns, given sufficient computing resources. This paper explores the problems of integrating MLS and RT requirements where neither of these two constraints is present (i.e., in complex, supervisory-control RT systems).

2   Multilevel Secure Real-Time Distributed Operating System

This paper describes our approach to integrating the requirements for high-assurance MLS and supervisory-control RT processing in a Secure Alpha [3] prototype. In the Secure Alpha study project, we addressed the general problem of applications that are inherently multilevel in nature but are also required to provide timely, appropriate responses in the event of emergency and overload situations, which cannot be avoided a priori, because the system must be able to respond correctly to an unpredictable environment.

Existing RT scheduling policies such as static priority, earliest deadline first, and best effort [5] do not consider the access classes of computations when selecting the computation that should execute next. As we describe in Section 3, such scheduling policies can be exploited to covertly pass information across access classes or to make inferences about the state of the system or

---

[2] An access class can be thought of as the classification of data (e.g., UNCLASSIFIED, CONFIDENTIAL, or SECRET), or the clearance of users or processes.

the application. Current security techniques may not allow the system designer to address RT requirements. For example, enhancing security by inserting random delays or isolating processing capability by access class to reduce covert-channel bandwidths [16] ignores RT requirements and may jeopardize the overall mission.

However, in a high-assurance MLS RT system, the existence of the scheduling covert channel (see Section 3) cannot be ignored. There are several possible approaches that can be adopted. This paper describes an important contribution to an approach that monitors and controls the bandwidth available through the scheduler covert channel. Other approaches might:

- require that all system and application software be examined carefully (or verified) to have high assurance that there are no Trojan horses present. This might involve a large amount of software and is not verified at runtime.
- demonstrate that the channel cannot be exploited effectively in the actual execution environment. Such a proof would almost certainly be application-dependent. It might include analysis like that presented in this paper, but it would have to cover a wider range.

2.1 Alpha Real-Time Distributed Operating System

Secure Alpha is based on the Alpha non-proprietary operating system for mission-critical real-time distributed systems [8]. Space limitations preclude a complete description of Alpha's features. Consequently, we will present the Alpha programming model and those features that are most relevant to real-time scheduling in a multilevel-secure system. (Readers can find material describing other aspects of the system in a series of technical reports and papers. For instance, [13] and [15] describe the high-level requirements, approaches, and programming model, including the manner in which various failures are reflected to and handled by the application and support for fault tolerance and advanced features like atomic transactions and object replication. [14] describes the implementation of Alpha on a dedicated-function, multiprocessing platform and provides a detailed account of the Communication Virtual Machine and the basic communication protocols used to support the programming model as well as the Scheduler Processor interface and a number of scheduling policies and mechanisms.)

Alpha's RT processing paradigm is based on Jensen's Benefit-Accrual Model [6] and it contains mechanisms to support a form of the model. The Alpha kernel provides a programming model with the following abstractions to support the development of RT distributed applications.

- *Objects*, which are passive abstract data types that contain data and code organized into data-access operations;[3]
- *Threads*, which are execution control points that move among objects via operation invocation;[4]

---

[3]The data associated with an object can only be accessed by a subject that is executing the object's code.
[4]The address space of a thread is limited to the code and data of the object of which it is currently invoking an operation, and a private, object-specific stack area.

- *Operation invocation*, which is an approach similar to procedure calling by which threads execute object operations.[5]

The Alpha kernel employs a capability-based design [9]. Objects and threads are identified by capabilities, and they can be accessed only via an appropriate capability. The creation and modification of capabilities is controlled by the kernel. Alpha's capabilities are globally unique over time, and their values are independent of network location. The capabilities are non-forgeable because they are maintained entirely within the TCB, and the application can refer to them only via a "handle" that is valid only within the context of a single Alpha object.

2.2   Computing Environment

We assume that the computer system is organized to support a single mission, where all portions of the application are cooperating toward a common goal. This assumption makes it possible to compare the relative benefit of accomplishing various portions of the application. This mission-oriented nature differs from that of most distributed operating systems, which consist of independent sites with autonomous users and goals. It differs as well from many centralized systems, which support a variety of independent applications. The system is expected to react appropriately to its environment, despite emergency or overload situations.

2.3   Multilevel Security

Secure Alpha is intended to support applications that require high assurance for MLS, such as that defined by Class B3 of the Trusted Computer System Evaluation Criteria (TCSEC) [12]. The security functionality required for this type of system includes the labeling of subjects[6] and storage objects,[7] mandatory[8] and discretionary access control,[9] controlled object reuse,[10] user

---

[5]When a thread invokes an operation of an object, it gets a new address-space, completely disjoint from its previous address-space. Thus, the only way to observe or modify an object's data is via its defined operation interface.

[6]A subject is an active entity that observes and/or modifies the data protected by the TCB. In Secure Alpha, a subject is a thread. Each subject (thread) in a Secure Alpha system is labeled with a security access class.

[7]A storage object is a passive repository for data. In Secure Alpha an object is an Alpha-object. Each Secure Alpha object is labeled with a security access class.

[8]Mandatory access control prevents subjects from gaining access to objects for which the subjects have insufficient clearance.

[9]Discretionary access control allows subjects to grant or deny permission for other subjects to gain access to objects (within the limits established by the mandatory access control).

[10]Object reuse mechanisms ensure that residual data is removed from system resources before a new subject gains access to them (e.g., a page of memory must contain no data from the previous subject when it is assigned to a new subject).

identification and authentication,[11] trusted path,[12] and auditing.[13] It must be demonstrated that a computer system of this class enforces its security policy with a high degree of assurance. Mandatory access control must be enforced by a mechanism that satisfies the reference-monitor requirements of mediating all accesses by subjects to objects, being tamperproof, and being small enough to be thoroughly analyzed. The trusted computing base (TCB), which contains the code needed to enforce the security policy, should be highly structured and should contain only security-critical code. Hardware protection must be used to protect the TCB from tampering by non-TCB code and to enforce mandatory access control. In addition, covert channels[14] must be identified, and the maximum bandwidth of each channel must be determined. According to the TCSEC, covert channels with bandwidths over one hundred bits per second must be eliminated (or reduced), and those with bandwidths between ten and one hundred bits per second must be audited, but covert channels with bandwidths less than ten bits per second may be tolerated.

3   Covert Channel Tradeoffs

Tradeoffs between maintaining security and accomplishing critical missions are common in the real world. For instance, after evaluating a situation and determining that the mission benefit outweighed the security risk, a commander might transmit sensitive information in the clear when no secure communication line is available. For Secure Alpha, we did not consider trading either mandatory or discretionary access control against the timeliness requirements to be an acceptable approach. That is, we consider access control checks to be required, not tradable, system functions. We believe that if the system is to be considered secure and evaluatable at Class B3, access control is always required. Moreover, because a predictable amount of time is dedicated to them, the access control checks need not interfere with the RT nature of the system.

However, there are other areas that can be traded. In particular, covert-channel[15] bandwidth may be varied as a function of the current success of the application in accomplishing its mission. Thus, when a mission is meeting its timeliness requirements relatively easily, security risk

---

[11]Identification and authentication mechanisms serve to provide assurance that only authorized users can use the system.

[12]A trusted path is a mechanism whereby the user can be assured that she is communicating directly with the TCB (and no application software can filter the communication).

[13]Auditing keeps records of a user's actions, in order that the user can be held accountable for them.

[14]Covert channels are discussed in Section 3.

[15]A covert channel is a communication mechanism that was not originally designed or intended to be used for communication that allows a high (classified) subject to signal high information to a low (less sensitive) subject. A simplified example could be a high subject that fills up an entire disk drive, and repeatedly releases and requests the last block. A low subject could repeatedly request a block. When the high subject wishes to transmit a one, it releases a block. The low subject is given the block, which it recognizes as a one. When the high subject wishes to transmit a zero, it does not release a block, and the request by the low subject to allocate a block fails, which the low subject recognizes as a zero.

can be limited by restricting the bandwidth of the covert channels, particularly the scheduling covert channel. If the mission is jeopardized, in part by covert-channel countermeasures, the countermeasures can be relaxed in a well-defined and controlled way to decrease the risk of overall mission failure. Our approach is to capture the relevant policy and encode it as resolution rules so that the system can automatically make an appropriate tradeoff when necessary.

In particular, the scheduling-covert-channel[16] bandwidth limits can be adjusted dynamically in accordance with mission goals. For example, if the destination of a flight is classified, but the classification is automatically reduced upon arrival, the resolution mechanism could be configured to ease restrictions on the scheduling covert channel bandwidth when the flight arrives within some threshold distance from the destination. These resolution rules will be protected and enforced by the TCB.

4   Traditional Covert Channel Reduction Mechanisms

Traditional solutions for removing or reducing the bandwidth of covert channels, such as virtualizing resources, slowing the system clock, and introducing noise into the system's operation, are not appropriate for RT supervisory-control applications because they may prevent the system from responding to its environment in a timely manner. The traditional practice is to reduce or eliminate covert channels by minimizing resource sharing between access classes in one or more of the following ways:

- *Static allocation* policies virtualize resources by statically allocating them to different access classes. While this eliminates the covert channels, it also interferes with the RT system's ability to maximize and control the sharing of resources in order to accomplish as much of the mission as possible.[17]
- *Delayed allocation* allows threads to borrow idle resources from lower access classes. If these idle resources are then requested by a low thread, however, they arc automatically retrieved from the high thread and reallocated. To disguise the resource sharing, all resource allocations are delayed to the longest time necessary to retrieve and allocate the resource. Although this approach can eliminate covert channels, it may allow a critical task at a high access class to fail because it cannot commandeer resources from less critical tasks at lower access classes.
- *Slowing the system clock* or reducing its granularity reduces the bandwidth of timing channels, but also reduces an RT system's ability to accomplish its mission by making the schedules less precise.

---

[16]Covert channels are introduced by the sharing of finite resources such as physical memory or operating system tables. One such resource that is important in an RT system is the processor. If a low subject can detect a difference in the availability of the processor because of the execution of a high subject, then the high subject can transmit information from a high object to the low subject in violation of the mandatory secrecy policy.

[17]Some RT systems are resource limited, because of space, power, or budgetary restrictions.

- *A randomizing scheduler* can reduce the bandwidth of timing channels by randomizing the actions of the processor scheduler, thereby introducing noise into the covert channel and making covert communications more difficult, but this approach's unpredictable behavior makes it inappropriate for an RT system because of the difficulty in scheduling computations effectively.

5   A Uniform Framework for Security and Timeliness

The existing methodology for providing assurance that MLS TCBs correctly enforce their security policy was extended to include a *timeliness component* in the Secure Alpha security policy [4, 2]. In the trusted systems development methodology, all crucial system requirements flow from the security policy, and all verification steps trace back to the policy. Timeliness must be a component of the security policy if it is necessary to make tradeoffs between "standard" security features and timeliness. It is also necessary to include timeliness as a component of the security policy if the verification effort is to provide assurance of the timeliness of the system.

Without such a coherent framework, either the system cannot be allowed to make tradeoffs, or the security policy must be ignored under certain circumstances. The Secure Alpha security policy explicitly addresses the possibility of conflicts between the secrecy and timeliness components of the security policy and defines a *resolution component* to specify the appropriate behavior when such conflicts occur.[18] To be effective, the *resolution rules* must be guided by application-specific knowledge. We motivate our approach by considering an example involving the scheduling covert channel.

Consider a land-based missile defense installation that must exhibit a high degree of RT behavior to be able to detect potential threats, perform evaluations, and destroy incoming missiles. The system contains a variety of classified material such as battle plans, weapon capabilities, and evaluation algorithms, and has users with a variety of clearances. Assume that a low subject is using the processor to monitor sensors when a high subject becomes ready and requests the processor so that it can perform a mid-course flight correction for a missile interceptor. One might expect that the flight-control subject should execute because its function is more important to the mission. However, the fact that the low subject would be able to observe a delay caused by a high subject would potentially allow covert information to be transmitted. In this example, if the scheduler executes the high subject it could violate the mandatory security policy (specifically, covert channel bandwidth limits), and if it does not execute the high subject it will violate the timeliness policy. The system must resolve these conflicting requirements, because some action must occur.

---

[18]Other approaches are possible. The critical issue is incorporating RT and MLS requirements in a single framework. Our choice allows us to use the existing security assurance methodology, to demonstrate with comparable assurance that the system provides a combined solution to the combined problem under study.

By including timeliness in the security policy for Secure Alpha, we are acknowledging that the timeliness needs of the mission (e.g., the ability to intercept a missile) are potentially as critical as the requirement to prevent leakage of classified information. By including tradeoff resolution in the security policy, we provide a means for the a priori determination of how such tradeoffs are to be made for a given situation; it is certainly possible to design the resolution rules so that they favor security over timeliness (or to favor timeliness over security), if desired. A system such as Secure Alpha, which must respond correctly in emergency and overload situations, would not be usable unless such tradeoffs can occur.

The resolution of conflicts between MLS and RT requirements depends on numerous factors, such as the amount and sensitivity of classified information in the application, the number and variety of subjects, the scrutiny and rigor under which the code executed by subjects was produced, the number of access classes and the distribution of subjects among access classes, the physical security of the system, the number and types of RT constraints, and the consequences of failing to satisfy the RT constraints.

5.1   Timeliness in Secure Alpha

The security policy for Secure Alpha states that activities will be scheduled and performed in accordance with application-specified timeliness requirements. Thus, it must be possible to evaluate and compare the requirements and relative importance of all computations so that they can be meaningfully scheduled and performed.

Secure Alpha uses a best-effort scheduling policy that is compliant with a variant of the Benefit-Accrual Model [6, 7] to complete the set of computations that results in the highest aggregate value to the mission. The Benefit-Accrual Model associates a value with a single computation that depends on when the computation completes. The value associated with a set of computations depends on when the individual computations complete, and on a function, which may be specified by the application, that indicates how to accrue value over a collection of computations.

As a thread executes, it performs a series of time-constrained tasks, each of which is represented by an object operation. Over its lifetime, the thread may invoke operations of many objects. The Secure Alpha scheduler bases its decisions on scheduling attributes associated with the tasks by the application. Time constraints, which are specified by special system calls placed in the code, are used by an application to associate scheduling attributes with tasks. The scheduling attributes consist of a *time-value function*, the functional *importance* of the task, and the expected duration of the task. Importance is an indication of how valuable a task is relative to the other tasks that form the mission. An example of a time-value function is shown in Figure 1.

Threads are scheduled for execution whenever a scheduling event occurs. Examples of schedul-

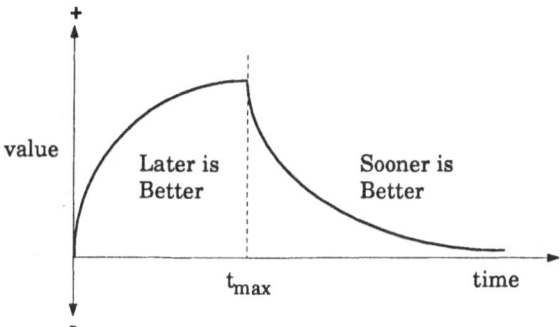

Figure 1: Abstract Time-Value Function

ing events that are recognized by Secure Alpha include completion of a task within its time constraint, blocking of a thread, failure of a time constraint, failure of an executing thread, creation of a thread, change of a thread's scheduling information, unblocking of a thread, and termination of a thread. When one of these occurs, the scheduler invokes the scheduling policy module to determine the threads that should be executed. The scheduling policy module evaluates the tasks competing for execution and, by assigning starting times to the tasks, forms a schedule that will lead to the greatest value for the mission. A task executes until it completes (or fails), or until more value can be obtained by executing a different task.

For example, given three tasks, A, B, and C, if task C could provide more mission benefit than either A or B, but A and B combine to provide more value than C, A and B may be executed while C starves. A best-effort schedule indicates more than just the order in which the threads execute. It also specifies precisely when each thread should begin to execute. For example, even if there is time to execute all three tasks, it is possible for the accrued benefit to be higher if the execution of A and B is timed precisely to maximize their value, while not leaving enough time to execute task C.

Each thread in Secure Alpha is labeled with an access class. The time at which threads are executed can be exploited as a covert channel between a high and a low thread. A schedule is defined to be secure if it does not cause any low thread to be preempted or have its execution delayed by the execution of a high thread (i.e., if no high threads interfere with the scheduling of any low threads). If a situation arises where greater mission benefit can be realized by executing a high thread instead of a low thread (which is ready and waiting to execute), then the schedule cannot be both timely and secure.

5.2   Resolution of Tradeoffs

For a given set of threads and their scheduling attributes, there may not exist a schedule that is considered both timely and secure by the Secure Alpha security policy. Other covert channels will exist with similar tensions between RT and MLS requirements, such as shared disks, communication networks, I/O channels, workstations, and printers. To resolve these conflicts, the security policy allows an authorized security officer to formulate application-specific tradeoff rules to be enforced by the TCB. The essential goal is to provide the security officer with the means to configure the TCB to enforce the resolution of tradeoffs in accordance with a mission-specific security policy. The exact expression of this resolution policy must be formulated individually for each mission. Thus, the TCB provides a combined solution to the problems of security and timeliness by strictly enforcing both, except when a pre-determined conflict occurs, in which case the resolution decisions made by the security officer are enforced. The TCB, with a high degree of assurance, solves the combined problem (given the security officer has configured the resolution mechanism correctly in accordance with the mission security policy).

A preliminary example of this type of configurable resolution mechanism was demonstrated with the Secure Alpha Rapid Prototype (SARP), described below. The approach we use in the SARP is to dynamically measure covert channel bandwidth to assess system security risk, and to allow an upper limit on the bandwith to be imposed by the security officer. This allows the system to adapt to the current situation (within limits). A low bandwidth indicates that the system would be able to tolerate an additional signal, while a high bandwidth indicates that the system would need a compelling reason to accept more security risk. We determine whether the task selected by the scheduler should be executed by comparing the current bandwidth with a threshold.

6   A Worked Example: The Secure Alpha Rapid Prototype

The *Secure Alpha Rapid Prototype (SARP)* is an operating system rapid prototype that contains a Secure Best-Effort Scheduler, as well as other security features, and upon which the Secure Alpha study's feasibility demonstration was built.

The SARP experiment demonstrated that a covert-channel-bandwidth limitation mechanism can be effectively incorporated into a best-effort scheduler to allow enforcement of the Secure Alpha security policy's resolution component. It was delivered and demonstrated to Rome Laboratory at the completion of the Secure Alpha study project in March 1993.[19]

SARP is based on a version of Alpha that runs on the Concurrent Computer Corporation 8000

---

[19]SARP is based on the Alpha software base as it existed in late February 1991.

Series machines (MIPS R3000-based multiprocessors). SARP uses two processors on each machine, one of which is dedicated to communication functions.

The SARP software architecture is largely the same as the Alpha architecture, rather than the Secure Alpha architecture that was developed during the Secure Alpha study. This was appropriate since the feasibility demonstration focused on the ability to monitor and limit the scheduling covert channel and a rapid prototype based on Alpha could readily support this work. Both the SARP and Secure Alpha architectures are layered, but the number of layers and the assignment of functions to layers differs for each. Alpha provides a general scheduler interface and a set of scheduling mechanisms[14] so that a broad range of scheduling policies can be supported. SARP includes a new Secure Best-Effort Scheduler policy that is built on that base.

SARP associates labels with the security subjects (SARP threads) and objects (SARP objects) in the system. Each label designates the access class of the corresponding subject or object. In addition to being necessary for mandatory access checks, these labels are also referred to by the Secure Best-Effort Scheduler to identify problematic execution schedules.

The Secure Best-Effort Scheduler is adapted from the earlier Alpha Best-Effort Scheduler. It has been enhanced to monitor and manage the scheduling covert channel. In addition, this scheduler executes on the application processor, which executes the application software, rather than on a dedicated-function processor of its own. This makes SARP less reliant on multiprocessor platforms than earlier versions of Alpha, making it potentially easier and less expensive to experiment with SARP and Alpha concepts. (This is the first incarnation of a best-effort scheduler on a uniprocessor machine.)

The Secure Best-Effort Scheduler in SARP monitors and manages the scheduling covert channel by examining the tentative execution schedule for the processor. This tentative schedule is always constructed by the best-effort scheduler; it permits a feasibility assessment to be performed for possible future schedules. The Secure Best-Effort Scheduler examines the tentative schedule, looking for

- *Preemption problems.* A preemption problem exists if the first subject in the tentative schedule, which is the subject selected for immediate execution, will preempt a subject that does not dominate it — that is if a high subject preempts a low subject.
- *Delayed execution problems.* A delayed execution problem exists if the first subject in the tentative schedule will delay the execution of a subject that does not dominate it — that is, if a low subject is delayed because a high subject is executing.

Both types of problems are timestamped and entered into a data structure that records all of the potential problems encountered in the last one-second time interval. Thus, the actual maximum bandwidth is monitored and controlled, regardless of how performance issues related to different hardware or software configurations might affect the theoretical maximum bandwidth.

The Secure Best-Effort Scheduler enforces a ceiling on the potential covert-channel bandwidth through the scheduler. In the feasibility demonstration, the experimenter sets this limit to some nonnegative value (setting it to zero is permissible). The scheduler will never allow more than that threshold number of bits to flow[20] through the scheduling covert channel in any one-second interval. Once that limit is reached, the covert channel is closed (i.e., no more bits are allowed to flow). It can again be opened when the measured bandwidth has been lowered, which happens automatically as time passes. The potential covert-channel bandwidth limit can be dynamically changed by the experimenter.

The scheduler closes the channel by executing only subjects that are dominated by all of the other ready subjects. If no such subject exists, then the processor will be idle.[21]

The Secure Best-Effort Scheduler can dynamically monitor and limit the potential covert-channel bandwidth through the scheduler and provides a foundation upon which higher-level control policies can be built. These policies, which would assess security and real-time application benefits and risks to set appropriate covert-channel bandwidths, will typically be application-dependent and were not explored as part of the feasibility demonstration because it focused on the lower-level monitoring and enforcement functions.

6.1   SARP Test Scenario

The feasibility demonstration studied the SARP's ability to monitor and control potential communication through the scheduling covert channel. The demonstration software also accommodated a mission workload so that the effects of the covert-channel countermeasures on all mission software could be assessed.[22] We concentrate here on the manner in which the

---

[20]Note that the "flow" of bits is only a potential flow (i.e., the scheduler can never know if the covert channel is actually being exploited).

[21]When at least two ready subjects are members of incomparable access classes and the bandwidth limit has been reached, the scheduler will not execute any of those subjects.

[22]SARP was designed and implemented to support the feasibility demonstration described in this paper in the short-term. In the longer term, SARP provides a testbed for the exploration of higher-level control policies, called *bandwidth regulation policies*, that set the covert-channel bandwidth limit based on security and real-time considerations for specific applications. By examining different mission workloads operating under different bandwidth regulation policies, we can assess the effects of covert-channel bandwidth management on both the covert communicators and the accomplishment of the system's mission. This is a moderately large problem that has no general solution. The SARP feasibility demonstration used a very simple bandwidth regulation policy: allow the channel to be used freely until a user-specified limit had been reached; then close the channel. Evaluating the effects of even this simple policy on the real-time behavior of the system involves the exploration of several variables, including: the number of subjects in the system, their access classes, the frequency and distribution of their time constraints, whether their time constraints are hard or soft, the processor load, the frequency and distribution of processor overloads, the number of cycles consumed by covert communicators, the relative importances of covert communicators and the other subjects in the system. These variables can be derived from specific applications or they can be studied in general. The effects on real-time performance can also be manifest in many ways, including, for instance, the total or relative number of time constraints satisfied, the degree

covert communicators exploit the scheduling channel.

The basic scenario studied included two covert communicators — a sender and a receiver — that were SARP threads executing within two separate objects. They never visited a common object during their attempted communication. The sender could change its functional importance to influence SARP's scheduling decisions. By setting a high importance value, the sender would be virtually assured of executing; by setting a low importance value, it would be virtually assured that the receiver would execute. Both the sender and receiver declared hard time constraints — that is, deadlines — and both declared that they needed most of the available processing cycles until their deadlines to successfully complete. Thus, the processor was most often overloaded when the covert communicators were attempting to communicate.

To obtain some degree of noise immunity, the receiver tried to "read" each value that the sender transmitted four times. Ideally, then, the receiver would receive four 1s when the sender transmitted a 1, and four 0s when a 0 was transmitted. Experiments showed that often three, four, or five 1s or 0s would be received, depending on the parameters selected for the experiment. By "averaging" the received bit stream, a relatively reliable covert communication channel could be formed, if no countermeasures were employed.

6.2   Description of SARP Experimental Software

Two separate software packages comprise the SARP feasibility demonstration: the software that runs in the Alpha testbed[23] (SARP and its application) and the software that runs on a Sun-3 workstation running SunOS 4.1 or later operating system (the communications multiplexor and demo control programs). The Unix-based software, referred to as the *MPX suite*, comprises three programs: *unixmpx*, *mlscmd*, and *simalpha*. 'unixmpx' provides a multiplexed communication path between threads executing on an Alpha system and programs running on a UNIX system. Using the EXT protocol, 'unixmpx' provides up to 254 independent, reliable communication links.

'mlscmd' provides a simple way to send a small set of commands to Alpha. The available commands include:

- *version* – returns the version identifier for the MPX suite.
- *set-parameter* – assigns a value to a designated parameter; some important parameters are listed below.

---

to which subjects are executed at optimal times, and the total application-specified benefit accrued[5, 17].

[23]The Alpha testbed consists of four Concurrent Computer Corporation 8000 Series machines (MIPS R3000-based multiprocessors) interconnected by an Ethernet local area network. Three of the machines run the Alpha Operating System, while the fourth runs Concurrent's RTU (Real-Time UNIX) Operating System and acts as a gateway to the Alpha OS nodes. A previous version of the Alpha testbed was based on custom multiprocessors based on the Motorola 680x0 processor family.

- *get-parameter* – returns the value of a designated parameter; some important parameters are listed below.
- *get-status* – returns a general system status snapshot; the specific information returned can be tailored for each experiment.

In addition, 'mlscmd' can also listen for unsolicited status generated by Alpha threads.

The third program, 'simalpha,' simulates the behavior of the MPX components of an Alpha system. It can be used to assist in the debugging of UNIX programs that wish to communicate via MPX. It is not needed during normal operation. Certain parameters are set by the experimenter, including:

- *sender-tc* – the critical time interval (deadline) specified by the covert sender, expressed in milliseconds[24]
- *sender-te* – the required computation time specified by the covert sender, expressed in milliseconds
- *sender-sigma-te* – the standard deviation expected in the sender's computation time, expressed in milliseconds
- *sender-hi-importance* – importance[25] of the sender when it wants to send a "1"
- *sender-lo-importance* – importance of the sender when it wants to send a "0"
- *receiver-importance* – importance of the receiver
- *pccbw-limit* – potential covert channel bandwidth that can be tolerated through scheduling channel, expressed in bits/second

Other parameters are measured by the scheduling policy framework or by the Secure Best-Effort Scheduler and can be queried by the experimenter at any time, including:

- *total-covert-flow* – total potential covert channel flow, expressed in bits
- *max-covert-flow-rate* – maximum observed potential covert channel bandwidth, expressed in bits/second
- *schedule-mods* – the number of times the tentative schedule was modified due to the threat of covert information flow
- *total-rescheds* – number of times the scheduler executed its selection algorithm
- *total-resched-time* – total time spent executing the scheduler's selection algorithm, expressed in ticks of the scheduler clock (one tick is 100 microseconds)

The covert receiver also sends the bit sequence that it "decoded" to the experimenter for examination.

---

[24]Hard deadlines were chosen as the type of time constraint because they allowed the covert communicators to achieve more throughput than did softer types of time constraints.

[25]The *importance* of a thread is used by the best effort scheduler to help determine the benefit that may be accrued by allowing the thread to complete it's computation in a timely fashion.

6.3 SARP Test Results

The basic attack mounted by the covert communicators is described in Section 6.1. To test the ability of SARP to monitor and control the scheduling covert channel, the communicators were tuned to a high degree of reliability when they had the system to themselves — that is, when there was no other load. SARP monitored the channel and reported on the measured potential bandwidth. To limit the available covert-channel bandwidth, the experimenter could set a bandwidth-control variable to any desired value (expressed as an integral number of bits per second). SARP's Secure Best-Effort Scheduler and the covert communicators were instrumented to record the effects of the bandwidth limitations.

6.3.1 Data from Individual Experiments

The following measurements were made on November 10, 1993, repeating earlier experiments so that the exact "received bit string" could be recorded. In the original experiment, the string was not recorded, but was characterized in the notes. These numbers are typical and repeatable, and they agree with those that were in the notes from the original experiments, and from the SARP feasibility demonstration at USAF Rome Laboratory in March, 1993.

Following is a discussion of the data from each of the experiments performed. Some parameter values were constant for all of the following experiments. The sender's importance was 1000000 (to send a "1") or 1 (to send a "0") The receiver's importance was always 1000. The data is summarized in Figures 2 and 3. In Figure 2, the headings are as follows:

- *Experiment* – identifies the experiment from which the numbers were drawn
- *tc* – the sender-tc (see Section 6.2)
- *te* – the sender-te (see Section 6.2)
- *sigma* – the sender-sigma-te (see Section 6.2)
- *pccbw* – the pccbw-limit (see Section 6.2)
- *bits* – the total-covert-flow (see Section 6.2)
- *max bw* – the max-covert-flow-rate (see Section 6.2)
- *sched mods* – the schedule-mods (see Section 6.2)
- *rescheds* – the total-rescheds (see Section 6.2)
- *reschedtime* – the total-resched-time (see Section 6.2)

Experiment 1 It takes a while for the receiver to get synchronized. Thereafter, communication is fairly reliable – as shown by the alternating strings of 0's and 1's in Figure 3.

Experiment 2 The receiver seems to get into sync. However, reception is subsequently disrupted when the sender is delayed to enforce the covert channel bandwidth limit.

The receiver perceives a longer string this time (see Figure 3) because the sender was delayed when the channel was throttled. (The receiver assumes the sender is always transmitting. It only stops when the control program for the experiment stops it.)

Experiment 3   The receiver perceives a longer string and never truly is able to get into sync with the sender.

Experiment 4   The channel is closed. During the period when sender/receiver synchronization usually occurs, the receiver does interpret a delay caused by other system activity as a "1."

The time required for each scheduling decision is less than under the previous case due to the fact that the sender is quickly shed, leaving only the receiver eligible to execute, and reducing the time required to make a scheduling decision (See the *rescheds* column of Figure 2 for experiment 4).

Experiment 5   Once this is synchronized it looks very reliable (see Figure 3).

Experiment 6   Once this is synchronized it looks very reliable (see Figure 3).

Experiment 7   This is typical of trying to pass 20 bits/sec through the covert channel. This looks very noisy and is not as repeatable as the results at lower covert communication rates. Section 6.3.2 explains the factors that impose this limit.

Experiment	tc	te	sigma	pccbw	bits	max bw	sched mods	rescheds	reschedtime
1	100	90	5	100	66	31	0	268	40.2
2	100	90	5	20	56	20	36	306	42.5
3	100	90	5	10	47	10	130	446	60.1
4	100	90	5	0	0	0	200	494	54.7
5	500	480	10	100	59	8	0	270	38.4
6	500	480	10	20	59	8	0	272	39.1
7	50	45	2	100	28	28	0	166	22.6

Figure 2: Numerical Results of Experiments

6.3.2   Test Conclusions

The following major conclusions were supported by the feasibility demonstration:

String Source	"Received" Bit String
Ideal String	000011110000111100001111000011110000111100001111000011110000111100001111000011110000111100001111
Experiment 1	101001000100010001000111110001111100011111000111110001111100011111000111110001111100011110001111
Experiment 2	110010001000100010001111100011000000000000000001000100010001000111110001110000000000000000010001 000100010001111
Experiment 3	100011000100010001000000000000000000000000000000001000100010001000010000000000000000000000000010 001000100010001000000000000000000000000010001000100010001000100010000000000000000000000010001000
Experiment 4	000001000000000000000000000000000000000000000000000000000000000000000000000000000000000000000000 000000000000000000000000000000000000000000000000000000000000000000000000000000000000000000000000 00000000000000000
Experiment 5	100001000000000000000000000000000000000000001000011110000111100001111000011110000111100001110000111
Experiment 6	100000000000000000000000000000000000000000000011110001111000111100011110001111000011110000111
Experiment 7	101010110101011011111111010111111010101011111101010110101010

Figure 3: "Received" Bit-Strings

- *Countermeasures for the scheduling covert channel can work.* A strict bandwidth limit can be imposed on the scheduling covert channel.
- *SARP's Secure Best-Effort Scheduler perceives more potential traffic than is actually present.* This occurs because the system cannot distinguish the intent of the threads declaring time-constrained computations. Not only does the system detect the preemption and delayed execution problems presented by the covert communicators, it can interpret harmless scheduling sequences as potential problems, too. The scheduler must assume that the channel is being exploited.
- *The countermeasures employed are very disruptive to covert communicators.* The scheme being employed allows the communicators to synchronize. However, each time the schedule is reordered to limit the bandwidth of the covert channel, the synchronization is jeopardized. Regaining synchronization is time-consuming.[26] For the covert communication scheme selected, the countermeasures essentially dictate that the covert communication will probably be most effective if the countermeasures can be avoided — that is, if the communicators do not try to transmit information at a rate that will exceed the currently established potential covert-channel bandwidth.
- *Scheduling overhead is increased.* The amount of time required to make a scheduling decision is increased. This is expected because the secure schedule is generated by modifying the schedule that the standard Best-Effort Scheduler would have produced. For this experiment, triggering and executing the covert-channel countermeasure software adds approximately 10% to the time required to make a scheduling decision. Specifically, on average, it takes about 110 μs to execute the basic scheduling algorithm for the feasibility demonstration. (The time required to execute the algorithm is a function of load and number of ready threads — as well as their access classes — and will be different for other situations.)

---

[26]At one point, the notion that text should be transmitted covertly and displayed was explored. A major drawback of that approach was that once synchronization was lost, gibberish would be displayed. The alternating strings of 0s and 1s that were chosen instead of text for transmission permit the experimenter to recognize quickly when synchronization has been lost, regained, and so forth.

- *Covert-channel bandwidth can be made zero.* The scheduler can ensure that no subject ever executes unless its access class is dominated by those of all other ready subjects. The covert receiver might still interpret delays as transmission. So in a system that is not completely quiet, to the extent that the receiver cannot account for where all of the processing time went, the receiver might still guess that the sender is transmitting 1s — when, in fact, the sender never has the opportunity to execute while the receiver is ready to execute.
- *Bandwidths on the order of 100 bits/s are possible if no limits are imposed on the covert-channel bandwidth.* With the scheme used, it is possible to use the channel to send approximately 10 to 15 bits/s using the feasibility demonstration. This is the actual amount of information that the sender is transmitting. Because the receiver is sampling at four times that rate and the bit stream is relatively noise-free when countermeasures are not used, the feasibility demonstration itself is effectively passing 40 to 60 bits/s through the channel. Although the feasibility demonstration can try to increase this rate, only another factor of two or three appears possible — a number of samples are missed at this point, making the channel appear to be much noisier.

To see why the scheduling channel bandwidth is on the order of 100 bits/s, consider the scheduling channel signaling method. As indicated above, a typical scheduling decision requires 110 μs. If one subject is to preempt the other for transmission to take place, which must happen fairly frequently under this scenario, another 100 μs or so must be used to switch context to another subject. If nothing else happens in the system, fewer than 5000 such exchanges can occur each second. For transmission to occur, the sender and receiver must each declare a time constraint, each must complete its time-constrained computation — either successfully or not — and the sender must adjust its functional importance to change transmission characters. This requires at least five scheduling decisions for each transmitted bit, implying that fewer than 1000 bits can be transferred each second even if no computation is performed by either the sender or receiver. However, the detection method used by the receiver is to attempt to consume some number of processor cycles, assuming that if they are not available it is due to the actions of the sender. When more cycles are consumed, the receiver can be more certain about whether or not the sender interfered with its processing request. Because all of the figures quoted above are averages, and there is some variation, it is certain that the 1000 bit/s rate is unobtainable. Under ideal circumstances, perhaps 500 bits/s could be obtained, but this rate has not been attained in experiments to date and might be unattainable.

The maximum attainable bandwidth is a function of the Secure Best-Effort Scheduler implementation. A more efficient implementation (and/or faster hardware) might permit a higher potential bandwidth. Moreover, other signaling schemes might be employed that can attain higher bandwidths. Both of these areas can be studied in the future. Nonetheless, even if higher bandwidths are attainable under other circumstances, they can still be limited dynamically by the scheduler.

7 Summary

In this paper we described an approach for addressing the inherent conflicts between multilevel security and real-time by considering them together within the context of a security policy model. The key idea is to use the security assurance methodology as a framework for providing assurance that other mission-critical system properties (in this case, timeliness) are also implemented correctly. This is crucial when different classes of equally important requirements conflict. In the case of multilevel security vs. real-time, the conflict is clear. There are times when preventing a security leak could cause the failure of the mission, and, if the mission is critical to national security, the resolution of this conflict must be handled correctly.

We described a rapid prototype system that was built to prove the concept, the Secure Alpha Rapid Prototype (SARP). The prototype was implemented with a modified best-effort scheduler, capable of limiting the covert-channel bandwidth of the multilevel scheduler channel. The SARP experiments show that the bandwidth-limiting mechanism is effective, and that the TCB could manage this mechanism at run-time to enforce the Secure Alpha security policy's resolution component.

References

[1] AFCEA. *SDI Battle Management/Command, Control, and Communications Technology Study.* Technical report, AFCEA, October 1987.

[2] R. Clark, D. Wells, and I. Greenberg. Secure Alpha: Software requirements specification for a class B3 multilevel secure real-time distributed operating system. SRI International, Menlo Park, CA, August 1992.

[3] I. Greenberg, P. Boucher, R. Clark, D. Jensen, T. Lunt, P. Neumann, and D. Wells. The Secure Alpha study - Final summary report. Computer Science Laboratory, SRI International, Menlo Park, CA, March 1993.

[4] I. Greenberg, T. F. Lunt, R. Clark, and D. Wells. Secure Alpha: Security policy and policy interpretation for a class B3 multilevel secure real-time distributed operating system. Computer Science Laboratory, SRI International, Menlo Park, CA, April 1992.

[5] E. D. Jensen. Asynchronous decentralized realtime computer systems. In *Proceedings of the NATO Advanced Study Institute on Real-Time Computing*, Saint Martin, October 1992.

[6] E. D. Jensen. A realtime scheduling model for asynchronous decentralized computing. In *International Symposium on Autonomous Decentralized Systems*, Kawasaki, Japan, March 1993.

[7] E. D. Jensen. A timeliness model for scaleable realtime computer systems. In *IDA Workshop On Large, Distributed, Parallel Architecture Real-Time Systems*, Alexandria, VA, March 1993.

[8] E. D. Jensen and J. D. Northcutt. Alpha: A non-proprietary operating system for mission-critical real-time distributed systems. In *Proceedings of the 1990 IEEE Workshop on Experimental Distributed Systems*, October 1990.

[9] R. Y. Kain and C. E. Landwehr. On access checking in capability-based systems. In *Proceedings of the 1986 IEEE Symposium on Security and Privacy*, April 1986.

[10] M. J. Lindemann, J. L. Grubbs, Jr., W. J. Eason, L. L. Lehmann, E. F. Shields, and T. H. Evers. *MODFCS Architecture Baseline: A Combat System Architecture*. Technical Report 79-357, Naval Surface Warfare Center, January 1980.

[11] D. R. Morgan. PAVE PACE: System avionics for the 21st century. In *Proceedings of NAECON*. IEEE, 1988.

[12] National Computer Security Center. Department of Defense Trusted Computer System Evaluation Criteria. DOD 5200.28-STD, Department of Defense, December 1985.

[13] J. D. Northcutt. The Alpha operating system: Requirements and rationale. Archons Project Technical Report TR #88011, Department of Computer Science, Carnegie Mellon University, Pittsburgh, PA, January 1988.

[14] J. D. Northcutt and R. K. Clark. The Alpha operating system: Kernel internals. Archons Project Technical Report TR #88051, Department of Computer Science, Carnegie Mellon University, Pittsburgh, PA, May 1988.

[15] J. D. Northcutt and R. K. Clark. The Alpha operating system: Programming model. Archons Project Technical Report TR #88021, Department of Computer Science, Carnegie Mellon University, Pittsburgh, PA, February 1988.

[16] N. Proctor and P. Neumann. Architectural implications of covert channels. In *Proc. 15th National Computer Security Conf., Baltimore, MD*, pages 28–43, 13-16 October 1992.

[17] J. E. Trull, J. D. Northcutt, R. K. Clark, and S. E. Shipman. *An evaluation of Alpha real-time scheduling policies*. Department of Computer Science, Carnegie-Mellon University, Pittsburgh, PA, December 1988.

Fault-Detecting Network Membership Protocols for Unknown Topologies

Klaus Echtle and Martin Leu
Universität Dortmund
Fachbereich Informatik
44221 Dortmund, Germany
echtle@ls4.informatik.uni-dortmund.de
leu@ls4.informatik.uni-dortmund.de

Abstract

Network membership determines the set of faultless nodes and links in a computer network with point-to-point links. Our protocol solves this problem under a general combination of assumptions, which goes beyond known approaches in the fields of network exploration, distributed system level diagnosis and group membership:

- Neither the topology nor a superset of the nodes are known in advance.

- Mutual dependencies in the initial information of nodes are excluded. Consequently, global authentication based on signatures is not provided.

- Faults may affect any number of nodes and cause nearly arbitrary behaviour.

The key issue of our solution is the application of special cryptographic functions instead of usual signatures for message authentication. According to the unlimited fault number the protocol is only fault-detecting, not tolerating.

Keywords: membership protocol, fault detection, distributed system, modified signature methods

1 Introduction

The development of computer networks leads to general distributed systems with sufficient structural redundancy to apply various fault tolerance techniques – checkpointing, reconfiguration, software-implemented TMR etc. However, in large general purpose distributed systems nodes are switched on and off, maintained, added and removed frequently and independently. A *network membership protocol* must make clear whether and where a fault-tolerant process system can be allocated and started. Therefore, it must detect the current redundant structure and determine the point in time, when the system can pass from normal to fault-tolerant operation. By solving this problem distributed systems or parts thereof can be made available for fault-tolerant applications.

A component is *present* iff it is operational (switched on) and not detected to be faulty. A network membership protocol must provide the information which components (nodes and links) are currently present. The difficulties in protocol design heavily depend on the following four, the points in *time* when the protocol has to be executed, the initial *knowledge* about the system structure, the *fault* assumption and the required *reaction* on fault occurence. The main alternatives to these "problem parameters" are as follows:

Choices in the points in *time* when to execute the network membership protocol:
T1 on system startup, where the initial presence of components must be detected,
T2 on restart in an unclear situation after ocurrence of a severe fault exceeding the fault tolerance capability and causing entire system failure,
T3 on insertion or removal of a component,
T4 on component failure,

An effective solution to *T3* and *T4* can be obtained straight-forward by insertion/removal protocols or system diagnosis protocols, respectively. In a fault-prone environment they can be mapped to consensus protocols [26] in principle. Here we concentrate on the start/restart cases *T1* and *T2*.

Choices in the assumption of initial *knowledge* available in all faultless nodes before the network membership protocol is executed:
K1 Type of links: bus, ring, point-to-point links, or mixture thereof,
K2 topology known or unknown,

K3 properties of the communication system: transparent routing, reliable message transfer between faultless nodes, reliable broadcast, synchronism,

K4 set of components (if known, only the "current" subset must be determined),

K5 individual information about components: identifier, digital signature check function, further individual functions, resources and properties.

Here we concentrate on point-to-point links (*K1*). No further assumption is made: The topology (*K2*) and the set of components are unknown (*K4*) – the network membership protocol must explore an arbitrary network of any size. Special properties like those listed in *K3* are not presupposed. Nodes do not need knowledge of individual properties of the components (*K5*) except the identity of their direct neighbours in the network. Consequently our solution must get along without a global signature system [22, 13]. Even simple signatures to protect against technical faults only [10] are not available.

Choices in the *fault* assumption [16, 4, 21, 9]:

F1 Arbitrary faults (any behaviour of faulty nodes, including Byzantine faults)

F2 restricted Byzantine faults only (almost the same as *F1*, however, particular types of faulty behaviour excluded),

F3 omission or timing faults only,

F4 fail-silent behaviour (similar to immediate and perfectly covering absolute tests),

F5 always faultfree behaviour.

In this paper we assume unintentional technical faults – not "intelligent" attacks or intrusions by adversaries. Therefore, *F2* can be viewed as a very general case, only excluding extremely unlikely fault types. On the other hand, *F3* or *F4* can be violated by a few technical faults. Our network membership protocol is based on *F2*: Faulty components may exhibit all types of wrong behaviour except destructive cooperation (multiple Byzantine faults which complement one another in their destructive effect). Refer to section 2.2 for a formal definition.

Choices in the required *reaction* on fault occurence during execution of the network membership protocol:

R1 detection,

R2 localization,

R3 tolerance.

Our solution provides fault detection (*R1*) rather than fault tolerance. Instead of wasting redundancy for some "startup fault tolerance", one needs a "green traffic light" to indicate that fault-tolerant operation with sufficient redundancy is guaranteed from now on. If it is not lit, conventional fault location and removal must be done, before the start is attempted again.

Some choices in the multi-dimensional field characterized by *T1* to *T4*, *K1* to *K5*, *R1* to *R3*, and *F1* to *F5* lead to well-known problems:

- Group membership [6] addresses the problem of collecting the information about the presence of nodes and/or processes in a fault-tolerant way. Transparent communication and some of the knowledge K1 to K5 are presupposed. Known solutions [7, 1, 14] only deal with the simpler fault types F3 and F4.

- *System level fault diagnosis* [20, 25, 3, 27] is a well-established research field, which addresses the question of checking the components of a *given* configuration, where most of the knowledge *K1* to*K5* presupposed. Diagnosis protocols are either based on neighbour tests [19, 2] or result comparisons [18, 8, 23]. Typically it is assumed that, once a node is diagnosed faultless all its actions throughout the rest of the protocol are faultless, too (assumption is even stronger than *F4*).

- *Exploration of network topology* [5, 24] is necessary whenever a network with unknown topology begins operation. Most of the known solutions are optimized for efficiently detecting the structure of nodes and links, but do not provide appropriate reactions to fault occurrence (*F5*).

The brief overview shows that the solutions in the related fields are not suitable for the network membership problem addressed in this paper. Here we aim at as few and as weak assumptions as meaningful in a general distributed system environment – no global a-priori knowledge about the system (*K2* to *K5*) and the general fault assumption *F2* only excluding unlikely destructive cooperation of faulty nodes. In other words, the topology of an unknown and possibly faulty point-to-point network (*K1*) must be explored. We require a so-called *network membership protocol* to detect both present *nodes* and present *links*. We focus on the start (*T1*) and restart (*T2*) where the exploration problem is most difficult.

The whole area of problems in starting a distributed system is not completely understood yet. In the past fault-tolerant systems have been started successfully, of course. The usual startup, however, is simplified by special conditions, as can be seen from the first of the following cases:

- When the topology is known in advance, the process allocation can be predetermined. After switching on, the presence of the nodes can be examined by the operator manually or by use of

a monitoring system without causing problems. Fault-tolerant operation is guaranteed immediately.

- When both the topology and the current fault state are unknown and the fault tolerance technique
  comprises *reconfiguration* in any sense, then it is not sufficient just to start the fault-tolerant
  processes anywhere in the system. In addition, one must make sure that sufficient faultless
  spare nodes and faultless spare links are available. Two problems are illustrated for a TMR process using primary and spare paths – a simple type of link reconfiguration:

  1) The TMR process may produce the correct output by masking a fail-silent node.
  However, one cannot distinguish whether this node is faulty (and hence covered by the fault
  model) or just switched off (see figures 1).

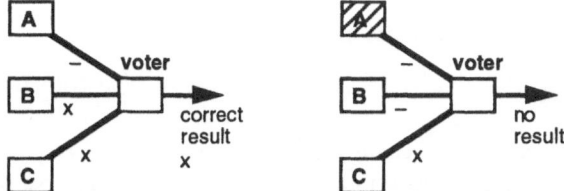

Fig. 1 Left: Node A switched on and fail-silent, Right: Node A switched off, node B switched on and fail-silent aftersome time (spare links not shown for simplicity).

In the latter case, according to the fault model, an additional node may become fail-silent and
violate the majority. As a countermeasure against problem (1) the TMR system must start as a 3-
out-of-3-system and then turn to a usual 2-out-of-3-system. Only an increased number of
proved faultless components at the beginning guarantees the minimum number of faultless
components during subsequent operation.

2) Even then, link reconfiguration may fail after a primary link was found to be faulty. The
3-out-of-3-System only checks whether all primary components are present. The set of all
possible paths is not tested and therefore can contain too less and/or faulty links.

Reconfiguration of processes, nodes, links and/or paths is a substantial part of nearly all efficient
fault tolerance techniques. Therefore, the presence of all possible spare components must be
checked at start or restart time – note, that this exactly expresses the specification of network membership.

We admit that there are exceptions for particular fault tolerance techniques. For an n-out-of-m-sys-
tem, as an example, the presence of $m - n + 1$ disjoint paths between any pair of nodes is
sufficient. However, when a set of n-out-of-m-systems is started, the redundant paths may over-
lap. Therefore, a general network membership protocol is the more efficient solution. In the

generality of the assumptions made above we solve the network membership problem for the first time.

Section 2 describes the system and the fault model formally. Section 3 presents a partial solution for a restricted fault model and demonstrates the arising difficulties. In section 4 the appropriate countermeasures are described and the complete network membership protocol is presented. A sketch of the correctness proof is deferred to the appendix. We conclude in section 5.

2 Models and system view

2.1 System model

We model a distributed system as consisting of a set of n *nodes* $\mathcal{N} = \{N_1, \dots, N_n\}$ and a set of m *links* $\mathcal{L} = \{L_1, \dots, L_m\}$, where $L_i \in \mathcal{N} \times \mathcal{N}$ for all $1 \leq i \leq m$. The links are point-to-point and bidirectional: $(x, y) \in \mathcal{L} \Rightarrow (y, x) \in L$.

Several independent executions of the network membership protocol with different initiating nodes are possible. The set of the initiating nodes is called *configuration manager* $C = \{C_1, \dots, C_c\} \subseteq \mathcal{N}$ consisting of c *manager nodes* $C_1, \dots, C_c$.

Initially a node N_i owns the following knowledge:

• its *identifier* N_i, which must be static and globally unique.

• its membership in the configuration manager. If $N_i \in C$, then the corresponding programs are locally available. These programs contain a list of all manager nodes for the purpose of future cooperation among them.

• its adjacent links $N_i.\mathcal{L} = \{(x, y) \in \mathcal{L} : x = N_i \vee y = N_i\}$. After a successful local test of its adjacent links the direct neighbours $N_i.\mathcal{N} = \{x \in \mathcal{N} : N_i.\mathcal{L} \cap x.L \neq \emptyset\}$ are also known.

No further initial knowledge of the nodes is presupposed. All nodes except the manager nodes have identical programs to participate in the network membership protocol. A manager node C_i

owns an individual signature s_i, which is an encryption function as defined in the following section, and a set of corresponding decryption functions $S^* = \{s_1^{-1}, \dots, s_c^{-1}\}$ initially distributed with the program.

2.2 Fault model

The sets of faultless nodes and links are denoted by $\mathcal{N}^c$ and $\mathcal{L}^c$, respectively, the sets of faulty ones by $\mathcal{N}^f$ and $\mathcal{L}^f$. In our *fault model* the numbers $f_{\mathcal{N}} = |\mathcal{N}^f|$ of faulty nodes and $f_L = |\mathcal{L}^f|$ of faulty links must not exceed upper bounds $F_{\mathcal{N}}$ and F_L, respectively: $f_{\mathcal{N}} \leq F_{\mathcal{N}}$ and $f_L \leq F_L$. Various fault tolerance techniques to be used simultaneously in the distributed system have individual upper bounds. The network membership protocol, however, does not induce constraints on the upper bounds $F_{\mathcal{N}} \leq n$ and $F_L \leq m$.

The behaviour of faulty nodes is characterized by fault assumption *F2*. It is defined precisely in this section according to usual fault models for highest reliability requirements. Our fault model covers all realistic types of technical faults occuring arbitrarily, but excludes intelligent attacks maliciously designed and applied to break protection systems. Faulty nodes do not have "criminal energy". However, the failure following from a technical fault can be worse than just an evenly distributed random alteration. Faulty nodes can access any internal data, do any computation by chance and send an arbitrary message. Since worst case failures may appear similar to malicious actions, we use terms like "malicious", "cryptographic function", "encryption function", "legal", "illegal" etc. as an analogy (for clarity, one could as well use terms like "pseudo-encryption function"). On the other hand it is well known that faults do not cause all types of failures with noticeable probability. Extremely unlikely failures are excluded to simplify the network membership protocol. Here we only exclude the most unlikely failure case where faulty nodes unintentionally (!) complement each other in their destructive effect, as is formally defined in the rest of this section.

Our solution heavily utilizes block check characters to achieve fault detection. Note that usual character generation functions like CRC are not sufficient according to our failure assumption *F2*. A faulty node may send a critically wrong message after adding the appropriate block check character to its wrong local data. Therefore, we use special cryptographic functions instead, to be defined in the following. With respect to these functions our fault model only excludes the case that one faulty node wrongly sends out its local (private) secret information and another faulty node receives this information and uses it to corrupt the protocol execution. If stochastical independence of faults in different nodes is assumed this is probably the slightest restriction of the general Byzantine fault model.

(Cryptographic functions for protecting data against intelligent attacks are computationally demanding [22, 13] whereas protection against "unintelligent" faults is not very costly to achieve [10]).

A one-to-one function f is called *encryption function* iff the following statements hold with high probability even for faulty network components:

- Deriving $f(x)$ from x implies the knowledge (i. e. local availability) of f.

- Deriving x from $f(x)$ implies the knowledge of f^{-1}.

- f is not derived from f^{-1} and vice versa.

The inverse function f^{-1} is called *decryption function*..

Let $E := \{f_1, f_2, \ldots , f_m\}$ be a set of encryption functions and $W := \{w_1, w_2, \ldots ,w_n\}$ be a set of words. Then (E, W) is a *cryptographic system*. The set

$$T(E,W) := \bigcup_{n=0}^{\infty} T_n$$

where $T_0 := W$ and $T_n := T_{n-1} \cup \{f(a_1\oplus a_2\oplus\ldots\oplus a_v)|\ v\in IN,\ a_1,a_2,\ldots,a_v\in T_{n-1},\ f\in E\}$

is called the *set of encryption trees upon (E, W)*. An element t of T(E, W) is called *encryption tree* ('$\oplus$' denotes the concatenation of words).

Let $t \in T(E, W)$ be an encryption tree and F be a set of encrytion and decryption functions ($f\in F \Rightarrow f\in E \vee f^{-1}\in E$). The set

$G(t, F) := D(t, F) \cup$

$\quad\quad \{x|x\in G(y,F) where\ y=f(a_1\oplus a_2\oplus\ldots\oplus a_v)\ and\{a_1, a_2, \ldots , a_v\} \subseteq D(t,F)\wedge f\in F\cap E\}$,

$D(t, F) := \{ t \} \cup$

$\quad\quad \{ x\ |x\in D(y, F)\ where\ t=f(a_1\oplus a_2\oplus\ldots\oplus a_v)\ and\ y\in \{a_1, a_2, \ldots , a_v\} \wedge f^{-1}\in F\backslash E\}$

consists of all encryption trees which can be generated from t using functions of F and is called *generation set of t upon F* (The length of concatenated words is considered to be static, of course).

The sets $N_i.E_{leg} \subseteq E$, $N_i.D_{leg} \subseteq \{f^{-1}|f \in E\}$ and $N_i.T_{leg} \subseteq T(E, W)$ define which encryption functions, decryption functions and encryption trees a node N_i can *legally* know (from initialization or from message reception according to the protocol specification). The set $A_i := G(N_i.T_{leg}, N_i.E_{leg} \cup N_i.D_{leg})$ consists of all encryption trees which N_i can yield by using legal information only.

The sets $N_i.E_{ill} \subseteq E$, $N_i.D_{ill} \subseteq \{f^{-1}|f \in E\}$ and $N_i.T_{ill} \subseteq T$, on the other hand, define encryption functions, decryption functions and encryption trees which are said to be *illegal* for N_i. The malicious use of this information can possibly result in undetectable faults.

Of course, only elements of the former sets are necessary to act according to the protocol specification. Elements of the latter sets are principally not provided.

For example, let $E = \{f_1, f_2\}$, $W = \{w_1, w_2, w_3\}$, $N_i.E_{leg} = \{f_1\}$, $N_i.D_{leg} = \{f_1^{-1}\}$, $N_i.D_{ill} = \{f_2^{-1}\}$, and $N_i.T_{leg} = \{ f_1(w_1 \oplus f_2(w_2)) \}$. Now a faulty node N_i receiving $f_1(w_1 \oplus f_2(w_2))$ may form an infinite number of values $x_1, x_2, \ldots$ thereof. Some example values are given in the following:

Apply f_1:	$x_1 = f_1(f_1(w_1 \approx f_2(w_2)))$
Apply f_1^{-1}:	$x_2 = w_1 \approx f_2(w_2)$
Concatenate:	$x_3 = f_1(w_1 \approx f_2(w_2)) \approx w_3$
Separate x_2:	$x_4 = w_1$
Separate x_2:	$x_5 = f_2(w_2)$
Apply f_1 to x_4:	$x_6 = f_1(w_1)$

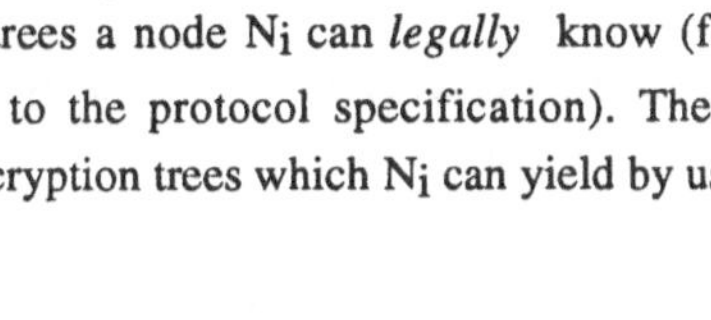

Fig. 2   Trees of the first three example values

However, N_i is never able to form w_2 because w_2 is protected by $f_2(w_2)$ and N_i does not know $f_2^{-1} \in N_i.D_{ill}$.

In all practically relevant cryptographic systems the sets of <u>legal</u> information of some node N_i intersect with the sets of <u>illegal</u> information of another node N_j. Consequently, if N_i and N_j are faulty than N_i can communicate illegal information to N_j or vice versa. Note, that the mere exchange of this illegal information not inevitably leads to undetectable faults. Both, the exchange of illegal information **and** the malicious use of it must coincide. A faulty node N_i uses illegal information maliciously if it generates encryption trees

$$t \in G(N_i.T_{leg} \cup N_i.T_{ill}, \ N_i.E_{leg} \cup N_i.D_{leg} \cup N_i.E_{ill} \cup N_i.D_{ill}) \setminus A_i.$$

> The failure model supposes that a faulty node N_i may behave arbitrary except that it never sends an encryption tree
>
> $$t \in G(N_i.T_{leg} \cup N_i.T_{ill}, \ N_i.E_{leg} \cup N_i.D_{leg} \cup N_i.E_{ill} \cup N_i.D_{ill}) \setminus A_i.$$

Of course, the smaller the sets $N_i.E_{ill}$, $N_i.D_{ill}$ and $N_i.T_{ill}$ can be made the more conservative the failure model is.

2.3 System view and principal accuracy limitations

The specification of the network membership protocol requires that each present manager node C_i obtains a *view* $\mathcal{V}_i = (\mathcal{N}_i, \mathcal{L}_i)$ of the whole currently present system. The view is defined as follows:

V1 The set $\mathcal{N}_i$ of nodes is a subset of the set of existing nodes: $\mathcal{N}_i \subseteq \mathcal{N}$

V2 In the case the set $\mathcal{L}_i$ of links contains non-existing links, they form $f_\mathcal{N}$ additional disjoint paths between any pair of faultless nodes at the most:

$$\forall \ N_j, N_k \in \mathcal{N}_i \cap \mathcal{N}^c, \quad \forall \ \Pi \in P^*(\mathcal{N}_i, \mathcal{L}_i, N_j, N_k) : \ \left| \Pi \setminus P(\mathcal{N}, \mathcal{L}, N_j, N_k) \right| \leq f_\mathcal{N}$$

V3 If no fault occurs during the protocol execution view V_i contains all nodes and links:

$$\mathcal{N}^f = \varnothing \ \wedge \ \mathcal{L}^f = \varnothing \quad \Rightarrow \quad \mathcal{N}_i = \mathcal{N} \ \wedge \ \mathcal{L}_i = \mathcal{L}$$

Obviously, *V1* to *V3* allow certain deviations of a view from reality. However, all of these deviations are unavoidable in a fault-prone system. Faultless nodes completely isolated by faulty ones are not guaranteed to be noticed. Non-existing links appear, if two faulty nodes can consistently "pretend" that they are linked. However, in the appendix it is shown that no more than $f_\mathcal{N}$ additional disjoint paths between two faultless nodes can be formed. On the other hand, up to $F_\mathcal{N}$ paths may contain faulty nodes. Since $f_\mathcal{N} \leq F_\mathcal{N}$ only the latter limitation is relevant in *V2*. From p disjoint paths in its view a manager node can conclude the existence of $p - F_\mathcal{N}$ faultless disjoint paths in the real system. Refer to figure 3 for an example.

Due to different distances and hence different communication delays between nodes the manager nodes obtain their view step by step. In the absence of faults the network membership protocol must terminate (with identical and correct views for all manager nodes) after the time for crossing the whole network has elapsed two times. The network membership protocol must work properly for both asynchronous and

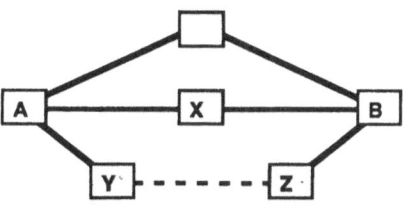

Fig. 3   Faulty node X in path A–X–B. Faulty nodes Y and Z pretend a non-existing link Y–Z (dashed line) reporting a non-existing disjoint path A–Y–Z–B.

synchronous networks. In the latter case termination is guaranteed after $2 \cdot d \cdot \Delta$, where d is the diameter of the network and Δ is an upper bound for a single message delay $\delta_{ij} \le \Delta$ between direct neighbour nodes N_i and N_j (including the corresponding local computations in N_i and N_j).

In the presence of faults the protocol may be aborted due to wrong behaviour of faulty nodes. The view reported to the manager nodes may be empty or violating *V1* to *V3*. However, only fault detection is required according to *R1*: The protocol terminates successfully after the initiating configuration manager states that the reported view satisfies a *termination criterion* depending on the goal to be achieved. E. g. if the goal is to allocate nodes and links for a TMR process system with one spare node for reconfiguration the termination criterion would require the resulting view to contain at least four nodes and at least two disjoint paths between any pair of the four nodes. Often, the termination criterion will be much more complex, of course. However, it is usually not as difficult as finding subgraphs, which is NP-complete, because a constrained amount of additional intermediate nodes can be tolerated.

If the termination criterion is not satisfied, the manager nodes remain waiting. Hence they do not start fault-tolerant application processes. Waiting eventually ends after the system is found without wrong behaviour in a subsequent run of the protocol (after maintainance or automatic passivation of faulty nodes).

A possible deviation in the views of different manager nodes cannot be avoided. A faulty node x may "hide" a further node y first and then propagate its existence at an arbitrary point in time. Due to arbitrary message delays different manager nodes will be aware of y at different times. The network membership protocol itself does not provide countermeasures against this type of inconsistency. Instead, the configuration manager has the choice between the following two:

• For special fault-tolerant application even inconsistent views may be sufficient provided that the intersection of the views contains the redundancy required by the given fault-tolerant application. Note that, depending on F_N and F_L, a manager node from its own view can determine a subset of the intersection of the views of the faultless manager nodes.

• For general fault-tolerant applications the manager nodes examine the portions of the view they receive step by step. As soon as they detect that the node and link redundancy enables an agreement protocol, they reach agreement on (sufficiently) identical views by the following protocols: Unless synchronized clocks are available an initial clock synchronization algorithm [15] is executed first (synchronous network presupposed). Then, one of various types of agreement protocols can be chosen to reach consensus upon the views. Even the fault detecting part of a distance agreement protocol [11, 12] with lowest communication expense is sufficient. In this approach the message number is kept minimum by comparing views (or signatures thereof) relatively to each other during protocol execution.

3 A partial solution of the network membership problem

This section outlines a simple network membership protocol for the fail-silent fault model (F4 in section 1). It will be extended for a more general fault model in section 4. The given protocol is based on the echo algorithm developed by Chang [5]. Basically, an echo algorithm distributes an inquiry from an initiator node to all other nodes in a network of unknown topology by flooding explorer messages. Each node receiving an explorer replies by sending an echo message on the reverse route. Chang shows that his algorithm is optimal in both message number and execution time.

The protocol NM0 is given as a collection of processes in a Pascal-like pseudo programming language. The *initiator* of the protocol executes the process Explore_NM0 while the other nodes execute the processes HandleExplorer_NM0 and HandleEcho_NM0. A message is specified as a tuple of values enclosed in square brackets. The first value specifies the message type, further values represent the data. All Send-operators are non-waiting and Receive-operators are blocking. The first parameter is the link on which the message is sent or received. The second parameter specifies the message to be sent or to be received. This enables selective receive by prespecifying the contents of message fields (especially the type field). '*' is used as a wildcard.

Process of initiator node N_i:

```
1   process Explore_NM0:
2   var  Link : LinkType,  L : ListType,
3            PathSet : set of ListType;
4   SendViaAllLinks([Explorer, [N_i]]);
5   PathSet := ∅;
6   repeat
7       Receive(Link, [*, L]);
8       PathSet := PathSet ∪ {L};
9   until TerminationCriterion(PathSet);
10  output PathSet.
```

Processes of an arbitrary node N_j:

```
11   var ReturnLink:     LinkType;

12   process HandleExplorers_NM0:
13   var  Link : LinkType,  L : ListType;
14   loop
15       Receive(Link, [Explorer, L]);
16       if "is first explorer" then
17           ReturnLink = Link;
18           SendViaAllExcept(Link,
                            [Explorer, L≈N_j]);
19       endif;
20       Send(Link, [Echo, L⊕N_j]);
21   endloop.

22   process HandleEchoes_NM0:
23   var  Link : LinkType,  L : ListType;
24   loop
25           Receive(Link, [Echo, L]);
26           Send(ReturnLink, [Echo, L]);
27   endloop.
```

Two types of messages, explorers and echoes, are used in this protocol. Explorers basically consist of node lists which initially only contain the initiator. The explorers are flooded (lines 4 and 16 – 19). During this process the node lists of the explorers grow because each relaying node adds its identifier. Furthermore an echo is returned to the initiator for each explorer received to carry back the collected node list (line 20). A received echo is just relaid using the link on which the first explorer was received (lines 25 – 26). The initiator continuously receives messages and adds their node list to its system view. The protocol terminates when the system view satisfies the termination criterion described in section 2.3.

Although this solution has no fault-detection mechanisms at all fail-stop behaviour is tolerated because this leads to the loss of topology information at most. However, this protocol is vulnerable to all types of more complicated faults like omitting to add an identifier, adding several identifiers (not necessariliy the own) or even changing a received node list. In the following extended solution faults of these kinds will be detectable.

4 A network membership protocol

Before describing our complete network membership protocol the fault model given in section 2.2 is specialised by defining the sets of legal and illegal functions and encryption trees. Our protocol uses a cryptographic system (E, W) as defined in section 2.2 with $E := \{f, g, h\} \gg S$ and $W := N$ to achieve fault detection. The following table presents the specialised fault model (R_i denotes the set of encryption trees which N_i receives during the current protocol execution according to its specification):

	$N_i \notin C$	$N_i \in C$	N_i is initiator
E_{leg}	$\{f, g, h\}$	$\{f, s_i, h\}$	$\{f, s_i, h\}$
D_{leg}	$\varnothing$	$\{f^{-1}, g^{-1}\} \cup S^*$	$\{f^{-1}, g^{-1}, h^{-1}\} \cup S^*$
T_{leg}	$\{N_i\} \cup f(\mathcal{N}) \cup R_i$	$\mathcal{N} \cup f(\mathcal{N}) \cup R_i$	$\mathcal{N} \cup f(\mathcal{N}) \cup R_i$
E_{ill}	S	$\{g\} \cup S\backslash\{s_i\}$	$\{g\} \cup S\backslash\{s_i\}$
D_{ill}	$\{f^{-1}, g^{-1}, h^{-1}\} \cup S^*$	$\{h^{-1}\}$	$\varnothing$
T_{ill}	$\mathcal{N}\backslash\{N_i\}$	$\varnothing$	$\varnothing$

The network membership protocol of section 3 is vulnerable to the following three fault cases: A faulty node N_i can
1 undetectably change a node list,
2 omit to add the node identifier,
3 add node identifiers other than its own.

The following countermeasures are appropriate for these fault cases:
1 A node list is protected by a special block check character C which reflects the number and order of the list entries. The block check character cannot be undetectably recomputed or changed.
2 The successor checks for the entry of the predecessor. For this purpose a node N_i must know the identity of its neighbours $N_i.\mathcal{N}$.
3 The validity of node identifiers is provable by adding information redundancy. Identifiers are never sent in plain text. Such, a node $N_i \notin C$ only knows its own identifier and cannot generate other valid identifiers.

Each node $N_i \in C$ gets to know identifiers of other nodes in plain text (from protocol executions of its own). Therefore, normal nodes need to have a different entry scheme for their node identifiers which cannot be forged by N_i.

The last case shows that configuration managers and normal nodes have to be distinguished carefully. To integrate these countermeasures consistently an explorer $M := [explorer, h, L, C]$ now consists of four message fields where the first field again specifies the message type, h is a so called entry protection function, L carries the node list and C is the mentioned block check character. The entry schemes for the two types of nodes are as follows:

$$N_i \notin C\colon M' := [explorer, \; h, \; L \oplus f(N_i), \; h(g(N_i \oplus C))] \tag{4.1}$$

$$N_i \in C\colon M' := [explorer, \; h, \; L \oplus f(N_i), \; h(s_i(C))] \tag{4.2}$$

The function f is to encrypt the secret node identifiers. Of course, f^{-1} is only known to configuration managers. The function h, which is chosen by the initiator of the protocol and distributed with the explorer messages, serves to encrypt the block check character C. Consequently, h^{-1} is only known to the initiator itself. Note, that h is specific for the current protocol execution and can also serve as a protocol identifier. Hence, C cannot be undetectably changed. Also, it cannot be recomputed because secret information of other nodes would be needed. The function g is not available to configuration managers. Therefore, the entry scheme for the normal nodes cannot be forged. The inverse g^{-1} can be available to all nodes but is only needed by configuration managers. Configuration managers use their signature function in their entry scheme. Note, that for checking the entry of a predecessor N_i only $f(N_i)$ needs to be known rather than the plain text identifier N_i.

	$N_i \notin C$	$N_i \in C$	N_i is initiator
initial knowledge	$f, g,$ $N_i, f(N_i.\mathcal{N})$	$f, f^{-1}, g^{-1}, s_i, s^*,$ probably[1] $\mathcal{N}, f(\mathcal{N})$	$f, f^{-1}, g^{-1}, s_i, s^*,$ $h, h^{-1},$ probably[1] $\mathcal{N}, f(\mathcal{N})$
knowledge gained by message reception	$h,$ $f(\mathcal{N})$	$h,$ $\mathcal{N}, f(\mathcal{N})$	$\mathcal{N}, f(\mathcal{N})$

[1] This information can originate from another protocol execution.

The following procedures can be implemented using the above functions:

MakeEntry([type, h, L, C])	enters the own node identifier into a node list L and the corresponding block check character C. Obviously, this procedure is different for normal nodes and configuration managers. It uses f and g or sigi, respectively.
MakeEcho([type, h, L, C])	changes an explorer into an echo by setting the appropriate type and once again applying h to C to prevent further manipulation.
CorrectNB(L, i)	checks for a message received via link i if the last entry in L matches the correct neighbour. It uses $f(N_i.\mathcal{N})$.
Consistent([type, h, L, C])	checks if all contained node identifiers are valid (using the information redundancy) and if L and C match. If type is echo the additional application of h on C is also checked. It uses f^{-1}, g^{-1}, h^{-1} and $\mathcal{S}^*$.

The fault-detecting network membership protocol NM1 is based on the protocol presented in section 3 but additionally uses the above procedures. The following additional lines are needed to change NM0 into NM1:

7.1 **if** "explorer" ∧ ¬ CorrectNB(L, Link) **or** Consistent([type, h, L, C]) **then**	15.1 **if** ¬ CorrectNB(L, Link) **continue**; 15.2 MakeEntry([Explorer, h, L, C]);
7.2 **output** "Failure"; **terminate**;	...
7.3 **endif**;	19.1 MakeEcho([Explorer, h, L, C]);

The initiator checks the consistency of all received messages and the validity of all received node identifiers. In case of an error the protocol is aborted (alternatively one can omit the erroneous message and proceed with the protocol execution). On reception of an explorer the other nodes check for the entry of the predecessor. In case of an error the message is omitted. The process HandleEchoes is not changed except for the different message format.

Figure 4 illustrates an example execution of the network membership protocol. Explorers are depicted by grey arcs and echoes by hatched ones. The encryption trees of the block check character of each message are given explicitly. The initator N_1 has just started the protocol. Only N_2 is already explored and has sent an echo back to the initiator. Hence, its current topology view $\mathcal{V}_1$ consists of the nodes N_1 and N_2 and the link (N_1, N_2).

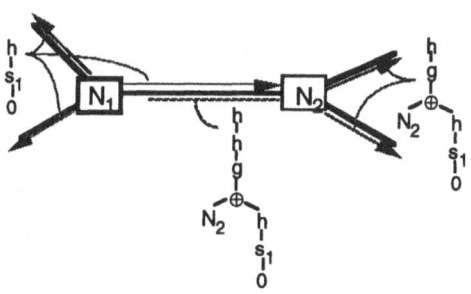

Fig. 4   Two nodes of a sample network with initiator N_1 just having started a protocol execution.

A sketch of the correctness proof is deferred to the appendix.

In the above solution for each received explorer one echo is generated and the length of the carried node list grows by one. Several optimizations thereof are possible. Generally, one is faced with a tradeoff between saving echoes, i. e. communication costs, and the robustness of the protocol against the loss of topology information. Simulation experiments were carried out to measure the efficiency improvements achieved by several optimized variants. We are not able to discuss these issues in detail due to space limitations.

5   Conclusion

The typical application area of a network membership protocol is a large and general distributed system with an unknown current topology, where fault-tolerant process systems must be allocated and started and, moreover, spare nodes and/or links are required from the beginning on for the purpose of any type of reconfiguration. The termination criterion of the network membership protocol defines the point in time when fault-tolerant operation can be guaranteed.

There are many startup cases where the network membership problem does not arise: Pure static redundancy may entirely exclude reconfiguration, or a reliable monitoring system is provided for a system where the topology is known in advance. However, in many existing systems, as we suppose, startup is simply based on the assumption that sufficient resources are present after switching them on. Then, the fault-tolerant process systems are started and their correct behaviour is proven by the operator. Confidence in this approach may be justified for small systems easy to survey. For large and complex systems not managed centrally, the network membership enables fully automated startup of fault-tolerant process systems without any assumptions about the presence of components or even the current topology. Thereby, the lots of large existing networks not designed with respect to dependability requirements are opened for (partly) fault-tolerant use – the termination criterion decides whether and where.

Since the network membership protocol is taken as the initial stage of various fault tolerance techniques, its own fault model comprises any number of faults and nearly any type of malicious behaviour (except the malicious cooperation among at least two faulty nodes).

Future research in this field may be directed to the following problems:
* Ring or bus links should be included in an efficient way.
* In extremely large networks the exploration should be aborted after a sufficient part of the network has been covered with respect to the termination criterion.

Remark of the authors. The crytographic functions employed in this protocols are tailored to their specific needs and partition the set of nodes in two classes: senders (normal nodes) and receivers (manager nodes). Meanwhile, we developed a new authentication concept called relative signatures [17] applicable within an a-priori unknown set of nodes. This general approach eliminates the need for partitioning the set of nodes .

References

[1]   G. Alari, A. Ciuffoletti: Group membership in a synchronous distributed system; *Fifth European workshop on dependable computing, EWDC-5*, Lissabon, 1993.

[2]   A. Bagchi, S. L. Hakimi: An optimal algorithm for distributed system level diagnosis; *FTCS-21 digest of papers*, IEEE, 1991, pp. 214 – 221.

[3]   R. Bianchini, R. Buskens: An adaptive distributed system-level diagnosis algorithm and its implementation; *FTCS-21, digest of papers*, IEEE, 1991, pp. 222 – 229.

[4]   F. Cristian, H. Aghili, R. Strong, D. Dolev: Atomic broadcast: from simple message diffusion to byzantine agreement; *FTCS-15, digest of papers*, IEEE, 1985, pp. 200 – 206.

[5] E. Chang: Echo algorithms: depth parallel operations on general graphs; *IEEE Transactions on software engineering*, vol. 8, no. 4, 1982, pp. 391 – 401.

[6] F. Cristian: Agreeing who is present and who is absent in a synchronous distributed system; *FTCS-18, digest of papers*, IEEE, 1988, pp. 206 – 211

[7] F. Cristian: Reaching agreement on processor-group membership in synchronous distributed systems; *Distributed Computing*, vol. 4, Springer, Heidelberg, 1991, pp. 175 – 187

[8] A. Dahbura, K. Sabnani, L. King: The comparison approach to multiprocessor fault diagnosis; *Transactions on Computers*, IEEE, vol. C-36, no. 3, march 1987, pp. 373 – 378.

[9] K. Echtle: Fehlermodellierung bei Simulation und Verifikation von Fehlertoleranz-Algorithmen für verteilte Systeme; *Informatik-Fachberichte 83*, Springer, Heidelberg, 1984, pp. 73 – 88.

[10] K. Echtle: Fault-masking with reduced redundant communication; *FTCS-16, digest of papers*, IEEE, 1986, pp. 178 – 183.

[11] K. Echtle: Fault masking and sequence agreement by a voting protocol with low message number; *6th symposium on reliability in distributed software and database systems, conf. proc.*, IEEE, 1987, pp. 149 – 160.

[12] K. Echtle: Distance agreement protocols; *FTCS-19, digest of papers*, IEEE, 1989, pp. 191 – 198.

[13] S. Goldwasser, S. Micali, R. Rivest: A digital signature scheme secure against adaptive chosen-message attacks; *SIAM J. Comput.*, vol. 17, no. 2, 1988, pp 281 – 308.

[14] H. Kopetz, G. Grünsteidl, J. Reisinger: Fault-tolerant membership service in a synchronous distributed real-time system; *Int. working conference on dependable computing for critical applications, conf. preprints*, UCSB, 1989, pp. 167 – 174.

[15] R. Kieckhafer, C. Walter, A. Finn, P. Thambidurai: The MAFT architecture for distributed fault tolerance; *IEEE Transactions on computers*, vol. 37, no. 4, 1988, pp. 398 – 405.

[16] J.-C. Laprie, J. Arliat, C. Beounes, K. Kanoun, C. Hourtolle: Hardware- and software-fault tolerance: definition and analysis of architectural solutions; *FTCS-17, digest of papers*, IEEE, 1987, pp. 116 – 121.

[17] M. Leu: Relative signatures and their implementation; *accepted paper for the First European Dependable Computing Conference EDCC-1*, Berlin, oct. 1994.

[18] J. Maeng, M. Malek: A comparison connection assignment for diagnosis of multiprocessor systems; *FTCS-11, digest of papers*, IEEE, 1981, pp. 173 – 175.

[19] E. Maehle, K. Moritzen, K. Wirl: A graph model for diagnosis and reconfiguration and its application to a fault-tolerant multiprocessor system; *FTCS-16, digest of papers*, IEEE, 1986, pp. 292 – 297.

[20] F. Preparata, G. Metze, R. Chien: On the connection assignment problem of diagnosable systems; *IEEE Transactions on Computers*, vol. 16, Dec. 1967, pp. 848 – 854.

[21] D. Powell: Error assumptions and evaluation of their influence on system dependability; in: *Second European workshop on dependable computing, EWDC-2*, Firenze, 1990.

[22] R. Rivest, A. Shamir, L. Adleman: A method for obtaining digital signatures and public-key cryptosystems; *Communications of the ACM*, vol. 21, no. 2, acm, 1978, pp. 120 – 126.

[23] A. Sengupta, A. Dahbura: On self-diagnosable multiprocessor systems: diagnosis by the comparison approach; *FTCS-19, digest of papers*, IEEE, 1989, pp. 54 – 61.

[24] A. Segall: Distributed network protocols; *IEEE Transactions on Information Theory*, vol. 29, no. 1, 1983, pp. 23 – 35.

[25] E. Schmeichel, S. L. Hakimi, M. Otsuka, G. Sullivan: On minimizing testing rounds for fault identification; *FTCS-18, digest of papers*, IEEE, 1988, pp. 266 – 271.

[26] H. Strong, D. Dolev: Byzantine agreement; *Compcon 83, conf. proc.*, IEEE, 1983, pp. 77 – 81.

[27] N. Vaidya, D. Pradhan: System level diagnosis: combining detection and location; *FTCS-21, digest of papers*, IEEE, 1991, pp. 488 – 495.5.

Appendix: Verification

Let $P(\mathcal{N}, L, N_i, N_j) \subseteq L$ denote the set of all acyclic paths between nodes $N_i \in \mathcal{N}$ and $N_j \in \mathcal{N}$ via links of L. A pair of paths $(p_q, p_r) \in P(\mathcal{N}, L, N_i, N_j) \times P(\mathcal{N}, L, N_i, N_j)$ is called *disjoint*, iff p_q and p_r do not share any nodes other than N_i and N_j. A subset $P_{dis} \subseteq P(\mathcal{N}, L, N_i, N_j)$ is called *path-disjoint* iff any pair of paths of P_{dis} is disjoint. A path-disjoint subset $P_{opt} \subseteq P(\mathcal{N}, L, N_i, N_j)$ with maximum cardinality $|P_{opt}|$ is called *path-optimal*. The set of all path-optimal subsets of $P(\mathcal{N}, L, N_i, N_j)$ is denoted by $P^*(\mathcal{N}, L, N_i, N_j)$.

The properties *V1*, *V2* and *V3* (section 2.3) of NM1 can be proven as follows:

Let $t \in T(E, W)$ be an encryption tree with $t = f(\ldots)$, $f \in E$ arbitrary. Any operation on t other than

- $f^{-1}(t)$

- $f'(t_1 \oplus t_2 \oplus \ldots \oplus t \oplus \ldots \oplus t_n)$, $t_i \in T(E, W)$, $f' \in E$

is called *random falsification*.

<u>Lemma 1:</u> A random falsification of a block check character is detectable.

<u>Proof:</u> An encryption function is a one-to-one function. Let y be a random falsification of f(x). Then $f^{-1}(y) \neq x$ holds and $f^{-1}(y)$ is a random falsification of x. Furthermore, a concatenation of words is randomly falsified iff at least one "part" is randomly falsified. A random falsification of a node identifier is detectable because of its information redundancy.

Because a block check character is an encryption tree with only node identifiers as leaves, Lemma 1 can be shown via structural induction using the above properties of encryption functions and the concatenation operation. ◆

<u>Lemma 2:</u> A node N_i cannot undetectably enter a node identifier N_x with $N_i \neq N_x$ to a received node list.

<u>Proof:</u> To undetectably enter a wrong identifier the resulting node list must be consistent. Lemma 1 requires all entries in the node list to be made according to either (4.1) or (4.2). Trivially, all incorporated node identifiers have to be valid.

<u>Case 1. $N_i \notin C$:</u> Then $N_x \in N_i.T_{ill}$. According to the fault model, N_i cannot have generated an encryption tree using N_x. Hence, a contradiction.

<u>Case 2. $N_i \in C$:</u> Then $g \in N_i.E_{ill}$. It follows that N_i cannot have added N_x to L using (4.1). This implies that (4.2) was used with signature function s_i because $s_j \in N_i.E_{ill}$ for $s_j \neq s_i$. This obviously would have been detected by the initiator. Hence, a contradiction. ◆

<u>Lemma 3:</u> A node N_i cannot undetectably modify the node list of an echo message.

<u>Proof:</u> The block check character of an echo message is protected by a two-fold application of the entry protection function h. This protection cannot be removed by any node N_i except the initiator because $h^{-1} \in N_i.D_{ill}$. Hence, any operation on a such protected block check character will be detectable by the initiator. ◆

The following theorems directly refer to the definition of the resulting system view $\mathcal{V}_i = (\mathcal{N}_i, \mathcal{L}_i)$ as given in *V1*, *V2* and *V3*.

Theorem V1: $\mathcal{N}_i \subseteq N$.

<u>Proof:</u> This follows immediately since the validity of node identifiers is provable. ◆

Theorem V2: $\forall\ N_j, N_k \in \mathcal{N}_i \cap \mathcal{N}^c,\ \forall\ P \in P^*(\mathcal{N}_i, \mathcal{L}_i, N_j, N_k)$:

$$\left| \Pi \setminus P(\mathcal{N}, \mathcal{L}, N_j, N_k) \right| \le f_{\mathcal{N}}$$

Proof: Let $N_j, N_k \in \mathcal{N}_i \cap \mathcal{N}^c$ be two arbitrary faultless nodes of $\mathcal{N}_i$ and let $\Pi \in P^*(N_i, L_i, N_j, N_k)$ be an arbitrary path-optimal subset of $P(\mathcal{N}_i, \mathcal{L}_i, N_j, N_k)$. Then for all $p \in \Pi \backslash P(\mathcal{N}, \mathcal{L}, N_j, N_k)$ the following holds:

$\quad \exists (N_\mu, N_\nu) \in p: \ (N_\mu, N_\nu) \notin \mathcal{L} \wedge (N_\mu, N_\nu) \in \mathcal{L}_i$

This implies that at least one of the received node lists contained either (N_μ, N_ν) or (N_ν, N_μ). Without loss of generality, say (N_μ, N_ν). Lemma 1 implies that the received node list was not randomly falsified. Lemma 2 guarantees that the message M which carried the information actually passed N_m and N_n. Because $(N_\mu, N_\nu) \notin \mathcal{L}$ M had to pass at least one intermediate node N_x to get from N_μ to N_ν. See figure 5 for an illustration. This implies that N_x behaved faulty because it did not add its identifier to the node list of M. From Lemma 3 follows that M was an explorer message when it reached N_ν because N_ν added its identifier successfully. This implies that N_ν behaved faulty because it did not check that the identifier of N_x was missing in the received node list. Note that $N_\nu \ne N_j$ and $N_\nu \ne N_k$ because $N_j, N_k \in \mathcal{N}^c$ and therefore is not part of any other path in $\Pi \backslash P(\mathcal{N}, \mathcal{L}, N_j, N_k)$ because $\Pi \backslash P(\mathcal{N}, \mathcal{L}, N_j, N_k)$ is path-disjoint. It follows that for each element of $\Pi \backslash P(\mathcal{V}, N_j, N_k)$ there must be at least one faulty node. ®

Figure 5 shows how the message M which was sent out (as an explorer) by the initiator N_i travelled through the network. It actually passed N_μ and N_ν before being returned to N_i (as an echo) because both N_μ and N_ν made correct entries in the respective node list. Since there is no direct link between N_μ and N_ν M had to pass at least one intermediate node N_x.

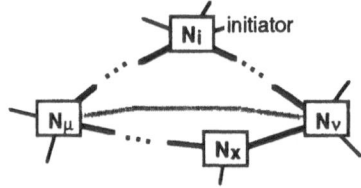

Fig. 5   The bold lines indicate the path M travelled. Because both, N_x and N_n, behaved faulty the node list of M is
$L = [f(N_i),\dots,f(N_m),f(N_n),\dots]$, pretending a link between N_m and N_n.

Theorem V3: $\quad \mathcal{N}^f = \varnothing \wedge \mathcal{L}^f = \varnothing$
$\quad \Rightarrow \quad \mathcal{N}_i = \mathcal{N} \quad \mathcal{L}_i = \mathcal{L}$

The proof of this theorem can be easily obtained by structural induction.

Secure Systems

Denial of Service: A Perspective

Jonathan K. Millen
The MITRE Corporation
Bedford, Massachusetts 01730
U.S.A.

Abstract

The scope of "denial-of-service protection" can be limited by comparing it and contrasting it with related concepts. The objectives and general concepts that drive current research have already been clarified to some extent by prior work. We summarize the general conclusions that have emerged, and assess their implications for the development of denial-of-service protection requirements and the guidance of future research.

1   Introduction

The three traditional concerns of computer security are confidentiality, integrity, and denial of service. Confidentiality and integrity have been addressed with fair success by designing operating systems that enforce various access-control models, together with trusted or semi-trusted software that provides special services and depends on the operating system to preserve its integrity. There are, as yet, no systems claiming to provide denial-of-service protection.

One possible factor contributing to the absence of systems with denial-of-service protection is the lack of a common understanding as to what "denial-of-service protection" means. This need not be an insurmountable obstacle, since there still does not exist complete agreement as to what "confidentiality" and "integrity" mean. A multitude of alternative access control policies have been proposed, and recent work has emphasized a desire for flexibility in accommodating users' needs [1]. In the area of label- based security controls, the intelligence community has fostered development of compartmented-mode workstations with floating "information labels" that represent control markings beyond the TCSEC levels [21]. There are proposals for more sophisticated discretionary policies, such as [20]. The needs of the commercial environment spawned the Clark-Wilson model for integrity [7] and the Chinese Wall policy [6].

Despite the intellectual turmoil in these areas, computer systems now exist that provide useful security services at varying assurance levels. The advances have been due to a willingness to formulate well-defined models of desirable features, and implement them. The strongest mechanisms and highest assurances are associated with features whose implementation relies to the greatest extent on hardware, but there has remained a need for trusted software to provide adequate user interfaces and application-dependent services.

To make similar progress in denial-of-service protection, we must begin by identifying and delineating denial-of-service protection objectives that are within the scope of computer security, and which are implementable. Models are needed, both for the purpose of guiding system design, and also for analyzing designs to prove that they have desirable properties. Although a 1985 workshop concluded that "No generic denial of service conditions could be identified which were independent of mission objectives," a research note on the subject had already identified shared resource allocation as at least one fundamental objective characteristic of denial-of-service protection [9]. Subsequent modelling work has continued to stress shared resource allocation as a central problem [22, 10, 19].

In this paper we wish to step back from detailed modelling issues, and focus on the objectives and general concepts that drive current research, and which have already been clarified to some extent by the prior research. We will try to limit the scope of what "denial-of-service protection" means by comparing it and contrasting it with related concepts. We will summarize the general conclusions that have come out of prior work, and assess their implications for the development of denial-of-service protection requirements and the guidance of future research.

2 Basic Ideas

2.1 Definitions

The following definitions are fundamental to subsequent discussion. They are based on concepts developed in the denial-of-service modelling study [19]:

- *Denial of Service:* failure of a guaranteed service due to malicious invocation.

- *Service:* an interface to computer system or computer network resources. It receives and satisfies requests to access a resource or related set of resources.

- *Denial-of-Service Protection Base (DPB):* a resource monitor satisfying a given waiting-time policy under specified user agreements.

- *Resource Monitor:* a collection of services with a common interface mechanism.

- *Waiting time policy:* a guarantee limiting the real time required to satisfy service requests.

- *User Agreement:* a constraint on service requests imposed to enable service guarantees.

- *Benign Process:* a process that observes user agreements.

- *Malicious Process:* a process that violates user agreements.

2.2 The Denial-of-Service Protection Base

The root definition for denial-of-service protection is that of a Denial-of-service Protection Base, or DPB. The difference between a DPB and a resource monitor is simply that a resource monitor may or may not provide denial-of-service protection. If it does, and the nature of the protection is specified, then it is a DPB.

A DPB is analogous to a Trusted Computing Base, or TCB. The objective of both is to define a perimeter separating the trusted software (and hardware) inside the perimeter from user software outside the perimeter that might be malicious. Some policy—a waiting time policy, for a DPB, or an access-control policy, for a TCB—is enforced at the perimeter, which is an interface to system services. In order to be successful at enforcing the policy, a DPB or TCB must be protected from unauthorized modification. Also, there should be no way to bypass it, to access or interfere with resources it controls. These are the properties of a reference monitor, and other authors have observed that a similar approach is appropriate for denial-of-service protection [2].

As noted in [19], if a DPB is possible at all, it can exist in a computer system that does not have a TCB providing confidentiality. On the other hand, if there is a TCB for a particular system, a DPB could be combined with it, to take advantage of the TCB's mechanism for protecting itself and other data from unauthorized modification. A DPB is also subject to access controls imposed by a TCB. The DPB cannot and need not provide unauthorized access.

If there is both a TCB and a DPB on the same system, it is natural to ask what the structural relationship is between them. One conclusion follows from the fact that the TCB maintains ultimate control over access to most, if not all, essential system resources. A DPB cannot guarantee access to those resources without the cooperation of the TCB. Any guarantee of service by the DPB is made not only on its own behalf but also on behalf of the TCB. In that sense, the TCB must be regarded as part of the DPB. The TCB might have to be reexamined and even redesigned in order to ensure that it supports service guarantees.

The scope of DPB protection consists of whatever resources are under its control. It is obvious that *the only services whose availability can be guaranteed by a DPB are those provided by the DPB itself.* Hence the DPB must offer those services whose loss would be viewed as a denial-of-service problem.

A service guarantee is really a contract between the DPB and its users. It has three parts, which will be discussed below:

- A service specification
- A waiting-time policy
- User agreements

2.3   Service Specifications

Denial of service has been defined as failure of a guaranteed service due to malicious invocation. There are a number of qualifications in that definition that should be clearly understood. First, a "service" has been defined as an interface to computer system or computer network resources. In a broader setting, this term would have to be qualified as a "computer system processing service" or something similar.

We must assume that there is some precise specification of exactly what the service is supposed to do, in the sense of system functionality. If the specification is fuzzy, there can be disagreement over whether a given behavior constitutes denial of service.

We have said that service is access to a resource, but this leaves open the question as to what the resource is, and what constitutes access to it. We are concerned only with shared resources, since there is no way for one user to deny another user access to a private resource. There are many possible shared resources that deserve protection. An operating system controls access to memory, devices, and processor time. Network software controls access to connections, virtual circuits, and related facilities.

All resources are abstractions, characterized by the language used to refer to them across the service interface. Some resources, like memory, may seem relatively concrete, but again this depends on the level of the interface. Are we referring to physical memory or virtual memory? Some examples of more abstract resources are information in a database—see, for example, Dobson's approach to expressing this [8]—and message delivery in a network.

2.4   Waiting-Time Policies

We also have to consider what kinds of "guarantees" might be made. It has been noted by many authors that indefinite delays in service constitute denial, so that a waiting time policy is part of any denial-of-service protection guarantee. There are different kinds of waiting-time policies.

Waiting time can be understood in terms of a process that is carried out cooperatively by user software and DPB software. The term "process" should be taken in a general sense that includes not only the usual operating system abstraction that exists within a single CPU, but also any defined sequence of tasks. In a network, we may consider the "process" of communication between parties at different hosts. The waiting time is the interval from the moment a service request is made, to the moment the service provides a response that enables processing to continue. Thus, a process experiences a waiting time with every service request it makes.

While the DPB takes some time to process requests itself, the real problem is the fact that the DPB may decide to schedule other processes before returning control to the process that originated the request. Thus the time used by other processes contributes to each process's waiting time.

It can be nontrivial to define the waiting time for particular services, and the definition will certainly be different for different sorts of services. The denial-of-service modelling study [19] required a fairly elaborate model to define waiting time for a conventional operating system resource allocation context. The literature on real-time systems is also a potential source of insight on how to specify time constraints. The "time-constrained reactive automata" model of Bestavros, for example, pays particular attention to realistic notions of causality and asynchrony, though it does not address resource allocation [3].

Since the detailed definition of waiting time for a service depends on the particular service, it should be left as a research objective to determine appropriate definitions for services of interest. Once a definition of waiting time is given, the next step is to prescribe a policy limiting it.

Two types of policies have been discussed in earlier literature: *maximum waiting time* (MWT), discussed by Gligor [9], and *finite waiting time* (FWT), studied by Yu and Gligor [22] and Haigh, et al [10]. The modelling study [19] also defined a third category of policy, *probabilistic waiting time* (PWT).

A maximum waiting time policy simply provides a specific bound on waiting times. More elaborate MWT policies might specify a number of different bounds that would be applicable in different system states or for different users, depending, perhaps, on their priority. In a real-time system, it can be as much of a problem for events to occur too early as too late. Interference leading to this kind of problem might more appropriately be regarded in the category of service functionality rather than waiting time.

A finite waiting time policy says only that waiting time cannot be infinite, i.e., that service will eventually be granted. It is a weaker property than MWT, since no specific upper bound is guaranteed. The attraction of FWT is that it can be expressed in temporal logic, which offers a conceptually pleasant environment in which to conduct proofs that the policy is satisfied. Also, it is hard to find a system that satisfies FWT without also, in fact, satisfying MWT, though the bound may be very large.

A probabilistic waiting time policy is one that specifies a statistical constraint on waiting time, such as a maximum expected waiting time. The advantage of a PWT policy is that if a system has probabilistic behavior, a MWT might not be possible, and the PWT policy is stronger and more informative than a FWT claim, and perhaps more optimistic than the MWT if it exists. Is it not more reassuring to be told that the average waiting time is one millisecond than to hear that the maximum is forty minutes, or that no maximum is promised? The techniques of stating and proving PWT policies have yet to be investigated, however.

With respect to PWT policies, it is important to keep in mind that ordinary performance analysis differs from denial-of-service needs because user behavior cannot be assumed to follow any otherwise plausible distribution when users may be malicious. Guaranteed probabilistic waiting time behavior arises only from a random element in the DPB resource allocation policy. Randomness may be built into a resource monitor as a way of ensuring fairness. An example of the conscious introduction of randomness into a scheduling policy is given in [12].

2.5   User Agreements

It is not possible, in general, for a DPB to support guaranteed service without some restrictions on that service, which would not be necessary if the resources were not shared. User agreements are constraints on the behavior of service users which must be obeyed in order to prevent denial of service. The notion of user agreement is relatively old, arising in the theory of operating systems. An example is the "ordered resource acquisition" constraint for preventing deadlock. Yu and Gligor called attention to the role of user agreements in denial-of-service protection [22].

What if some users do not obey the user agreements? Those users are called "malicious," and the DPB is not required to serve them, though it still has to provide the guaranteed service to "benign" users, the ones who observe the user agreements. The DPB must also be capable of recognizing malicious usage and revoking resources, if necessary, from malicious users who attempt to hold onto some resource beyond agreed limits.

An example of a user agreement in a simple operating system setting was given in [19]. One can expect that any service providing access to shared resources will require users to forbear requests for excessive quantities of a resource, and to release resources after some specified reasonable time. The DPB can revoke resources, of course, and will if necessary, but the user agreement specification permits users to plan their resource usage in such a way as to avoid disruption.

Networks already have mechanisms for congestion control that require cooperation from hosts, so these concerns should not have as extreme an impact in that context. Nevertheless, the identification and handling of malicious nodes is a new problem.

2.6 Policy Conflicts

It has been suggested that the objective of denial-of-service protection may conflict with confidentiality or integrity protection. One concern, brought up by Dobson, is the question of expectations; a user may feel entitled to more extensive service or better functionality than the system permits [8]. A DPB will impose user agreements and other resource restrictions that will seem burdensome in comparison to systems that do not afford denial-of-service protection. Dobson's remedy is to formalize the guarantees and conditions as a contract between the service provider (the DPB) and the service consumer, using a sufficiently expressive language, suited to the nature of the resource.

An example of a apparent conflict between denial-of-service protection and integrity is the following. Message integrity may be supported by appending a checksum, encrypted or otherwise, to a message. Encryption of the message itself also serves as an integrity protection, at least to the extent of detecting modifications. But a checksum provides a new opportunity to deny service, since modifying bits in the checksum will cause the whole message to be thrown away, while modifying the same number of bits in the message itself may hardly be noticeable due to redundancy in the message. Similarly, a one-bit change in an encrypted message block may be enough to totally garble the decrypted message.

While integrity protection seems to be detrimental to service, the effects cited above are only significant for random errors, not to malicious software attacks. A malicious attack on network service will cause whole messages to be intercepted, not a few bits. Even from the point of view of random errors, the extra vulnerability is not quantitatively significant compared to the level of error handling that networks are routinely equipped to manage, with such techniques as error-correcting codes and higher-layer protocols.

There is some apprehension that denial-of-service protection might conflict with confidentiality because its management of shared resources could introduce covert channels or make it difficult to close existing channels [10]. It is difficult to see what might happen without consideration of specific services and channels, but there are some general observations that suggest that this problem might not be as serious as anticipated. First, while denial-of-service deals with shared resources, the resources are already shared, so the potential for covert channels already exists in the system. Second, the way a DPB manages sharing is by introducing user agreements. User agreements are typically unilateral; they are rules that each individual user process must satisfy, regardless of what other users happen to be doing. They do not require new communication between users. In that sense they tend to isolate users from one another and decrease opportunities for covert channels. Even timing channels, would be lessened, since the influence of a DPB would be to give the users less discretion over how long they could retain a resource.

2.7 Assurance

If the service fails to meet its specification even though all users have acted benignly, the problem is not denial of service, but rather a design or implementation defect that should be addressed by other software engineering approaches.

If a service meets its waiting-time obligation but it is incorrect, and the failure is traceable to some previous malicious request, the problem must be considered a kind of denial of service. No specific policy need be stated to assert that this should not happen, since it is implicit in the specification of the service itself. When evaluating a particular DPB, however, one can still ask whether sufficient attention has been paid to this possibility. It is an assurance concern. If it is a design problem, it could be due to inadequate theoretical analysis of the model. If it is an implementation problem, it has implications for the design of adequate test suites.

Assurance should be distinguished from fault tolerance, the ability of a system to maintain normal service despite random failures in some components. Fault tolerance is a kind of functional specification, which may be implemented with more or less assurance. Fault tolerance has limits; only certain patterns of failure permit recovery. Denial-of- service guarantees do not, by definition, address hardware failures, even though they must be considered as part of "system availability" in the broad sense.

Even though assurance is not a denial-of-service problem per se, there is precedent for imposing assurance requirements on systems that are being evaluated for security properties. The TCSEC has elaborate assurance requirements for higher-class TCB's, and similar requirements were carried over into the ITSEC and would apply to availability functionality. Since assurance is the underpinning for any security guarantees, it is not unreasonable to apply whatever software engineering approaches are believed effective to DPBs.

3   Related Concepts

In order to place the above definitions into perspective, we need to look at some related concepts. Certain terms, such as "reliability" and "availability," which have been used in a denial-of-service context, have been borrowed from a broader system setting. The standard military definitions of these and a few related terms are given below, taken from MIL-STD-721C. Deeper discussions of these terms are given in [15].

3.1 Definitions

- *Availability:* a measure of the degree to which an item is in an operable and commitable state at the start of a mission when the mission is called for at an unknown (random) time.

- *Reliability:* (1) the duration or probability of failure-free performance under stated conditions. (2) the probability that an item can perform its intended function for a specified interval under stated conditions.

- *Dependability:* a measure of the degree to which an item is operable and capable of performing its required function at any (random) time during a specified mission profile, given item availability at the start of the mission.

- *Failure:* the event, or inoperable state, in which any item or part of an item does not, or would not, perform as previously specified.

- *Fault:* Immediate cause of failure (e.g., maladjustment, misalignment, defect, etc.).

3.2 Discussion

One drawback of the term "denial of service" is that it is an undesirable property. When stating requirements and technical goals, it is natural to look for a term expressing some corresponding positive property. Terms such as "availability" are attractive for that reason, but they already have precise technical meaning in related contexts. Merely stating that they mean something different for software does not entirely eliminate the confusion that can result from their connotations.

Sometimes attempts are made to retain the original definitions. For example, availability is often expressed mathematically as:

$$\frac{MTBF}{MTBF + MTTR}$$

where MTBF is "mean time between failures" and MTTR is "mean time to repair." There is a long history to the notion that software failures can be regarded as random, although the probabilities are subjective. Randomness in the time between failures comes from lack of knowledge about how many bugs might remain in the program, plus uncertainty about when a user will happen to supply inputs that trigger a given bug.

There are pitfalls in adopting statistical definitions in a software context, however. Littlewood notes, for example, that MTBF actually does not exist for a program [16]. For, there is some non-zero a priori probability that the program is actually perfect, in which case it will never fail, i.e., the time between failures is infinite. No matter how small the probability of perfection is, the resulting mean time between failures is therefore also infinite. Littlewood's suggestion is to use other measures that still make sense, such as the failure rate.

Caution should be exercised, therefore, in using these terms in a software context. Nevertheless, they are reasonably safe when used in a qualitative sense, keeping in mind that "measures" of them are either not possible or very different from those based on random hardware failures. The difference between "reliability" and "dependability" is that dependability is task-oriented, asking whether the mission will be completed, while reliability is time-oriented, asking whether the system is still running after a given time. The phrase "software reliability" is commonly used in software engineering when studying statistical aspects of software quality, and "software dependability" is often used in connection with techniques to improve correctness, particularly formal methods such as program verification.

"Availability" might make sense if the "mission" is just a service request. Denial of service causes an unacceptable delay in providing the requested service, as though the necessary subroutine were unavailable due to some maintenance problem. For denial of service, the maintenance problem is a malicious attack. Thus, it is probably not unreasonable to consider availability as the opposite of denial of service, as long as the context is clear and no attempts are made to use unjustified probabilistic measures.

The standard definitions of "fault" and "failure" seem to work well for software, given the appropriate categories of faults and failures. Fault tolerance also applies to software, either when software is used to compensate for faults in other parts of the system, or when self-checking software detects and recovers from its own bugs.

In his book on program fault tolerance [18], Mili distinguishes between errors and faults. An error is an observed discrepancy between the expected and actual state of the program, while a fault is a property of the program that leads to an error. An error might not cause a failure, because it could be an implementation detail. That is, the state "expectation" is based on lower-level design details that are not implied by the specification. Mili's approach to fault tolerance is to design software to detect errors as they happen and perform error recovery (repairing the state).

Leveson notes that: "Fault tolerance techniques attempt to ensure robustness, i.e., to deliver a certain minimum of services even when faced with an unexpected or hostile environment. However, fault tolerance techniques usually assume that a correct answer can be obtained—that faults can be tolerated—and do not consider what to do when it cannot, i.e. what happens when the fault tolerance techniques fail or when they successfully implement incorrect specifications." [17]

In connection with a hostile environment, we should also mention "survivability," referring to the ability of a system to remain operative after physical attacks that destroy components. Survivability is to hardware fault tolerance as denial-of-service protection is to software fault tolerance.

Safety is distinguished from reliability by its focus on the human consequences of system failures, as opposed to the mere fact of failure. Leveson says: "In general, reliability attempts to ensure no failures whereas safety attempts to ensure no accidents." [17] A computer failure may or may not result in an accident, just as an error does not necessarily result in a failure.

Whether applying these ideas to hardware or software, or whether statistical measures are attempted, the principal discriminator affecting the approach one takes to remedy or compensate for them is whether the failures are random or due to malicious attack.

4 The Impact of Layered Architectures

A service is an interface, and it seems natural to expect service interfaces to correspond to some of the usual interfaces that exist between the layers of a computer system or network. Since layering is significant architecturally for the design of a system and the relationships among its services, layering is equally significant for the specification and implementation of denial-of-service protection for those services.

Looking inside an operating system, for example, the process management layer is implemented above a memory management layer, making use of it as a service. Process management shares this service with other operating system layers that depend on it, such as file management. In order to provide a maximum waiting time for process management services, there would also have to be a maximum waiting time for memory management services.

We can make a distinction between a DPB interface, which might be exposed to malicious users, and an internal service interface, which will be used only by higher layers within a DPB. It may appear easier to make service guarantees for an internal service, since its users are trusted not to be malicious. But it is not that much easier. Not only must a waiting-time policy be stated and observed for the internal service, but appropriate user agreements must be identified and proved for the higher layers that use it.

The waiting-time policy and user agreements for internal services will be different in detail from the ones applied at the DPB interface. The user agreements constitute a new proof obligation to be satisfied by the higher DPB layers. If a higher layer is expected to be fault-tolerant with respect to lower-level failures, the burden of proof obligations is changed somewhat. An interesting discussion of this concern is given in [10].

Finally, the design and implementation of the internal service must be examined for conformance to its waiting-time policy. This last step is more like performance analysis or real-time computing verification than denial-of-service analysis, however.

4.1 Networks

Network denial-of-service analysis is complicated by the existence of vulnerabilities that have no parallel for stand-alone systems. First, network communication can be disrupted by hardware failures and physical attacks that render nodes and links inoperative. What differentiates networks from stand-alone systems in this regard is the fact that they can, and are expected to, recover from a limited number of failures of this kind. This is the realm of survivability and hardware fault-tolerance. These do not truly constitute a denial-of-service problem as we have defined it, since malicious software is not responsible, but they must be taken into consideration when stating service guarantees.

Second, particular switches or hosts may be taken over by hostile parties, and their software replaced by malicious software. This makes possible a more sophisticated threat than simply bringing the node down, since the malicious node can generate false traffic. The network can still operate with some nodes and links down, but network software can be tricked into disastrous mistakes by false messages. Finally, we should also mention active wiretapping as a related concern, since it is another way of introducing false traffic.

There are at least two ways that false traffic can create denial-of-service problems. One is simply through an attempt to flood the network, thereby delaying legitimate messages. Packet-switching networks respond to unusual loads through various congestion control techniques. The problem is that congestion control algorithms do not attempt to determine whether a user is malicious, and benign users may suffer. Research is needed to identify special approaches to congestion control that specify user agreements and discriminate against nodes that violate them.

Another way to deny service is to misappropriate network control functions or subvert operating system functions by generating spurious messages that are trusted by naively written DPB software. The answer to the latter problem is stronger authentication techniques, such as those based on encryption. Many authentication techniques exist and more are being developed; what needs to be done is to apply them where needed.

Link encryption is a way to defeat active wiretapping, but it is ineffective against the threat of subverting an existing node.

5   Requirements

One reason for trying to pin down what denial of service is, and attempting to distinguish it from other security or functionality concerns, is so that denial of service protection requirements can be added, where appropriate, to system acquisitions. The current DoD approach to computer security requirements has been to separate product evaluation from system requirements specifications, by "factoring" the process through an evaluation phase. Specialized evaluation teams work with standard "criteria" and assign an evaluation class to a computer system. When requirements for specific installations or systems are written, they can simply call for a particular evaluation class, and thus avoid the effort of developing a complete, customized set of computer security requirements.

To fit in with the current process, it would be desirable to develop a standard set of denial- of-service protection criteria. They could be implemented in the form of a revision to the TCSEC, or they could be incorporated into the planned Federal Criteria. Because denial- of-service protection is a new security objective, its criteria cannot be regarded as just an interpretation of existing requirements, such as those developed for networks and database systems. Some insight into the task of developing new criteria for a different security objective may be obtained by looking at the effort to develop integrity criteria.

5.1 Analogy with Integrity

Integrity, like denial of service, has long been recognized as a security concern that is not adequately addressed in the confidentiality-oriented TCSEC, even though it requires system integrity, and some TCSEC-specified features or options contribute to user data integrity. The incentive to work on new criteria was spurred not only by the recognition that integrity needs were not being met, but also by research on models that offered some chance of meeting significant needs successfully, or at least significantly better than previous models. The Clark-Wilson model [7] and the Boebert-Kain type- enforcement mechanism [5] stand out as advances beyond the first lattice-based models given by Biba [4].

Moving from models and even implementations of new concepts into standardized

criteria is difficult. In the case of integrity, an effort to revise the TCSEC has so far generated proposed integrity-oriented control objectives [13], and a report surveying the breadth of integrity concerns and models [14]. Actual criteria are still in the future. However, the importance of modelling is clear, as is the importance of working out the proper wording and coverage of specific criteria.

5.2 ITSEC Availability Criteria

The European Harmonized Criteria, or ITSEC [11], states "availability" as a security objective, defined as: "information and other IT (Information Technology) resources can be accessed by authorized users when needed." The availability objective shows up in the organization of the requirements in the form of a heading, "Reliability of Service," which "should cover any functions intended to insure consistency of service and availability of service." There is a predefined functionality class F7 that "sets high requirements for the availability of a complete system or special functions of a system."

The actual criteria in F7, under the Reliability of Service heading, are brief: a paragraph on error detection and error recovery, and one on continuity of service. The error detection and recovery paragraph requires the ability to maintain operation in the face of "a failure of certain individual hardware components," and requires a statement of maximum time for the reintegration of a repaired or replaced component. The continuity of service requirement guarantees a maximum response time for "certain specified actions" and protection against deadlock.

The coverage of class F7 is broader than just denial of service. It makes sense that users who are concerned about system availability will be interested in requirements that address all possible causes of failure, either malicious software or other problems such as hardware component failure. It is still moot whether hardware failures and corresponding remedies such as fault tolerance—whether implemented in hardware or software—ought to be addressed in "Computer Security" requirements.

In another sense, the coverage of class F7 is narrower than denial of service, because it does not recognize the scope of protection that would be required to address this category of threat. The requirement to protect against deadlock is a good example of how it misses the point. Deadlock does, indeed, deny a resource to a process that needs it. However, if a malicious process wants to deny a resource, it need not contrive a deadlock situation; it can simply acquire and hold the resource.

By failing to discriminate between denial-of-service protection and other kinds of availability concerns, the ITSEC availability criteria are unlikely to stimulate development of solutions to problems that are traceable to the action of malicious software. There should be separate availability-related criteria for various kinds of functionality, such as fault tolerance, and others specifically addressing denial of service. The denial-of-service functions should be categorized further according to the diversity of technical problems that are represented and technical approaches that might be brought to bear on them.

References

[1] M. D. Abrams, K. E. Eggers, L. J. La Padula, and I. M. Olson, "A Generalized Framework for Access Control: An Informal Description," *Proc. 1990 National Computer Security Conference*, October, 1990.

[2] E. M. Bacic and M. Kuchta, "Considerations in the Preparation of a Set of Availability Criteria," *Third Annual Canadian Computer Security Conference*, Ottawa, Canada, May 1991, 283-292.

[3] A. Bestavros, "Time-Constrained Reactive Automata," *Proc. Real-Time Systems Symposium*, IEEE Computer Society, 1991, 244-253.

[4] K. J. Biba, "Integrity Considerations for Secure Computer Systems," ESD- TR-76, NTIS AD-A039324, Electronic Systems Division, Air Force Systems Command, April, 1977.

[5] W. Boebert and R. Kain, "A Practical Alternative to Hierarchical Integrity Policies," *Proc. 8th National Computer Security Conference,* 18-27.

[6] D. Brewer and M. Nash, "The Chinese Wall Security Policy," *Proc. 1989 Security and Privacy Symposium,* IEEE Computer Society, 206-214.

[7] D. D. Clark and D. R. Wilson, "Comparison of Commercial and Military Computer Security Policies," *1987 Symposium on Security and Privacy,* IEEE Computer Society, 184-194.

[8] J. Dobson, "Information and Denial of Service," *Database Security V: Status and Prospects,* IFIP Transactions A-6, 1992, 21-46.

[9] V. Gligor, "A Note on the Denial-of-Service Problem," *Proc. 1983 Symposium on Security and Privacy,* IEEE Computer Society, 139-149.

[10] J. T. Haigh, R. C. O'Brien, W. T. Wood, T. G. Fine, M. J. Endrizzi, "Assured Service Concepts and Models. Volume 3. Availability in Distributed MLS Systems," Secure Computing Technology Corp., Arden Hills, MN, January 1992.

[11] *Information Technology Security Evaluation Criteria (ITSEC),* Der Bundesminister des Innern, Bonn, May 1990.

[12] W. Hu, "Lattice Scheduling and Covert Channels," *Proc. 1992 Symposium on Security and Privacy,* IEEE Computer Society, 52-61.

[13] National Computer Security Center, "Integrity-Oriented Control Objectives," C Technical Report 111-91, October, 1991.

[14] National Computer Security Center, "Integrity in Automated Information Systems," C Technical Report 79-91, September, 1991.

[15] J. C. LaPrie (ed.), *Dependability: Basic Concepts and Terminology,* Springer-Verlag, 1992.

[16] B. Littlewood, "How to Measure Reliability and How Not To," *IEEE Trans. on Reliability,* Vol. R-28, No. 2, June 1979, 103-110.

[17] N. G. Leveson, "Verification of Safety," *Safety of Computer Control Systems 1983 (SAFECOMP '83),* IFAC, Pergamon Press, New York, 1983, 167-174.

[18] A. Mili, *An Introduction to Program Fault Tolerance,* Prentice Hall, New York, 1990.

[19] J. K. Millen, "A Resource Allocation Model for Denial of Service," *Proc. 1992 Symposium on Security and Privacy,* IEEE Computer Society, 137-147.

[20] R. S. Sandhu, "Expressive Power of the Schematic Protection Model," *J. Computer Security,* Vol. 1, No. 1, 1992, 59-98.

[21] *Department of Defense Trusted Computer System Evaluation Criteria,* DOD 5200.28-STD, December, 1985.

[22] C-F. Yu and V. D. Gligor, "A Specification and Verification Method for Preventing Denial of Service," *IEEE Trans. on Software Engineering,* Vol. 16, No. 6, June 1990, 581-592.

Reasoning about Message Integrity

Rajashekar Kailar*, Virgil D. Gligor*, Stuart G. Stubblebine†
*Electrical Engineering Department
University of Maryland, College Park, Maryland 20742
†USC Information Sciences Institute
Marina del Rey, California 90292
U.S.A.

Abstract

We propose an approach for reasoning about message integrity protection in cryptographic protocols. The set of axioms presented herein relate design parameters and assumptions of message integrity protection mechanisms to generic message integrity threats. Comparison of threat properties derived using these axioms with the policy goals for integrity protection aids in assessing the strength (or lack thereof) of message integrity protection mechanisms. We provide examples to illustrate the use of our approach in examining the weaknesses of message integrity protection mechanisms, and also in suggesting modifications in their design parameters.

Categories and Subject Descriptors: C.2.4 [**Computer-Communication Networks**]: General - *Security and Protection*, Distributed Systems; D.4.6 [**Operating Systems**]: Security and Protection - *cryptographic controls*; E.3 [**Data**]: Data Encryption.

Key Words and Phrases: Message integrity, integrity threshold, effective threshold, block membership, order, cardinality, plaintext, ciphertext.

1   Introduction

Cryptographic protocols, and in particular, authentication protocols, rely on message integrity protection. However, mechanisms for protecting message integrity have been shown to be error-prone ([3], [4], [14], [18], [20]). Past studies of message integrity have focused primarily on finding attack scenarios in message integrity protection mechanisms, suggesting solutions to eliminate vulnerabilities, and proposing new algorithms for message integrity protection [2], [3], [4], [5], [14], [15], [18], [21]. Recently, an operational model for message integrity, and a general method for designing message integrity protection mechanisms was also proposed [21]. However, to date, analysis of message integrity protection mechanisms has been done in an ad-hoc manner. A method which considers the types of threats that the environment is exposed to and analyzes protection mechanisms to determine whether they achieve their goals[1] has not been proposed. An analysis method that relates design attributes and assumptions of message integrity protection mechanisms to message integrity goals is useful in (1) analyzing extant protection mechanisms for their threat resistance properties, (2) designing new protection mechanisms, and (3) to help gain insight into properties of message integrity protection mechanisms.

In this paper, we propose a set of axioms which model the threats to message integrity in a given environment. These axioms relate the design parameters and assumptions of message integrity protection mechanisms to generic message integrity threats. Comparison of threat properties derived using these axioms with policy goals for integrity protection allows one to assess the strength (or lack thereof) of protection mechanisms. The analysis of protection mechanisms using our axioms has shown that some of the previously flawed mechanisms [18], [19], [20] do not achieve their policy goals. The analysis of protection mechanisms has also suggested modifications in their design parameters to remove existing vulnerabilities. Although the axioms presented herein mainly concern shared key cryptosystems, this reasoning can be extended to public-key cryptosystems (i.e., those that use RSA scheme [17] or variants thereof) as well. We provide some guidelines on how this may be done in Appendix A.

In our model of computation (section 2), we assume that messages are broken up into blocks and that the encryption algorithm is applied to each of these blocks. Hence, the axioms presented herein refer to block enciphering, and cannot be applied to stream encryption. The axioms presented herein model the threats assumed in our model. Hence, these axioms are applicable in analyzing message integrity protection mechanisms only against the types of threats assumed in our model.

The balance of this paper is organized as follows. In the next section, we briefly discuss our model of computation and introduce the notation that we use to reason about message integrity. In section 3 we briefly list some of our assumptions about the operating environment. In section 5, we discuss the intuition behind the axioms which relate design parameters and assumptions to the goals of protection mechanisms. We use the axioms to analyze some well known (flawed) protection mechanisms and show how the analysis suggests a change in design parameters in order to achieve the specified message integrity goals.

---

[1] Message integrity protection mechanism goals are usually stated in terms of a probability threshold; i.e., in the form *Probability with which the protection mechanism is vulnerable to message integrity compromise is less than a specified threshold* [20].

2 Model of Computation

Our model of computation is derived from the operational model for message integrity of [19]. This model consists of a set of principals (e.g., users, processes or machines), a distinguished principal (X) called the *attacker* and an open network consisting of buffers to/from which messages are sent/received by all principals, including X. Messages sent over the network may or may not be encrypted, depending on whether their application requires confidentiality or not. Encryption algorithms use keys (random numbers) as arguments, and when applied to plaintext, produce ciphertext.

The plaintext message is broken up by the encryption algorithm into a sequence of blocks and the encryption algorithm is applied to each of those blocks. At the receiving end, the ciphertext blocks are decrypted to obtain the original message. If the plaintext message is viewed as an ordered set of plaintext blocks, then the integrity of a received message is said to be preserved if the received (decrypted) message contains all (and only) the plaintext message blocks that are sent, in the proper order (as intended by the sender). In the terminology of [18], these set properties are called *membership* (M), *order* (O), and *cardinality* (C). The M, O and C properties of plaintext messages are mapped onto checksum functions. We refer to these functions collectively as "MOC" functions, and the checksums that they compute over all message blocks, as MOC-values.

The goal of an attacker is to construct a valid message representation (i.e., a message representation which decrypts properly and passes the message integrity checks at the receiving end). The goal of the system is to preserve message integrity by keeping the probability of a successful message integrity compromise below a specified threshold. The attacker is capable of recording all message exchanges that occur over the network. Further, the attacker can distinguish between ciphertext blocks that are encrypted with two different keys, without necessarily knowing either key. This is a reasonable assumption, since the knowledge of the session identifier of a message is often indicative of the use of a (particular type of) key.

Whenever a key which is used as an argument in a function (e.g., encryption key in encryption) is not assumed to be secret, the attacker is assumed to know the value of the key with certainty. Otherwise (if the key is secret), the attacker can only guess the value of the key randomly from the space of the key. If a random key which is secret is used as an argument in the domain of the encryption function the ciphertext can be predicted with a very small probability by X, given that X has chosen the plaintext values.

To construct a valid message representation, the intruder must find (1) the matching encrypted MOC value for the plaintext message or a plaintext message for an encrypted MOC value, and (2) find the ciphertext corresponding to the plaintext that has the proper MOC value. The second objective may not be relevant in cryptographic protocols using public-key cryptosystems (i.e., those using the RSA scheme [17] or variants thereof), since finding the ciphertext for plaintext amounts to finding the encryption key which is publicly known.

In cryptographic protocols using conventional (shared-key) cryptosystems, the attacker can find the ciphertext for a plaintext message in three ways, namely, by (1) finding the encryption key, (2) by matching the plaintext with the plaintext blocks for which the corresponding ciphertext blocks are known (i.e., by collecting known plaintext-ciphertext block pairs), and (3) by using the knowledge of domain-range pairs of the encryption function (i.e., plaintext-ciphertext pairs) to make a "calculated guess" of the ciphertext representation.

The attacker can find the encrypted MOC value for a plaintext message in two ways: (1) by finding both the encryption key and MOC function key, or (2) by using the knowledge of the plaintext message and MOC value pairs to make a calculated guess of the encrypted MOC value. An attacker can find a plaintext message for a given encrypted MOC value in three ways: (1) by finding the encryption key and the MOC function key, (2) by using the knowledge of the plaintext and MOC value pairs in addition to the *many to one* mapping from the plaintext messages to the MOC values[2], and (3) by using the knowledge of the plaintext-ciphertext block pairs to construct a plaintext message for an encrypted MOC value.

The attacker action using known plaintext-ciphertext block pairs is made less effective by using random blocks in the domain of the encryption and MOC functions, in addition to the keys. These random blocks are commonly referred to as confounders. In addition, some encryption algorithms use an initial permutation of the plaintext blocks before encrypting them. The permutation is often referred to as an *Initialization Vector,* (IV). We model the use of confounders and Initialization Vectors (IV's) collectively, and call them *confounding text.* The effect of adding them is to multiply the (range) space of the ciphertext message representation for a given plaintext message (domain) by the space of the confounding block(s). However, the effect of adding confounding components to the domain of the MOC function on the resulting MOC space is not as straight forward. This is because the size of the MOC block is fixed, and in the worst case, for any message, the MOC value can be randomly guessed with probability of success at least equal to the reciprocal of the MOC block space. We discuss this case in greater detail in our axioms.

2.1   Notation and Definitions

Run : An epoch during which a set of principals (e.g., users, machines or programs) share a communication channel (e.g., a set of cryptographic keys) to exchange information. In particular, the principals which participate in this type of communication, and the messages that they send/receive thereby, are said to "belong" to the *run.*

X : A principal which does not belong to the session (i.e., an illegitimate member of the session, or an attacker).

---

[2]Given any message, the probability that its MOC value is equal to some C is at least equal to the reciprocal of the MOC space.

P_i: A plaintext block (with index i denoting its position in the message), the size of which is a constant defined in the cryptographic protocol.

P_i': A *modified* plaintext block. P_i' is the result of applying some function on P_i; e.g., in Cipher-Block-Chaining, the modified plaintext block is obtained by Exclusive-OR-ing the plaintext block with the previous ciphertext block.

$\{P_i'\}_{tkey}$: Ciphertext block obtained by encrypting a *modified* plaintext block P_i' using the encryption key (tkey). In the case of unencrypted messages (i.e., $tkey$ is null) $\{P_i'\}_{tkey} = P_i'$.

MOC($mkey, M$): *Membership, order* and *cardinality* value computed over message M (consisting of a set of b plaintext or ciphertext blocks), using the MOC function key ($mkey$). The MOC function key refers to the key used as an argument by the function to compute the checksum.

X **has** $(n, tkey)$: Principal X has n unique plaintext and ciphertext block pairs encrypted with $tkey$. In particular, X can *have* a plaintext-ciphertext pair either by encrypting known plaintext, or by decrypting a ciphertext, or by observing ciphertext messages whose (plaintext) contents are known.

X **has** (m_{tkey}^{mkey}): Principal X has m unique (plaintext) messages and their corresponding MOC-values computed with $mkey$, and (optionally,) encrypted with $tkey$. X can *have* plaintext message and encrypted MOC-value pairs either by sending/receiving known messages and observing the encrypted MOC-values, or by observing ciphertext messages whose plaintext contents are known, and noting their encrypted MOC-values. In cryptographic protocols that do not encrypt the moc-function values, the $tkey$ is treated as a known constant (say zero), and the $tkey$ space is unity.

X **finds** F: Principal X finds information F, either by deriving F from the information it possesses or by randomly guessing F.

X **finds** F **for** G: Principal P finds F such that F matches G in accordance with some matching criteria. In our axioms, G and F are used to denote either a plaintext and its corresponding ciphertext, or a text string and its corresponding MOC value (or vice-versa), or a plaintext and its corresponding *confounded* text (formed by an initial permutation or by the addition of a confounding block).

Secret(F, Run): Information F is secret in the Run. In particular, members of the session who legitimately share F are assumed not to reveal F to anyone outside of the Run.

$|F|$: The number of all possible values that F can take; i.e., size of the space of F.

T_F : The threshold of F. This is defined as the reciprocal of the space of F.

T_{tkey}: The reciprocal of the space of the encryption key. In the case of an plaintext message, the key is null, and the key space is unity.

T_{mkey}: The reciprocal of the MOC key space. In the case of un-keyed MOC function, the $mkey$ is null, and the $mkey$ space is unity.

T_{block}: The reciprocal of the ciphertext block space.

T_{moc}: The reciprocal of the MOC block space.

T_{conf}: The reciprocal of the confounder block space.

T_{int}: The policy threshold for message integrity.

$|(P', \{P'\}_{tkey})|$: The number of all possible (modified) plaintext and ciphertext block pairs that can be generated using the $tkey$.

ET_F The *effective threshold* of F. This is defined to be the reciprocal of the *effective search space* of F if F is a secret and F has been used as an argument in some function n times. In particular, if the threshold of F is T_F, then the effective threshold

$$ET_F \equiv \frac{T_F}{1-\sigma(n)},$$

where $\sigma(n) > 0$ iff $n > 0$, and $\sigma(n+1) \geq \sigma(n)$ if $n \geq 0$. Intuitively, $\sigma(n)$ is a measure of the amount of information that the knowledge of n domain-range pairs of a function which uses F as an argument provides in "guessing" F. When n equals the size of the space of the domain-range pairs, ET_F equals 1.

ET_{tkey} The *effective* encryption key threshold if $tkey$ is assumed to be a secret and the $tkey$ has been used n times in plaintext block encryption.

$$ET_{tkey} \equiv \frac{T_{tkey}}{1-\delta(n)},$$

where $\delta(n) > 0$ iff $n > 0$, and $\delta(n+1) \geq \delta(n)$ if $n \geq 0$.

Intuitively, $\delta(n)$ is a measure of the amount of information that the knowledge of n plaintext-ciphertext block pairs provides in "guessing" $tkey$.

ET_{mkey} The *effective* MOC key threshold if $mkey$ is a secret and has been used m times in message checksum computation.

$$ET_{mkey} \equiv \frac{T_{mkey}}{1-\phi(m)},$$

where $\phi(m) > 0$ iff $m > 0$, and $\phi(m+1) \geq \phi(m)$ if $m \geq 0$.

Intuitively, $\phi(m)$ is a measure of the amount of information that the knowledge of m message-MOC-value pairs provides in "guessing" $mkey$.

ET_{block} The *effective* ciphertext block threshold after n plaintext and ciphertext pairs have been generated.

$$ET_{block} \equiv \frac{T_{block}}{1-\epsilon(n)},$$

where $\epsilon(n) > 0$ iff $n > 0$, and $\epsilon(n+1) \geq \epsilon(n)$ if $n \geq 0$.

Intuitively, $\epsilon(n)$ is a measure of the amount of information that the knowledge of n plaintext-ciphertext block pairs provides in "guessing" the ciphertext block for a given plaintext block.

ET_{moc} The *effective* MOC block threshold after m message and encrypted MOC-value pairs have been generated.

$$ET_{moc} \equiv \frac{T_{moc}}{1-\chi(m)},$$

where $\chi(m) > 0$ iff $m > 0$ and $\chi(m+1) \geq \chi(m)$ if $m \geq 0$.

Intuitively, $\chi(m)$ is a measure of the amount of information that the knowledge of m plaintext message and MOC-value pairs provides in "guessing" the MOC-value for a given plaintext message.

ET_{cmoc} The *effective* MOC block threshold after m message and encrypted MOC-value pairs have been generated in message types which use confounders.

$$ET_{cmoc} \equiv \frac{T_{moc}}{1-\chi(m) \times T_{conf}},$$

where $\chi(m) > 0$ iff $m > 0$ and $\chi(m+1) \geq \chi(m)$ if $m \geq 0$.

Intuitively, $\chi(m)$ is a measure of the amount of information that the knowledge of m plaintext message and MOC-value pairs provides in "guessing" the MOC-value for a given plaintext message.

ET_{pair} The probability of success in finding a given plaintext-ciphertext pair among n such pairs.

If the number of all possible pairs is $|(P', \{P'\}_{tkey})|$, and the number of pairs that have already been generated is n, then, assuming a uniform distribution of blocks,

$$ET_{pair} \equiv \frac{(n+1)}{|(P',\{P'\}_{tkey})|}.$$

In the analysis presented in this paper, *effective thresholds* are used for establishing that if ET_F is a function of n (the number of available domain-range pairs of a function), then $ET_F > T_F$ for $n > 0$. In other words, the effective threshold is larger than the real threshold for non-zero values of n. Further, ET_{pair} is useful in deriving a limit on the number of pairs that can be generated. We will discuss this in some detail in section 5.

3   Environment Assumptions

In reasoning about message integrity protection mechanisms, we make several assumptions about the operating environment and about the properties of the underlying system functions. We refer to these assumptions collectively as the *environment assumptions*. The axioms that we present, and hence, the results derived using them are based on these assumptions.

- The encryption/decryption functions of block ciphers break a plaintext message M into b blocks $P_1, P_2, ..., P_b$ and apply a specific function which transforms the b plaintext blocks into $P'_1, P'_2, ..., P'_b$, (e.g., in cipher-block chaining, P'_i is a function of P_i and $\{P'_{i-1}\}_{tkey}$) and encipher each block with the same key; i.e.,

$$\{M\}_{tkey} = \{P_1, ..., P_b\}_{tkey} = \{P'_1\}_{tkey}, \{P'_2\}_{tkey}, ..., \{P'_b\}_{tkey}$$

- The block size of the cipher is a constant defined in the design of the protection mechanism. Pair $(P', \{P'\}_{tkey})$ denotes the modified plaintext P' and its corresponding ciphertext block $\{P'\}_{tkey}$. While the correlation between a plaintext block P and its corresponding ciphertext block $\{P'\}_{tkey}$ cannot necessarily be determined, there is a one-to-one mapping between a modified plaintext block P' and its corresponding ciphertext block $\{P'\}_{tkey}$. In the rest of the paper, we refer to *modified* plaintext and ciphertext block pairs simply as plaintext-ciphertext block pairs.

- In message types which do not use checksums, to construct a ciphertext representation for the plaintext message $P_1, P_2, ..., P_b$, X must find the *modified plaintext*[3] $P'_1, P'_2, ..., P'_b$, and

  - encrypt each modified block with $tkey$ to produce the ciphertext representation, or
  - find the corresponding ciphertext blocks $\{P'_1\}_{tkey}, \{P'_2\}_{tkey}, ..., \{P'_b\}_{tkey}$, from among the known plaintext and ciphertext block pairs.

  Possession of plaintext blocks $P_1, P_2, ..., P_b$ does not necessarily allow X to find the *modified plaintext blocks* $P'_1, P'_2, ..., P'_b$ with certainty. However, there are examples ([18]) where it is possible to obtain P'_i from P_i with certainty (e.g., all cryptographic protocols which use DES-CBC [6] encryption mode).

- In message types which use checksums, the MOC function is computed over the entire message (all blocks), and may be encrypted with the $tkey$. In such message types, the possession of ciphertext blocks $\{P'_1\}_{tkey}, ..., \{P'_b\}_{tkey}$ corresponding to plaintext blocks $P_1, ..., P_b$ does not allow X to construct a valid message representation unless he can also find the corresponding *modified* MOC value $\{MOC(mkey, (P_1, P_2, ..., P_b))'\}_{tkey}$ for the plaintext message $P_1, P_2, ..., P_b$. The valid message representation is idealized as:

---

[3]The modified plaintext blocks are a function of the plaintext blocks, and may not always be determined with certainty from the knowledge of the plaintext blocks.

$$\{P_1'\}_{tkey}, \{P_2'\}_{tkey}, ..., \{P_b'\}_{tkey} \{MOC(mkey, (P_1, P_2, ..., P_b))'\}_{tkey}.$$

Note that the placement of the MOC block in the message representation is a function of the message type, and is not considered in this analysis. The probability with which X can find a valid message representation $\{M'\}_{tkey} = \{P_1, P_2, ..., P_b, MOC(mkey, M')\}_{tkey}$ is less than or equal to the probability of finding the ciphertext blocks $\{P_1\}_{tkey}, \{P_2\}_{tkey}, ..., \{P_b\}_{tkey}$ and the moc-value $\{MOC(mkey, M)\}_{tkey}$.

- The encryption and checksum (MOC) functions are common knowledge. The only unknown arguments in these functions are the keys that they use.

- In any collection of modified plaintext and ciphertext block pairs, modified plaintext blocks occur with uniform probability. This assumption is used to derive the probability of finding a modified plaintext block and its corresponding ciphertext block in a group of plaintext-ciphertext block pairs. In examples where some blocks are more probable than others, this assumption can be relaxed to give a more detailed analysis.

- At any given instance in the run, the number of all possible valid message representations using a $tkey, mkey$ pair is much larger than the number of valid message representations that are generated during the lifetime of these keys. In particular, the ratio of the maximum number of messages generated to the total number of possible messages is smaller than T_{int}.

- Legitimate principals maintain the privacy of the secret components of the keys that they possess.

- The encryption key, moc-key and confounding blocks are chosen at random from their candidate spaces.

Preliminary Observations

- The number of all possible (modified) plaintext-ciphertext block pairs that can be generated using the $tkey$ is not larger than the number of all possible values that a ciphertext block can take. While this limit is quite obvious, it is significant because the inequality also supports properties of cryptographic protocols in which not all (modified) plaintext and ciphertext block pairs can exist, since some plaintext message blocks are not considered valid.

- The number of all possible message representations and their corresponding MOC-values is much larger than the number of all possible values that a moc-block can take. This is because the message may consist of several blocks (which can be permuted) and each block can have $|block|$ values. For instance, if we consider a message containing 5 or more blocks, where each block is 64 bits in length, the space $|M, \{MOC(mkey, M)\}_{tkey}|$

would be much larger than 2^{256}, which is the space of the largest MOC block used in practice.

- In this analysis, we do not consider the vulnerabilities due to replay of old messages. This type of threat does not compromise message integrity of individual messages. However, our approach can be used to analyze message integrity attacks that are due to the use of old message contents (or blocks) to construct new messages.

4 Axioms

We present a set of axioms which relate message integrity protection mechanism design parameters and assumptions to the message integrity goals. The axioms follow from our model of computation and environment assumptions. Hence, the axioms derive message integrity properties in the context of the set of threats that are assumed within our model of computation.

4.1 Secrecy

This axiom states that if F is some information that is assumed to be secret throughout the session run, then a principal X can find F with probability less than or equal to ET_F.

$$A_1: \quad \begin{array}{l} \textbf{Secret}(F, Run) \Rightarrow Prob[X \ \textbf{finds} \ F] \ \leq \ ET_F \\ \neg \ \textbf{Secret}(F, Run) \Rightarrow Prob[X \ \textbf{finds} \ F] \ = \ 1 \end{array}$$

where ET_F is the *effective threshold* of F (section 1).

That is, if F is assumed to be a secret, then the choice of F by some principal X is uniformly random in the "effective" space of F. If F is not a secret, then it can be found with certainty by the intruder X. We use this axiom in the analysis of protection mechanisms to define the probability with which keys can be predicted. If a key is assumed to be secret, we define the probability with which X finds the key to be at most equal to the effective threshold of the key. In the application of this axiom, unkeyed functions (e.g., unkeyed MOC) are considered to be equivalent to keyed functions which use known constants as their keys. This is because in both these types of functions, the value of the range can be predicted with certainty for a given domain, if the function does not add any unknown component to its domain. In such mechanisms, the key can be found with certainty.

4.2 Finding a Ciphertext Representation for a Given Plaintext

In this axiom, we compare the three search processes using which X may be able to find a matching ciphertext representation $\{M\}_{tkey}$ for a given message $M = P_1, .., P_b$.

$$A_2: \quad \frac{X \textbf{ has } (n, tkey); \ Prob[X \textbf{ finds } tkey] = S}{Prob[X \textbf{ finds } \{M\}_{tkey} \textbf{ for } M] \leq max(S, (ET_{pair})^b, (ET_{block})^b)}$$

where $n \leq |(P', \{P'\}_{tkey})|$, and b is the number of blocks in message M. If $tkey$ is a secret, S is ET_{tkey}. Otherwise, S is 1. By definition, ET_{pair} is the probability of finding a given plaintext-ciphertext pair among n pairs. The intruder would require b such pairs, each of which would have to be searched independently. Hence, we obtain the second term $(ET_{pair})^b$. By definition, ET_{block} is the effective threshold of a ciphertext block. The intruder would need b such blocks to compose a message. Hence, the term $(ET_{block})^b$.

The probability with which X finds $\{M\}_{tkey}$ for a given M is at most as large as the maximum of these three probabilities (i.e., the one which is most likely to be successful). In this axiom, we do not take into consideration the probability with which X can also find a matching checksum for the plaintext (We describe this in another axiom, namely, A_3). In the analysis of message integrity protection mechanisms, we use this axiom in conjunction with the property that the ciphertext block space is no smaller than the plaintext-ciphertext block pair space. That is, ET_{pair} is greater than $(n + 1) \times T_{block}$. In this axiom, we do not take into account any *confounding text* that the layer may add to the domain of the encryption function. This omission is intentional, since we treat this feature separately in axiom A_4.

An interesting result of axiom A_2 is that if $(ET_{pair})^b > T_{int}$ or if $(ET_{block})^b > T_{int}$ then $Prob[X$ **finds** $\{M\}_{tkey}$ **for** $M]$ is not limited by T_{int}. >From the definitions of these effective thresholds, it follows that if $n > |(P', \{P'\}_{tkey})| \times (T_{int})^{\frac{1}{b}}$ or if $\epsilon(n) > 1 - \frac{T_{block}}{(T_{int})^{\frac{1}{b}}}$, then $Prob[X$ **finds** $\{M\}_{tkey}$ **for** $M]$ is not limited by T_{int}. This is interesting because we can define a lifetime restriction on the $tkey$ based on this condition. >From our observation that the ciphertext block space is larger than the plaintext and ciphertext block pair space, this result also implies that if $n > \frac{(T_{int})^{\frac{1}{b}}}{(T_{block})}$, then $Prob[X$ **finds** $\{M\}_{tkey}$ **for** $M]$ is not limited by T_{int}.

4.3 Checking the Validity of Messages

In this axiom, we describe the process by which X may find either the MOC-value for a given plaintext message, or a plaintext message that corresponds to a given MOC-value.

A_3: $\dfrac{X \text{ has } (m^{mkey}_{tkey}) \, ; \, X \text{ has } (n, tkey);}{Prob[X \text{ finds } tkey] = S; \; Prob[X \text{ finds } mkey] = T}$
$\overline{Prob[X \text{ finds } \{MOC(mkey, M)\}_{tkey} \text{ for } M] \leq max(S \times T, ET_{moc})}$
$Prob[X \text{ finds } M \text{ for } \{MOC(mkey, M)\}_{tkey}] \leq max(S \times T, ET_{moc}, (ET_{block})^b)$

That is, if principal X has m pairs of plaintext messages M and their corresponding MOC values computed with key $mkey$ and encrypted with $tkey$, and X has n plaintext-ciphertext block pairs, then the probability with which X finds $tkey$ is S, and the probability with which X finds $mkey$ is T, then X can find $\{MOC(mkey, M)\}_{tkey}$ for M with probability less than or equal to the maximum of $S \times T$, and ET_{moc}. X can find M for $\{MOC(mkey, M)\}_{tkey}$ with a probability which is not greater than the maximum of $S \times T$, ET_{moc}, and $(ET_{block})^b$.

The reasoning behind this axiom is similar to that of A_2. X may be able to find the MOC-value corresponding to a given text in two ways, namely:

1. By finding both the $mkey$ (in the case of a keyed MOC function) and the $tkey$ (in the case of an encrypted MOC-value), The probability of this is $S \times T$ if $tkey$ and $mkey$ are unrelated. In some cryptographic protocols, this may not be the case, as $mkey$ and $tkey$ may be identical (e.g., Kerberos V4), or the discovery of $tkey$ may aid in finding $mkey$.

2. By computing a MOC-value for the message M using the knowledge of the m message and MOC-value pairs. X can find the MOC-value for the message by searching the *effective moc-block space*. This search succeeds with probability ET_{moc}.

X may be able to find the plaintext for a given (possibly encrypted) MOC-value in three ways, namely

1. By finding both $mkey$ and $tkey$, and using the knowledge of the moc-function to construct a message M which has the required MOC-value. The probability of this is at most $S \times T$ (discussed above).

2. By making use of the many to one mapping from messages to MOC-values. Since any given message has a particular MOC-value with probability T_{moc}, the probability with which any chosen message M will have the MOC-value $\{MOC(mkey, M)\}_{tkey}$, given that X has m message and MOC-value pairs is at most equal to ET_{moc}, the *effective moc threshold* (in this axiom, non-existence of a confounder is assumed), using the same arguments as before.

3. Lastly, X can choose a plaintext message representation M by randomly choosing b blocks to match the message by searching the *effective block space*. Using the environment assumption we have that the probability of success in this type of search is limited by the *effective block threshold*.

The probability of finding $\{MOC(mkey, M)\}_{tkey}$ for a given M (or vice-versa) is at most as large as the maximum of the probability of success in the corresponding search options. In this axiom, like in A_2, we assume that the MOC function does not add any *confounding text* to the text M.

This assumption will be relaxed in A_4 and A_5. $ET_{moc} > T_{int}$ or $(ET_{block})^b > T_{int}$ then $Prob[X$ **finds** $\{MOC(mkey, M)\}_{tkey}$ **for** $M]$ or $Prob[X$ **finds** M **for** $\{MOC(mkey,M)\}_{tkey}]$ are not limited by T_{int}. From the definitions of the effective thresholds, this implies that if $\chi(m) > (1 - \frac{T_{moc}}{T_{int}})$ or $\epsilon(n) > 1 - \frac{T_{block}}{(T_{int})^{\frac{1}{b}}}$, then the $Prob[X$ **finds** $\{MOC(mkey, M)\}_{tkey}$ **for** $M]$ or $Prob[X$ **finds** M **for** $\{MOC(mkey, M)\}_{tkey}]$ are not limited by T_{int}.

4.4 Confounding the Domain of Encryption Function

In practice, encryption and MOC functions can add *confounding text* to their domain in order to counter known plaintext attacks. In the previous axioms (A_2 and A_3), we have not taken into consideration the effect of *confounding text* on the probability of the success of attacker actions. In this axiom, we take this into account and relate the probabilities that ensue due to adding a confounding component to the domain of an encryption function. The effect is to multiply the search space as derived in the previous axioms (A_2 and A_3) by a constant factor, namely the search space of the *confounding text*.

$$A_4: \quad \frac{Prob[X \text{ \textbf{finds} } \{M'\}_{tkey} \text{ \textbf{for} } M'] \le Prob[X \text{ \textbf{finds} } \{M\}_{tkey} \text{ \textbf{for} } M] \le W}{Prob[X \text{ \textbf{finds} } M' \text{ \textbf{for} } M] \le T_{conf}}$$
$$Prob[X \text{ \textbf{finds} } \{M'\}_{tkey} \text{ \textbf{for} } M] \le W \times T_{conf}$$

where $W = max(S, (ET_{pair})^b, (ET_{block})^b)$, is derived using axioms A_2. S is the probability with which $tkey$ is compromised. ET_{pair} and ET_{block} are defined in section 2.1.

That is, if the probability with which X finds $\{M'\}_{tkey}$ for M' is not greater than the probability with which X finds $\{M\}_{tkey}$ for M, which in turn is not greater than W, and if M' is obtained from M by the application of some layer function (i.e., a confounding function) with a range space of $\frac{1}{T_{conf}}$, then the probability with which X finds $\{M'\}_{tkey}$ for M is less than or equal to the product of W and T_{conf}.

The rationale behind this axiom is the following: We treat $[X$ **finds** $\{M'\}_{tkey}$**for** $M]$ as the simultaneous occurence of two events, namely (1) $[X$ **finds** M' **for** $M]$, and (2) $[X$ **finds** $\{M'\}_{tkey}$ **for** $M']$. If X finds both these matching pairs, X can find $\{M'\}_{tkey}$ for M. The probability of this joint occurence would then be equal to the product of their individual probabilities. The probability with which X finds M' for M is at most equal to T_{conf}. The joint probability of the events (1) and (2) above is at most $W \times T_{conf}$. The statement $Prob[X$ **finds**

$\{M'\}_{tkey}$ **for** $M'] \le Prob[X$ **finds** $\{M\}_{tkey}$ **for** $M]$ holds for any message type in which the space of M' is not smaller than the space of M. Since this is always the case, we do not explicitly write this as an assumption in our analysis. The reason for including this in our axiom is to allow the reader to understand the implication.

4.5   Confounding the Domain of MOC Function

In this axiom, we describe the effect of adding a confounding component to the domain of the MOC function.

$$
\begin{array}{l}
Prob[X \text{ \bf finds } \{MOC(mkey, M')\}_{tkey} \text{ \bf for } M'] \le \\
\qquad\qquad Prob[X \text{ \bf finds } \{MOC(mkey, M)\}_{tkey} \text{ \bf for } M] \le Z_1; \\
Prob[X \text{ \bf finds } M' \text{ \bf for } \{MOC(mkey, M')\}_{tkey}] \le \\
\qquad\qquad Prob[X \text{ \bf finds } M \text{ \bf for } \{MOC(mkey, M)\}_{tkey}] \le Z_2; \\
\underline{Prob[X \text{ \bf finds } M' \text{ \bf for } M] = Prob[X \text{ \bf finds } M \text{ \bf for } M'] \le T_{conf}} \\
Prob[X \text{ \bf finds } \{MOC(mkey, M')\}_{tkey} \text{ \bf for } M] \le max(Z_1 \times T_{conf}, ET_{cmoc}) \\
Prob[X \text{ \bf finds } M \text{ \bf for } \{MOC(mkey, M')\}_{tkey}] \le max(Z_2 \times T_{conf}, ET_{cmoc})
\end{array}
$$

with label **A5:** at the left.

where $Z_1 = max(S \times T, ET_{moc})$, and

$$Z_2 = max(S \times T, ET_{moc}, (ET_{block})^b)$$

are derived using axiom A_3. S and T are the probability with which X can find $tkey$ and $mkey$ respectively. T_{conf}, ET_{moc} and ET_{cmoc} are defined in section 2.1.

That is, if

- the probability with which X finds $\{MOC(mkey, M')\}_{tkey}$ for M' is not greater than the probability with which X finds $\{MOC(mkey, M)\}_{tkey}$ for M, which in turn is not greater than Z_1 (Z_1 is derived using axiom A_3),

- the probability with which X finds M' for $\{MOC(mkey, M')\}_{tkey}$ is not greater than the probability with which X finds M for $\{MOC(mkey, M)\}_{tkey}$, which in turn is not greater than Z_2 (Z_2 is derived using axiom A_3), and

- the probability with which X finds M' (which is the confounded version of M) for M is at most T_{conf},

then the probability with which X finds $\{MOC(mkey, M')\}_{tkey}$ for M is at most equal to the maximum of two quantities $Z_1 \times T_{conf}$ and ET_{cmoc}. The reasoning about the term $Z_1 \times T_{conf}$ is similar to that of A_4. The second term, which is the *effective moc threshold* is not a linear function of T_{conf}. This term represents the "calculated" choice of the MOC-value for a given message from the knowledge of the m message and MOC-value pairs. The presence of a confounder with a large space (and hence, a small T_{conf}) has the effect of bringing the *effective moc threshold* close to T_{moc}. The statements (1) and (2) above are true in any message type in which the space of M' is not less than the space of M. These statements are included in the axioms to enable the reader to understand the implication. The probability of finding M for a given $\{MOC(mkey, M')\}_{tkey}$ is limited by the maximum of two quantities $Z_2 \times T_{conf}$, and ET_{cmoc}. The reasoning here is similar.

An important consequence of axioms A_2 through A_5 is that if X **has** $(n, tkey)$ where n is not zero, and if T_{conf} is 1, then X can mount a verifiable plaintext attack, subject to X's ability to find a suitable MOC-value. If n is equal to the plaintext-ciphertext pair space, and if T_{conf} is one, then X can mount a chosen plaintext attack, again, subject to the same moc-constraint. If, however, X has known plaintext-ciphertext pairs $(n > 0)$, and T_{conf} is less than one, then possession of known pairs does not allow X to mount known or chosen plaintext attacks. Attacks in this case would be probabilistic in nature. The same reasoning holds for text and MOC value pairs (i.e., when X has (m_{tkey}^{mkey})). Attacks using the vulnerability of the moc function would be successful, subject to the ability of X to compromise the encryption scheme, if one is used.

5   Goals of Analysis

Message integrity is said to be established if the probability with which X (who is assumed not to have the encryption key) can find a suitable ciphertext representation for a given message, is less than the assumed integrity threshold. This is stated below as goal [G_1]. In addition, message types which make use of checksums for detection of modification of messages may want to ensure that X can neither compute a checksum for a given message [G_2], nor find a message corresponding to a given checksum [G_3] with a probability that is greater than the threshold.

$$Prob[X \textbf{ finds } \{M'\}_{tkey} \textbf{ for } M] \leq T_{int} \dots\dots\dots[G_1]$$

$$Prob[X \textbf{ finds } \{MOC(mkey, M')\}_{tkey} \textbf{ for } M] \leq T_{int} \dots\dots[G_2]$$

$$Prob[X \textbf{ finds } M \textbf{ for } \{MOC(mkey, M')\}_{tkey}] \leq T_{int} \dots\dots[G_3]$$

Definition: *A cryptographic protocol is said to satisfy message integrity conditions if it satisfies the condition* [$G_1 \vee (G_2 \wedge G_3)$].

The condition G_1 implies that X can not find the ciphertext $\{M'\}_{tkey}$ corresponding to plaintext M with a probability greater than T_{int}. G_2 implies that no attacker X can find a $\{MOC(mkey,$

$M')\}_{tkey}$ for some M with a probability greater than T_{int}, and G_3 implies that X cannot find a M' for $\{MOC(mkey, M')\}_{tkey}$ with a probability greater than T_{int}. Here, M' denotes the message string that is obtained when the layer function for encryption and MOC-value computation adds some unknown component to M. This *confounding text* is assumed to be chosen at random from a uniform space of size T_{conf}^{-1}. If there is no such unknown component in M', then $M' = M$.

Mechanisms which use only the properties of encryption algorithm to enforce message integrity protection without having any redundancy (using checksums) may aim to achieve only goal G_1. Goal G_1 may be relevant only to cryptographic protocols which use encryption for preserving secrecy. In cryptographic protocols which do not use encryption for secrecy (i.e., use non-confidential messages), G_1 would reduce to $Prob[X \textbf{ finds } M' \textbf{ for } M] \leq T_{int}$. This is precisely the definition of *confounding text*. Hence, cryptographic protocols which do not use encryption would need to use confounders in the domain to achieve goal G_1.

5.1   Lifetime Constraints on the Encryption Key

The goals of message integrity analysis translate to lifetime constraints on the encryption key, since the ease with which an attacker can find the $tkey$, or construct the ciphertext representation by having known plaintext and ciphertext block pairs increases with the lifetime of the key for a fixed encryption rate and computational resources available to the attacker. In this section, we briefly discuss the interpretation of the goals of integrity analysis in terms of lifetime constraints of the $tkey$.

- In axioms A_2 and A_4, using $(ET_{pair})^b < T_{int}$ and our assumption that $\frac{1}{|(P',\{P'\}_{tkey})|} \leq T_{block}$, we have that the total number of message blocks that are encrypted using $tkey$ during its lifetime must be less than $\frac{(T_{int})^{\frac{1}{b}}}{(T_{block} \times T_{conf})}$ if the goal G_1 should be achieved. If the maximum number of ciphertext blocks that are encrypted in any given time is R, then

$$lifetime(tkey) < \frac{1}{R} \times \frac{(T_{int})^{\frac{1}{b}}}{(T_{block} \times T_{conf})}.$$

  where $lifetime(tkey)$ should be read as "lifetime of $tkey$."

- >From A_2 and A_4, using $(ET_{block})^b < T_{int}$, we derive a lifetime restriction (i.e., limit on n) based on the limiting value of $\epsilon(n)$. That is,

$$lifetime(tkey) < \frac{1}{R} \times \epsilon^{-1}(1 - \frac{T_{block}}{(\frac{T_{int}}{T_{conf}})^b}).$$

  A precise calculation of this limit would require the analysis of the encryption function to find the function $\epsilon(n)$, and is beyond the scope of our work.

- If the maximum rate at which an attacker can exhaustively try $tkey$ (possibly off-line) to verify a certain computed ciphertext block with known plaintext and ciphertext block pairs is C, then

$$lifetime(tkey) < \tfrac{1}{C} \times (\tfrac{T_{int}}{T_{tkey}})$$

The lifetime of $tkey$ must not exceed the minimum of the above three quantities in order to ensure that the message integrity goals are established. That is,

$$lifetime(tkey) < min(\tfrac{1}{R} \times \tfrac{(T_{int})^{\frac{1}{b}}}{(T_{block} \times T_{conf})}, \quad \tfrac{1}{R} \times \epsilon^{-1}(1 - \tfrac{T_{block}}{(\frac{T_{int}}{T_{conf}})^b}), \quad \tfrac{1}{C} \times (\tfrac{T_{int}}{T_{tkey}}))$$

In this analysis, we do not explicitly consider the lifetime limits, but analyze the constraints on n, the number of available plaintext-ciphertext block pairs.

5.2 Lifetime Constraints on MOC-Function Key

The goals of message integrity analysis can also be translated to moc-key lifetime constraints, similar to those of encryption keys. In what follows, we interpret the goals in terms of necessary conditions on the lifetime of moc-keys.

- >From A_3 and A_5, using $ET_{moc} < T_{int}$, we have that the value of m must be limited such that

$$\chi(m) \leq 1 - (\tfrac{T_{moc} \times T_{conf}}{T_{int}})$$

This gives us a lifetime restriction on $mkey$. If K is the maximum number of valid message representations that can be constructed per given time,

$$lifetime(mkey) < \tfrac{1}{K} \times \chi^{-1}(1 - \tfrac{T_{moc} \times T_{conf}}{T_{int}})$$

A precise calculation of this limit would require the analysis of the MOC-function to find the function $\chi(m)$, and is beyond the scope of our work.

- If the maximum rate at which an attacker can exhaustively try $mkey$ (possibly off-line) is D, then

$$lifetime(mkey) < \tfrac{1}{D} \times \tfrac{T_{int}}{T_{tkey}}$$

The lifetime of $mkey$ must not exceed the minimum of the above two quantities in order to ensure that the message integrity goals are established.

6 Examples of Message Integrity Analysis

Message integrity protection weaknesses can be subtle, and can escape the scrutiny of designers if there are no guidelines on relating the design parameters to the message integrity goals of the protection mechanisms. In this section, we provide examples to show that using our approach can aid in explaining these weaknesses and also suggest modifications in the design parameters. Our approach to the analysis of message integrity protection properties involves the following steps.

1. Explicitly state the premises and design parameters of the message integrity protection mechanisms using semantics of the model of computation. The results derived are based on these premises. In particular, our approach cannot determine the soundness of these premises.

2. Apply the axioms on the premises and design parameters to derive expressions of the form $Prob$[message integrity compromise] $\leq P$.

3. Compare the probability limit (P) derived using the axioms with the policy threshold for message integrity. Depending on the choice of the integrity threshold, the analysis may (not) indicate a vulnerability. In our analysis, unless otherwise stated, we assume an integrity threshold of $\frac{1}{2^{64}}$. In other words, we assume that the level of message integrity protection is adequate if the probability of a successful integrity attack is at most equal to $\frac{1}{2^{64}}$.

4. Analyze the implications of the policy goals on the lifetimes of the keys. If these constraints are strong, then the protection mechanism is vulnerable due to the availability of known domain-range pairs of encryption or moc-functions.

5. Suggest modifications in protection mechanism design parameters to remedy existing vulnerabilities.

We use three protocol examples to illustrate the use of the axioms presented in this paper. The examples are chosen to illustrate different types of weaknesses that can be detected using this method. Additional analysis examples are given in [8].

6.1   Analysis of Kerberos V4 Protocol [2]

In this section, we analyze the Kerberos V4 protocol which uses DES (Data Encryption Standard [6]) for encryption. The Kerberos V4 was previously shown to be vulnerable in [2]. We show that this protocol does not achieve its goals, namely, that of keeping the probability of a successful message integrity compromise below the specified threshold. The Kerberos V4 protocol has the property whereby the same key is used both for the encryption function as well as for the moc function. This key is 7 bytes long, and hence, has a space of size 2^{56}. Hence, the key threshold is equal to 2^{-56}. The ciphertext blocks in this protocol are 8 bytes long. The *moc* value occupies a whole block (i.e., 8 bytes). The protocol does not use confounding text.

Hence, the confounder threshold is 1.[4]

Protocol Premises and Design Parameters

Secret($tkey$, Run)

Secret($mkey$, Run)

$mkey = tkey = key$

$T_{key} = \frac{1}{2^{56}}$

$T_{block} = T_{moc} = T_{int} = \frac{1}{2^{64}}$

$T_{conf} = 1$

The first two premises are about the secrecy of the encryption key ($tkey$) and the moc-key ($mkey$) in the session (Run). These premises are based on the fact that this protocol uses keyed encryption and moc-functions, the keys of which are secret within a session (Run). The remaining assumptions are about the design parameters of the protocol. The $mkey$ and the $tkey$ are the same, and their threshold is $\frac{1}{2^{56}}$. The ciphertext block, MOC-value, and the assumed policy integrity threshold are all 2^{-64}. The protocol uses no confounders (i.e., confounder space is 1).

Analysis

>From the assumptions about *secrecy*, and about $tkey$ and $mkey$ being identical (both equal to key), using axiom A_1, we have

$Prob[X$ **finds** $tkey] = ET_{key}$.

$Prob[X$ **finds** $mkey] = ET_{key}$.

Application of axioms A_2 and A_3 yield

$Prob[X$ **finds** $\{M\}_{tkey}$ **for** M $] \leq max(ET_{tkey}, (ET_{pair})^b, (ET_{block})^b) = W$

$Prob[X$ **finds** $\{MOC(mkey, M)\}_{tkey}$ **for** M $] \leq max(ET_{key}, ET_{moc}) = Z_1$

$Prob[X$ **finds** M **for** $\{MOC(mkey, M)\}_{tkey}] \leq max(ET_{key}, ET_{moc}, (ET_{block})^b) = Z_2$

Note that since $tkey = mkey = key$, in the application of A_3, the joint probability of finding both $tkey$ and $mkey$ is equal to the probability of finding one of them (i.e., key).

---

[4]We view this message type as being equivalent to a message type which uses a known constant as a confounder. The use of a known constant as a confounder does not serve any purpose. This view allows us to analyze both these message types with the same notation.

We now apply axioms A_4 and A_5, with the assumption that $Prob[X \textbf{ finds } M' \textbf{ for } M] \le T_{conf} = 1$. Using this, we obtain,

$Prob[X \textbf{ finds } \{M'\}_{key} \textbf{ for } M] \le W$ .... [R_1]

$Prob[X \textbf{ finds } \{MOC(key, M')\}_{key} \textbf{ for } M] \le max(Z_1, ET_{moc})$ .... [R_2]

$Prob[X \textbf{ finds } M' \textbf{ for } \{MOC(key, M')\}_{key}] \le max(Z_2, ET_{moc})$ .... [R_3]

Observations

In result R_1, the term W is at least as large as T_{key} (because $W = max(ET_{tkey}, (ET_{pair})^b, (ET_{block})^b))$. From the design parameters, $T_{key} > T_{int}$. Hence, $W > T_{int}$. Using this in result R_1, we can see that this protocol does not achieve goal G_1. Consider goals G_2 and G_3. In results R_1 through R_3, terms Z_1 and Z_2 are at least as large as T_{key}, which in turn is greater than T_{int}. Hence, the protocol does not achieve G_2 and G_3.

We now reason about these results. On first glance, it appears that the goal G_1 is not achieved due to the relatively small size of the key space. We will now show that this intuition is not entirely correct. Consider a variant of this protocols' design parameters; i.e., the same protocol with $T_{key} \le \frac{1}{2^{64}}$. Hence, $T_{key} \le T_{int}$. However, the term $W = max(ET_{key}, (ET_{pair})^b, (ET_{block})^b)$ can be less than or equal to T_{int} only if

$$(ET_{pair})^b \le T_{int}$$

>From the definition of ET_{pair},

$$\left(\frac{(n+1)}{|(P',\{P'\}_{key})|}\right)^b \le T_{int} \text{ .... (1)}$$

Using our assumption that the number of plaintext-ciphertext pairs that can be generated is not larger than the number of all possible ciphertext blocks, we can derive the condition $((n + 1) \times T_{block})^b \le T_{int}$ from (1). This means that

$$n < \frac{(T_{int})^{\frac{1}{b}}}{T_{block}} \le 1$$

The limit on n derived above suggests that even if T_{key} is decreased from 2^{-56} to 2^{-64}, all other design parameters being the same, this protocol can still not achieve the analysis goals for non-zero values of n. Choosing a large key size for this protocol does not turn out to be useful in achieving the goal G_1, since the ciphertext block size is still equal to the reciprocal of the assumed integrity threshold, making the search of the ciphertext block space easier than searching the key space when $n > 0$. This is the case because this protocol does not use confounding text, and hence, makes known plaintext-ciphertext block pairs available (hence, reducing the *effective* search space for unknown ciphertext blocks).

Again, consider the goals G_2 and G_3. On first glance, the reason for this protocol not being able to achieve goals G_2 and G_3 seems to be because the $mkey$ is the same as the $tkey$. However, this intuition is also not entirely true. Consider the variant of this protocol which uses an $mkey$ which is different from $tkey$. The terms Z_1 and Z_2 would then be

$$Z_1 = max(ET_{tkey} \times ET_{mkey}, ET_{moc}), \text{ and}$$

$$Z_2 = max(ET_{tkey} \times ET_{mkey}, ET_{moc}, (ET_{block})^b)$$

Since this protection mechanism has $T_{moc} = T_{int}$, the *effective moc threshold* (i.e., ET_{moc}) is larger than T_{int} for non-zero values of m. Hence, both Z_1 and Z_2 are greater than T_{int} for non-zero values of m. Hence, (from results R_2 and R_3) this protocol will not achieve the goals ($G_2 \wedge G_3$) if $m > 0$, in spite of having unique $mkey$ and $tkeys$.

Suggested Solution for Remedying Vulnerability
The vulnerability of this protection mechanism can be remedied by choosing a larger block size (i.e., $T_{block} < T_{int}$), in addition to a larger encryption key size (i.e., $T_{key} < T_{int}$) or adopting one of the following solutions:

- Choosing a larger moc block space. By doing this, the *effective moc threshold* is reduced to less than T_{int}.

- Using a confounder which has a large space, such that $\frac{T_{moc}}{1-T_{conf}}$ is at most equal to the assumed message integrity threshold.

- Choosing a moc function and the lifetime constraints on the $mkey$ such that the value of $\epsilon(m)$ is much smaller than 1 for the maximum number that m can attain during the lifetime of the $mkey$. By this choice, the term $\frac{T_{moc}}{1-\epsilon(m)}$ in Z_1 and Z_2 can be kept close to T_{moc}.

The rationale for these suggestions is that by increasing the size of both the encryption key space (such that $T_{key} > T_{int}$) as well as the ciphertext block space (such that the term $\frac{(T_{int})^{\frac{1}{b}}}{T_{block}}$ is larger than the number of plaintext-ciphertext block pairs that will be generated during the lifetime of the key), the goal G_1 can be achieved. To achieve G_1 or $G_2 \wedge G_3$, the design parameters can also be chosen such that the goals G_2 and G_3 hold. This can be done using the three alternative solutions presented above.

This example shows that vulnerabilities of message integrity protection mechanisms can be subtle and that adhoc and intuitive reasoning about them can be erroneous. Relating protection mechanism design parameters to the integrity goals and the environment threats allowed us to reason about the weaknesses of this protocol, and to discover some weaknesses that were not obvious on intuitive reasoning. In the following section, we provide an example in which we consider a protocol which uses two message types, and consider the effects of having a moc-function key that can be common to both types of messages.

6.2 Analysis of Privacy Enhanced Email

The Privacy Enhancement for Internet Electronic Mail (PEM) [11,12,13] supports confidentiality, integrity, and authentication for electronic mail transfer in the Internet. An earlier version of PEM [10] was shown to be vulnerable to integrity attacks against messages using the DES-MAC integrity check when these messages are sent to more than one receiver [16]. The PEM services support both single- and multiple-receiver messages.

The moc-key ($mkey$) of the single recipient session is known to members of the multiple receiver session, since the same $mkey$ can be used in both types of sessions. In the analysis that follows, we consider the effect of having the $mkey$ of the single receiver session known to the members of the multiple receiver session. In this protocol analysis, a principal who belongs to the multiple receiver session (and hence, knows the $mkey$ of the multiple receiver session as well as the single receiver session) and uses this knowledge of the $mkey$ to construct a single receiver session message between two other parties which belong to the multiple receiver session, is viewed as the attacker. It should be noted that the secrecy assumptions in the analysis are made from this viewpoint, and hence, may not apply in general, to any attacker.

The PEM protocol uses ciphertext blocks which are 8 bytes long, and hence, have a space of size 2^{64}. The moc block size is equal to the ciphertext block size. The protocol does not use confounding text. Hence, the confounder threshold is 1. The encryption key is 7 bytes long. Hence, its space is 2^{56} in size.

Protocol Premises and Design Parameters

Secret($tkey, Run$)

¬**Secret**($mkey, Run$)[5]

$T_{block} = T_{moc} = T_{int} = \frac{1}{2^{64}}$

$T_{tkey} = \frac{1}{2^{56}}$

$T_{conf} = 1$

(i.e., $T_{block} = T_{moc} = T_{int} < T_{tkey} < T_{conf} = 1$)

The protocol uses encrypted messages, applying a keyed encryption function on plaintext blocks. The encryption key is assumed to be secret in the protocol run. For the reasons discussed at the start of this section, the $mkey$ of the single receiver session is not assumed to be secret within the multiple user session. Although the protection mechanism uses a $mkey$

---

[5]The $mkey$ of the single user session is not a secret to the members of a multiple session which contains the members of the single user session. For example, if parties $A,B,$ and C belong to a multiple receiver session, then the $mkey$ of this session can be used by A to send (single receiver session) messages to B or to C.

which has space $= 2^{56}$, the assumption about the lack of secrecy of this key makes $T_{mkey} = 1$. The remaining premises listed above are about design parameters.

Analysis
>From the assumptions about secrecy, using A_1, we have that

$$Prob[X \text{ finds } tkey] = ET_{tkey}$$

$$Prob[X \text{ finds } mkey] = 1$$

Application of the axioms A_2, A_3 and the results derived above yields

$$Prob[X \text{ finds } \{M\}_{tkey} \text{ for } M] \leq max(ET_{tkey}, (ET_{pair})^b, (ET_{block})^b) = W$$

$$Prob[X \text{ finds } \{MOC(mkey, M)\}_{tkey} \text{ for } M] \leq max(ET_{tkey}, ET_{moc}) = Z_1$$

$$Prob[X \text{ finds } M \text{ for } \{MOC(mkey, M)\}_{tkey}] \leq max(ET_{tkey}, ET_{moc}, (ET_{block})^b) = Z_2$$

We now apply axioms A_4 and A_5, with the assumption that $Prob[X \text{ finds } M' \text{ for } M] \leq T_{conf} = 1$, and obtain,

$$Prob[X \text{ finds } \{M'\}_{tkey} \text{ for } M] \leq W \text{ .... } [R_1]$$

$$Prob[X \text{ finds } \{MOC(mkey, M')\}_{tkey} \text{ for } M] \leq max(Z_1, ET_{moc}) \text{ .... } [R_2]$$

$$Prob[X \text{ finds } M' \text{ for } \{MOC(mkey, M')\}_{tkey}] \leq max(Z_2, ET_{moc}) \text{ .... } [R_3]$$

Observations
The term $W = max(ET_{tkey}, (ET_{pair})^b, (ET_{block})^b)$ on the right hand side of result $[R_1]$ is larger than T_{int}, since $ET_{tkey} > T_{int}$. Hence, goal G_1 does not hold. Similarly, $Z_1 = max(ET_{tkey} \times T_{mkey}, ET_{moc}) > T_{int}$ and $Z_2 = max(ET_{tkey} \times ET_{mkey}, ET_{moc}, (ET_{block})^b) > T_{int}$ because $ET_{tkey} > T_{int}$ and $ET_{mkey} = 1$. Hence, $G_2 \wedge G_3$ does not hold.

Since neither G_1 nor ($G_2 \wedge G_3$) hold, this protocol does not achieve the goals of analysis. Although the analysis seems to indicate that the protocol is unable to achieve goal G_1 due to the relatively small space of the $tkey$, we observe that increasing the size of the encryption key space alone would not suffice for this protocol to achieve goal G_1. This is because the ciphertext block space in this protocol is equal to the reciprocal of the assumed integrity threshold, and hence, the *effective block space* would be smaller than the reciprocal of the integrity threshold for non-zero values of n. Hence, goal G_1 can be achieved if the key space as well as the ciphertext block space are increased such that $ET_{tkey} < T_{int}$, and the *effective block threshold* is smaller than T_{int} for the largest value of n possible during the lifetime of the encryption key. Another alternative to remedy the protocol weakness is to use a confounder which has a sufficiently large space such that the *effective moc threshold* is close enough to T_{int}.

This protocol falls short of goals G_2 and G_3 because of three reasons: (1) because it uses the same $mkey$ for the single as well as the multiple recipient sessions (i.e., the key for a single recipient session is not a secret) (2) the moc-block space is equal to the reciprocal of the assumed integrity threshold, and hence, the *effective moc threshold* is larger than the assumed integrity threshold for non-zero values of m, the number of available plaintext message and MOC-value pairs. (3) the ciphertext block size is equal to the reciprocal of the message integrity threshold, hence making the *effective block threshold* larger than the integrity threshold, for non-zero values of n, the number of available plaintext-ciphertext block pairs available. The goals G_2 and G_3 can be achieved if the three conditions above which cause the protocol weakness, are negated.

The solution to this problem suggested in [20] is to use two moc blocks, so that the resulting moc-space is the product of the individual spaces. This solution does not establish goal G_1, but achieves goals G_2 and G_3, hence satisfying the condition $G_1 \lor (G_2 \land G_3)$. Below, we analyze the proposed solution of [20] and show that it satisfies the message integrity goals. In [8], we present the analysis of this proposed solution and show that it does not exhibit the weakness of the original protocol.

6.3   Analysis of Message Digest of KRB_SAFE in Kerberos V5

In this protocol example, we show that choosing the ciphertext block space equal to or smaller than the reciprocal of the integrity threshold may reduce the effectiveness of choosing a large moc-block space to enforce the assumed message integrity threshold. The KRB_SAFE protocol in Kerberos V5 uses a checksum which is the DES-CBC encryption of the 128-bit RSA-MD4 digest. The ciphertext block size is 8 bytes. The encryption key and the $mkey$ are 7 bytes long. The protocol does not use confounders. By using a MOC-value which has a space 2^{128}, the protocol aims to keep the message integrity threshold anywhere between $\frac{1}{2^{128}}$ and $\frac{1}{2^{64}}$. In the analysis that follows, we show that this goal is not actually achieved by the protocol, owing to the fact that it uses a ciphertext block which has a space much smaller than the space of the MOC-value.

Protocol Premises and Design Parameters

Secret$(tkey, Run)$

Secret$(mkey, Run)$

$T_{moc} = \frac{1}{2^{128}}; T_{block} = \frac{1}{2^{64}}$

$T_{tkey} = T_{mkey} = \frac{1}{2^{56}}; T_{conf} = 1$

$T_{int} \in (\frac{1}{2^{128}}, \frac{1}{2^{64}})$

(i.e., $T_{moc} \le T_{int} < T_{block} < T_{tkey} = T_{mkey} < T_{conf} = 1$)

This protocol does not use encrypted messages. However, in this example, we are concerned with the DES-CBC encryption of the RSA-MD4 checksum, for which the protocol uses a $tkey$ which is assumed to be secret. The RSA-MD4 is a keyed moc-function, with the $mkey$ being secret. The digest space is 2^{128}, since it is 16 byte long. However, the digest is encrypted (using cipher-block-chaining) into two blocks of size 8 bytes each.

Analysis
>From the assumption that the $tkey$ and $mkey$ are secret, using A_1, we have,

$Prob[X$ **finds** $tkey] = ET_{tkey}$

$Prob[X$ **finds** $mkey] = ET_{mkey}$

Since this protocol does not use encryption for secrecy, goal G_1 is not relevant. We focus on finding the encrypted MOC-value for a given message (or vice-versa). Application of the axiom A_3 yields

$Prob[X$ **finds** $\{MOC(mkey, M)\}_{tkey}$ **for** M $] \le max(ET_{tkey} \times ET_{mkey}, ET_{moc}) = Z_1$

$Prob[X$ **finds** M **for** $\{MOC(mkey, M)\}_{tkey}]$

$$\le max(ET_{tkey} \times ET_{mkey}, ET_{moc}, (ET_{block})^b) = Z_2$$

We now apply axiom A_5, with the assumption that $Prob[X$ **finds** M' **for** $M] \le T_{conf} = 1$, and obtain

$Prob[X$ **finds** $\{MOC(mkey, M')\}_{tkey}$ **for** M $] \le max(Z_1, ET_{moc})$ ....[R_2]

$Prob[X$ **finds** M **for** $\{MOC(mkey, M')\}_{tkey}$ $M] \le max(Z_2, ET_{moc}, (ET_{block})^b)$ ...[R_3]

Observations
For this protocol to achieve the goal G_3, it is necessary that $ET_{block} \le T_{int}$ for a one block message. That is,

$$(\tfrac{T_{block}}{1-\epsilon(n)}) \le T_{int}$$

since the term Z_2 is at least as large as $\frac{T_{block}}{1-\epsilon(n)}$. But $\frac{T_{block}}{1-\epsilon(n)} > T_{block} > T_{int}$ for values of $n > 0$ in this protocol. The ciphertext space in this protocol is much smaller than the MOC block space. Hence, the protocol can achieve only a weaker threshold, namely $\frac{1}{2^{64}}$, although it uses a 128 bit moc block. The protocol fails to achieve the desired result, namely that of keeping the integrity threshold in the range $(\frac{1}{2^{128}}, \frac{1}{2^{64}})$.

The protocol is rendered weak because it uses DES-CBC to encrypt the 128 bit RSA-MD4 digest. The use of a 64 bit cipher-block makes the search space for the encrypted checksum smaller than 2^{128}. This is because the protocol does not use confounders, and hence, makes available known plaintext-ciphertext block pairs. Having a known plaintext-ciphertext pair which corresponds to one of the encrypted digest blocks allows the attacker to exhaustively search for the other encrypted digest block, from a space of size 2^{64}.

One possible solution to remedy this weakness is to use a confounding block of space 2^{64} which makes known plaintext-ciphertext block pairs available with a probability $\frac{1}{2^{64}}$, and hence, allows this protocol to achieve its message integrity goals. Another alternative solution (although probably less practical) is to use 16 byte long ciphertext blocks and 16 byte long encryption keys to encrypt the digest, instead of the 8 byte blocks and keys, respectively.

7   Conclusions

Our model of computation allows reasoning about integrity design parameters of cryptographic protocols. The reasoning provides a set of axioms that relate protocol design parameters and protocol premises to the goals of message integrity protocols, considering the set of threats that the system is exposed to (as assumed in our model). As shown in these examples, using simple reasoning, we have accounted for the vulnerabilities in protocols which have been previously discovered to be flawed, and have suggested modifications in design parameters to strengthen these protocols.

Acknowledgements

The first two authors are grateful to Tom Tamburo and Marty Simmons of IBM Federal Systems Company for their continued support and encouragement.

References

[1]   M.Burrows, M.Abadi and R.Needham, "A Logic of Authentication," *ACM Transactions on Computer Systems,* Vol.8, No.1, February 1990.

[2]   S.M.Bellovin and M.Merritt, "Limitations of the Kerberos Authentication System," *Computer Communications Review,* Vol.20, No.5, October 1990.

[3] G.I.Davida, "Chosen Signature Cryptanalysis of the RSA (MIT) Public-Key Cryptosystem," Tech. Report TR-82-2, Dept. of Electrical Engineering and Computer Science, University of Wisconsin, Milwaukee, WI, October 1982.

[4] D.E.Denning, "Digital Signatures with RSA and Other Public-Key Cryptosystems," *Communications of the ACM,* Vol. 27, No. 4, April 1984.

[5] D.Dolev and A.C.Yao, "On the Security of Public Key Protocols," *IEEE Transactions on Information Theory,* Vol. IT-29, No. 2, March 1983.

[6] Federal Information Processing Standard, "DES Modes of Operation," National Bureau of Standards, December, 1980.

[7] L.Gong, R.Needham and R.Yahalom, "Reasoning about Beliefs in Cryptographic Protocols," *Proceedings of the IEEE Symposium on Research and Privacy,* May 1990.

[8] R.Kailar, V.D.Gligor, S.G.Stubblebine, "Reasoning about Message Integrity," Technical Report (93-065), Electrical Engineering Department, University of Maryland, College Park, MD 20742.

[9] J.Kohl and B.C.Neuman, "Kerberos Version 5, RFC, Draft #4", Project Athena, MIT, December 20, 1990.

[10] J.Linn, "Privacy Enhancement for Internet Electronic Mail: Part I - Message Encipherment and Authentication Procedures," Internet Working Group, RFC-989, February, 1987.

[11] J.Linn, "Privacy Enhancement for Internet Electronic Mail: Part I - Message Encipherment and Authentication Procedures," Internet Working Group, RFC-1113, August 1989.

[12] J.Linn, "Privacy Enhancement for Internet Electronic Mail: Part II - Certificate-Based Key Management," Internet Working Group, RFC-1114, August 1989.

[13] J.Linn, "Privacy Enhancement for Internet Electronic Mail: Part III - Algorithms, Modes, and Identifiers," Internet Working Group, RFC-1115, August 1989.

[14] J.M.Moore, "Protocol Failures in Cryptosystems," *Proceedings of the IEEE,* Vol. 76, No. 5, May 1988.

[15] R.C.Merkle, "One Way Hash Functions and DES," in Advances of Cryptology, *Proceedings of Crypto '89* Santa Barbara, California, 1989.

[16] C.Mitchell and M.Walker, "Solutions to the Multidestination Secure Electronic Mail Problem," *Computers and Security,* Vol. 7, No. 5, 1988.

[17] R.L.Rivest, A.Shamir, and L.Adleman, "A Method for Obtaining Digital Signatures and Public Key Cryptosystems," *Communications of the ACM,* Vol. 21, No. 2, February 1978.

[18] S.G.Stubblebine and V.D.Gligor, "On Message Integrity in Cryptographic Protocols,"
     TR.No:2843, University of Maryland, College Park, Maryland 20742.

[19] S.G.Stubblebine and V.D.Gligor, "On Message Integrity in Cryptographic Protocols,"
     Proceedings of the IEEE Symposium on Research in Privacy, May 1992.

[20] S.G.Stubblebine and V.D.Gligor, "Protecting the Integrity of Privacy-Enhanced Elec-
     tronic Mail with DES-Based Authentication Codes," PSRG Workshop on Network and
     Distributed System Security, San Diego, CA, February 1993.

[21] S.G.Stubblebine and V.D.Gligor, "Protocol Design for Integrity Protection," *Proceedings
     of the IEEE Symposium on Research in Privacy*, May 1993.

A Reasoning about Message Integrity of Public-Key Cryptosystems

In this section, we provide some guidelines on extending our approach to the analysis of public-key cryptosystems. The set of axioms presented here is not necessarily complete. Other axioms may be necessary to model threats to the integrity of digital signatures.

In public-key cryptosystems (i.e., those using the RSA scheme [17] or variants thereof), the goals of message integrity protection are somewhat different from those of shared-key cryptosystems. Finding the ciphertext $\{M\}_{tkey}$ for a given plaintext M is not an issue here, since the $tkey$ is publicly known (by definition). Hence, the goals are simply G_2 and G_3. Axiom A_2 is relevant only for computing the probability of finding the encrypted signatures. The *membership*, *order*, and *cardinality* properties are mapped onto digital signatures using a secret key (i.e., the secret counterpart of the public-key). Hence, finding the signature for a plaintext message is equivalent to finding the MOC-value for a plaintext message, and vice-versa.

If the plaintext message $M = P_1, P_2, ..., P_b$, where each P_i is the integer value corresponding to a block, then the digital signature using the RSA scheme is

$$D(M) = P_1^d(\text{mod } pq) \, || \, P_2^d(\text{mod } pq) \, || \, ... \, || \, P_b^d(\text{mod } pq),$$

where d is the secret key of the party which signs the message, and p and q are large prime numbers. The symbol $||$ denotes signature block concatenation.

In [4], it was shown that if P_i can be factorized, and the signature blocks of the factors can be obtained, then the signature block corresponding to the ith block can be computed. Possession of b such computed blocks can enable an attacker to compute the signature. In [14], it was shown that possession of signatures to b arbitrary message blocks (where each plaintext block is some constant times P_i) enables an attacker to compute the plaintext message for a chosen signature.

With this introduction, we proceed to present a modified set of axioms for public-key cryptosystems. The axiom A_1 is still the same, and applies primarily to the secret counterpart of the public-key. As observed before, axiom A_2 is not relevant since obtaining ciphertext for plaintext is not a problem in public-key cryptosystems. Axiom A_3 for a b block message with the signature computed as above becomes:

$$
\text{A3'} \quad \frac{X \textbf{ has } (n, d) \; ; Prob[X \textbf{ finds } d] \leq T_d}{Prob[X \textbf{ finds } D(M) \textbf{ for } M] \leq max(T_d, \overset{b}{\prod} \frac{t_b}{|P_i, D(P_i)|}, (\frac{T_{block}}{1-\epsilon(n)})^b)}
$$

where b is the number of blocks in the message and in the signature, and t_b is the number of factors for each block which maximizes the probability of finding the signatures of all the factors, and $|P_i, D(P_i)|$ is the number of all possible message blocks and their corresponding signatures.

The reasoning here is as follows: the signature can be computed by (1) finding the key, or (2) by computing each of the signature blocks, or (3) by making a calculated guess at the signature representation using the knowledge of message blocks and their corresponding signatures. The first search has a probability T of success. Each block in the signature can be factored into t_b numbers, where the factors are chosen such that the probability of finding their corresponding signatures is maximum. Hence, the probability of finding the signature for a block by computing it from known message block- signature pairs is $\frac{t_b}{|P_i, D(P_i)|}$. The probability of finding the entire signature in this manner is at most equal to the product of the probability of finding the signature blocks of individual message blocks. Finally, a calculated guess at the signature may be done with a probability of success defined by $\frac{T_{block}}{1-\epsilon(n)}$, in a manner similar to guessing a ciphertext block for a plaintext block. The function $\epsilon(n)$ is monotonically non-decreasing, and is non-zero for values of n greater than zero.

Axioms A_4 and A_5 are identical to the ones presented here, except that the term *confounding text* now refers to the effect of hashing a plaintext message prior to encrypting it.

On the Security Effectiveness of Cryptographic Protocols

Rajashekar Kailar*, Virgil D. Gligor*, Li Gong†
*Electrical Engineering Department
University of Maryland
College Park, Maryland 20742
†SRI International
Computer Science Laboratory
333 Ravenswood Avenue
Menlo Park, California 94025
U.S.A.

Abstract

We introduce the notion of security effectiveness, illustrate its use in the context of cryptographic protocol analysis, and argue that it requires analysis of protocol property dependencies. We provide examples to show that, without dependency analysis, the use of some logics for cryptographic protocol analysis yields results that are inconsistent or unrealistic in practice. We identify several types of property dependencies whose use in protocol analysis methods can yield realistic analyses.

Categories and Subject Descriptors: C.2.4 [**Computer-Communication Networks**]: General- *Security and Protection*, Distributed Systems; D.4.6 [**Operating Systems**]: Security and Protection - *authentication, cryptographic controls*; K.6.5 [**Management of Computing and Information Systems**]: Security and Protection - *authentication*; E.3 [**Data**]: Data Encryption.

Key Words and Phrases: Security effectiveness, cryptographic protocol, cryptographic assumption, logic, dependency, threat countermeasure.

1 Introduction

Cryptographic protocol analysis using logics has conclusively shown that subtle security vulnerabilities which may escape designers' attention can be detected using formal methods. In the
last few years, several logics have been proposed and have been used in the analysis of protocols
to show that the protocols can (not) achieve certain goals ([2], [10], [11]). The application of
logics[1] allows one to derive properties which imply that certain threats are infeasible in the
system. By doing this, one attempts to show that a protocol is *effective* in countering a specified
(set of) threat(s). For instance, protocol properties, such as *session-key freshness, jurisdiction*
(i.e., established authority over a session key), and *session-key confidentiality,* are generally
considered sufficient to establish the effectiveness of a cryptographic protocol with respect to
threats of *unauthorized re-use of old session keys, unauthorized session-key generation,* and
unauthorized session-key disclosure.

However, logics for the analysis of cryptographic protocols derive properties based on cryptographic assumptions about the underlying system. The results of such analyses are dependent
not only on the validity of stated assumptions but also on that of often unstated ones. For
example, extant logics for the analysis of cryptographic protocols do not take into consideration the dependency of message confidentiality on the lifetime of encryption keys while
deriving properties of confidential message contents (e.g., session-key freshness). The analysis
of this dependency is important in determining whether current session key, old session keys,
and long-term keys (e.g., key-encrypting keys) are breakable (i.e., can be read by an active
intruder).

Missing cryptographic assumptions in analyzing cryptographic protocols may yield inconsistent
or unrealistic results. For example, the analysis and the logic itself may assume that a key is
unbreakable during the key lifetime, yet the protocol itself may generate a very large number
of known (or even chosen) plaintext-ciphertext block pairs, which can lead to the breaking of
that key. Or, the analysis and the logic itself may assume that an authenticated source of time is
available for an authentication protocol to establish message freshness. Missing cryptographic
assumptions in analyzing cryptographic protocols is particularly ironic because the relationship
between these protocols, logics, and cryptography is limited precisely to making assumptions
about the properties of underlying cryptographic systems.

In this paper, we argue that analyzing the dependencies of the derived protocol properties on
the underlying assumed properties is a necessary part of establishing security effectiveness.
We provide examples to show that the use of logics for cryptographic protocol analysis without
this dependency analysis may yield results that are inconsistent or unrealistic in practice. We
present a *dependency graph* and illustrate how it can be used in protocol analysis.

This paper contains five sections. In section 2, we briefly introduce the notion of security

---

[1]Here, and in the rest of the paper, we will use the term *logics* and *protocol analysis methods* to refer primarily
to the logics presented in references [2], [10] and [11], or other similar logics.

effectiveness and illustrate its use; i.e., how a specific threat function can be rendered infeasible by the properties of system functions. In section 3, we provide examples to illustrate the consequence of disregarding dependencies among derived and assumed properties, on the results of protocol analysis. In section 4, we present a *dependency graph* that underlies much of the analyses and illustrate its use. Section 5 concludes this paper. Additional examples of protocol dependencies appear in [12].

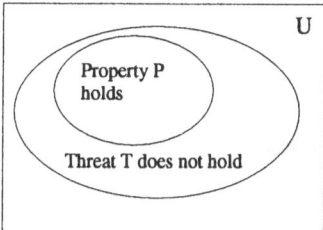

Fig 1. Security Effectiveness

2 Security Effectiveness

In figure 1, we illustrate the intuitive notion of *security effectiveness* of a system. The figure shows a set of reachable system states U. Threats are properties of intruder actions in a given environment. System properties are introduced to counter a specific set of threats. If P is a property that counters a threat T (i.e., $P \Rightarrow \neg T$), then the set of system states in which T does not hold contains the set of system states in which property P holds.

Intuitively, we say that a system security function is *effective* if its (correct) operation counters one or more identified threats. To reason about the effectiveness of a security function, it is useful to think of threats as threat functions that operate in the environment of system use, and to assume that the effects of these functions can be specified in the same way as that of the security functions; i.e., by specifying their properties. Then, *to demonstrate the effectiveness of a security function, we need only show that the properties of the system security functions and those of the threat functions cannot coexist.*

Properties of security functions have been derived in the past in various areas of computer system security. For example, functions implementing an information flow policy may have the property that "if information flows from variable x to variable y, the security level of y dominates that of x." Functions exhibiting this property could counter the threat properties of "unauthorized downgrading of classified data" or "unauthorized flow of sensitive information," which characterize functions implemented by Trojan Horses in unprivileged (malicious) application programs. Thus the effectiveness of the information-flow property can be established with respect to the threat of confidentiality violations by Trojan Horses.

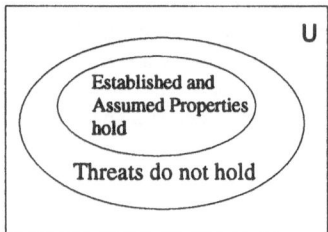

Fig 2. Threat Countermeasures - Established
and assumed properties

2.1   Property Dependencies

Cryptographic protocols and the logics used to derive their properties typically make several
assumptions about the system and about the operating environments. For example, the logics
for the analysis of cryptographic protocols commonly assume that (1) message integrity is
preserved, (2) clocks are synchronous, (3) timestamps are monotonic, (4) encryption keys are
unbreakable, (5) message sender can distinguish his own message from those that are sent by
others, (6) encryption keys are generated randomly, in a manner which does not allow their
prediction from the knowledge of old keys, and so on. The results derived using logics, are,
hence, dependent on these assumptions. These assumptions, if justified, imply that certain
threats cannot hold. For example, in cryptographic protocols which use long-term keys for
encryption, the assumption about the strength (complexity) of the encryption mechanism implies
that some threats of cryptanalysis cannot be successful over the lifetime of the long-term key.
Hence, system states in which assumptions hold are a subset of those in which the *assumed
threats* do not hold. That is,

Assumed Properties $\rightarrow \neg Threats_{\text{ASSUMED}}$.

Properties of the system, which are established by system analysis, imply that another set of
threats do not hold. That is,

Established Properties $\rightarrow \neg Threats_{\text{COUNTERED}}$

The nature of the established and assumed properties differs only in the sense that the former is
justified, whereas the latter may not necessarily be justified. Together, the established protocol
properties and the assumed properties imply that a certain set of threats cannot materialize.
Figure 2 shows a set of reachable system states, and the subset of states in which established
and assumed properties hold. This subset of states is contained in the set of states in which the
threats are negated by the assumed and established properties.

2.2 Effectiveness Analysis

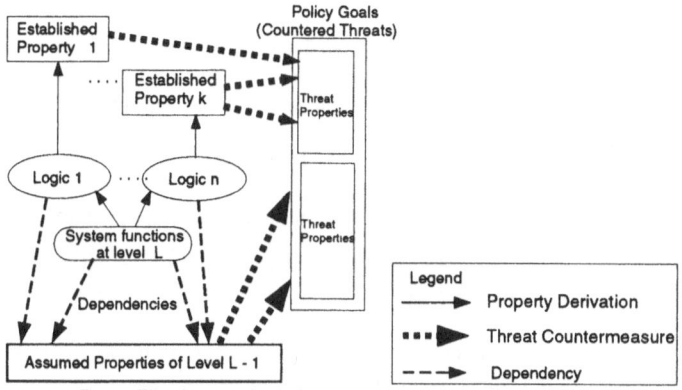

Fig. 3: Effectiveness Analysis.

Figure 3 illustrates the general approach to security effectiveness analysis. The figure shows three types of arcs - straight, dashed, and bold, denoting property derivation, property dependencies and threat countermeasures, respectively.

System function properties (represented as boxes labeled Established Property 1 through k) are derived by analyzing the system functions at some level L (represented as the oval in the center of the figure) using logics (represented as ellipses labeled Logic 1 through n). The system functions, as well as the logics used to derive the properties, make assumptions about the protocols at level $L - 1$ or lower (represented as the rectangle at the bottom of the figure). The validity of derived results is checked by taking into account the property dependencies, and the assumptions about the environment of operation. The properties so derived are analyzed for their ability to invalidate a set of threats (represented as rectangles on the right). If the set of threats that are rendered infeasible by the system properties (both established and assumed) corresponds to those that are required to be countered by the system, then one can say that the system is *security effective*. In figure 3, we note that the set of threats that are rendered infeasible by assumed properties is disjoint from the set of threats that are countered by properties which are established using logics. This follows from the fact that if a property that counters a threat is established, then it need not be assumed.

The correspondence between assumed/established properties and threats, however, is not necessarily one to one. Often, a threat property may be expressible as conjunctions and/or disjunctions of component threat properties. In order to counter the threat, the system must support properties which are at least as strong as the negation of the threat property. For example, a threat property may be of the form $T \equiv T_1 \vee T_2 \vee T_3$. If *Properties of system* $\Rightarrow (\neg T_1)$ $\wedge (\neg T_2) \wedge (\neg T_3)$, then the system is said to be effective in resisting the threat T.

2.3 An Example

Consider the threat of plaintext disclosure by cryptanalysis; i.e., a message encrypted with key K becomes known to a principal who does not have key K. Suppose that principals in the system can find the plaintext message contents without knowing the decryption key in only one of the following ways, namely

T_1 Obtain the plaintext message from some other principal (i.e., the plaintext message is leaked);

T_2 Obtain the ciphertext of a message and,

$\quad$ T_{2a} obtain the key K from some principal in the session (i.e., the key is leaked), and use it to decrypt the ciphertext message;

$\quad$ T_{2b} derive the plaintext by conducting cryptographic analysis on some N_1 or more known plaintext and ciphertext block pairs without possessing or deriving the key; or

$\quad$ T_{2c} derive the key K from the ciphertext by conducting cryptographic analysis on some N_2 or more plaintext and ciphertext block pairs encrypted with the key, and then decrypt the ciphertext.

Threat T_2 typically can not be assumed away in open, distributed systems, since any principal with access to the network can obtain ciphertext messages. However, this need not affect the effectiveness of the system, since other properties may render the possession of ciphertext useless in finding the plaintext. For example, if the system is known to support the properties P_1, P_2, P_3 and P_4 below, then threat T_2 may not pose a problem.

P_1 Principals maintain the secrecy of the key and of the message contents (i.e., the principals handling the encryption key and plaintext message do not use untrusted code).

P_2 The plaintext cannot be derived from the ciphertext without the key using the knowledge of fewer than N_1 plaintext-ciphertext block pairs.

P_3 The key K cannot be derived from the knowledge of less than N_2 plaintext and ciphertext block pairs.

P_4 The lifetime of key K and encryption throughput are limited such that the total number of plaintext and ciphertext block pairs generated using K is less than the minimum of N_1 and N_2.

The threat can be denoted as

$$T \equiv [T_1 \vee (T_2 \wedge (T_{2a} \vee T_{2b} \vee T_{2c}))]$$

>From the properties of the system, we can deduce that

$$P_1 \Rightarrow \neg(T_1)$$

$$P_1 \Rightarrow \neg(T_{2a})$$

$$(P_2 \wedge P_4) \Rightarrow \neg(T_{2b})$$

$$(P_3 \wedge P_4) \Rightarrow \neg(T_{2c}) \text{ Hence,}$$

$$P_1 \wedge P_2 \wedge P_3 \wedge P_4 \Rightarrow$$

$$\neg(T_1) \wedge \neg(T_{2a}) \wedge \neg(T_{2b}) \wedge \neg(T_{2c})$$

$$\Rightarrow \neg(T)$$

3   Modeling Threats

To argue that properties of system functions counter properties of security threats, it is essential to model system functions and threats. The invariants (P) obtained by analyzing the system functions imply that specific threat properties (T) are infeasible (i.e., $P \Rightarrow \neg T$) if the functions are implemented correctly.

For example, the specification and verification of cryptographic protocols of [3] uses a state transition model whose semantics allow the specification of system function properties as well as properties of threats. In the formal policy model of [3], the notion of an *intruder process* formally models external threats. Intruder actions (i.e., security threats) are modeled as state transforms which can be initiated by untrusted processes or intruder processes. The policy properties (P) are restrictions enforced on state transforms. If these restrictions are correctly enforced, the transforms that describe intruder actions are not permissible, and hence, the threat(s) due to these "illegal" state transforms (i.e., which result in unsafe states) are countered.

An example of implicit threat modeling can be found in [13], where the properties of message integrity intruder actions are axiomatized in relation to the design parameters of message integrity protection mechanisms. This axiomatization aids in comparing threat properties to those of the message integrity protection goals.

4   Examples of Protocol Dependencies

In earlier work [8], we noted that the logics for cryptographic protocols must ensure that whenever the required assumptions or policies are not satisfied by a protocol, the use of the logic's inference rules should fail to achieve desired, but unsupported, conclusions. In this section, we provide examples of protocol analyses using a fairly well understood logic [2] to illustrate the consequence of disregarding property dependencies on specific cryptographic properties.

4.1   Kerberos IV

In the analysis of the Kerberos protocol using [2], the protocol is shown to achieve its goal, namely that of securely distributing session keys to the two parties A and B in the presence of a hostile environment. At the end of the protocol, both parties are said to obtain "beliefs" in the origin of the key, namely the authentication server, and on the *freshness* of the key, meaning that it has never been used prior to the session. However, the results derived are dependent on the lifetime constraints of the long-term keys, and in particular, can be false if the number of available plaintext and ciphertext pairs are not limited by limiting the lifetime of those keys. We reproduce the simplified protocol description used in the analysis [2].

Protocol Description

1. $A \to S : A, B$

2. $S \to A : \{T_s, L, B, K_{ab}, \{T_s, L, Kab, A\}_{K_{bs}}\}_{K_{as}}$

3. $A \to B : \{T_s, L, K_{ab}, A\}_{K_{bs}}, \ \{A, T_a\}_{K_{ab}}$

4. $B \to A : \{T_a + 1\}_{K_{ab}}$

The notation $\{M\}_k$ denotes plaintext message M encrypted with key k. T_s is a timestamp that the server attaches to its messages. The analysis shows that at the end of the protocol run, both A and B obtain beliefs about the source that generated the key K_{ab}, and that K_{ab} is a key that was not used prior to the start of the session [2]. This is represented as:

A **believes** $A \xleftrightarrow{K_{ab}} B$

B believes $A \xleftrightarrow{K_{ab}} B$.

The protocol, however, can be vulnerable to at least two types of attacks. For example,

1. If encryption is performed using DES [7] (in Cipher Block Chaining mode since the text exceeds one block), party A can use all 2^{56} values of keys K to encrypt known text $\{T_s, L, K_{ab}, A\}$. Then A can compare each $\{T_s, L, K_{ab}, A\}_K$ with $\{T_s, L, K_{ab}, A\}_{K_{bs}}$. If they match, then A can conclude that $K = K_{bs}$.

2. Party A can obtain known plaintext-ciphertext block pairs by repeatedly sending requests to the authentication server S for a key to communicate with B. If A can obtain sufficient known plaintext-ciphertext block pairs, then A may exploit the knowledge of those pairs and the DES function to derive the key K_{bs}, since the only unknown component in the encryption is the key (the DES function itself is not a secret).

The first attack can be countered by using confounding text within the encrypted text, hence decreasing the probability of success in the known plaintext attack. However, the presence of confounding text cannot prevent the intruder from obtaining known plaintext and ciphertext block pairs in the second attack. This is true because the presence of a confounder in the plaintext only affects its neighboring ciphertext block in Cipher Block Chaining mode [17]. However, from the complexity of the encryption algorithm and the size of the block cipher (i.e., those of DES in this case), a limit may be derived on N, the number of known plaintext-ciphertext block pairs required to uniquely determine the encryption key. If the protocol does not limit the lifetime of the long-term key, K_{bs}, such that the total number of plaintext-ciphertext block pairs is less than N, then the results derived using logics are not justifiable, since A could uniquely derive K_{bs}, and send a key K'_{ab} to B.

While it seems essential that the lifetimes of all long-term keys must be limited, the dependency of message secrecy on their non-derivability from the plaintext-ciphertext pairs may not exist in all protocols. For instance, the protocol proposed by Davis and Swick [4], and implemented as the ENC_TKT_IN_SKEY option of the Ticket Granting Server exchange in [14] eliminates this dependency by using limited lifetime session keys instead of the long-term keys, thereby reducing the amount of ciphertext that can be obtained using any one key (viz., [12]). However, that protocol still depends on the strength of the ticket-granting server's long-term key; i.e., K_{TGS}. The lifetime of this key can be limited appropriately.)

4.2 Needham-Schroeder Shared Key Protocol

The analysis of the Needham-Schroeder protocol [16] using the BAN logic [2] indicates that the protocol falls short of its goals, namely that of distributing a fresh session key K_{ab} to A and B, and in particular, causing the two parties to "believe" that the key is fresh and was created by the right source (i.e., the authentication server). The simplified protocol description is given below:

Protocol Description

1. $A \rightarrow S : A, B, N_a$

2. $S \rightarrow A : \{N_a, B, K_{ab}, \{K_{ab}, A\}_{K_{bs}}\}_{K_{as}}$

3. $A \rightarrow B : \{K_{ab}, A\}_{K_{bs}}$

4. $B \rightarrow A : \{N_b\}_{K_{ab}}$

5. $A \rightarrow B : \{N_b - 1\}_{K_{ab}}$

At the end of the protocol run, according to the analysis, party B does not have evidence to believe that K_{ab} has been issued during the current session by the server, since the key is delivered to B in message 3, which has no component which B believes to be fresh. While this result is reasonable, it does not necessarily mean that the protocol is vulnerable to replay attacks, since B can still verify whether A knows K_{ab} by the handshake message exchange (i.e., messages 4 and 5). Replay of message 3 from a previous session would be successful only if the party which replays the message has also compromised the expired session key $K_{ab_{(expired)}}$ [5]. Replay of message 3 without the knowledge of $K_{ab_{(expired)}}$ would only allow the intruder (say X) to obtain $\{N_b\}_{K_{ab_{(expired)}}}$, but would not enable X to reply with $\{N_b - 1\}_{K_{ab_{(expired)}}}$. The ciphertext $\{N_b\}_{K_{ab_{(expired)}}}$ would not be of much use to X in deriving $K_{ab_{(expired)}}$, since the corresponding plaintext nonce N_b is unknown to X. The protocol itself does not produce plaintext-ciphertext pairs for the session key. If suitable confounders would be added to plaintext and if the lifetime of the session keys would be limited such that the maximum number of known plaintext-ciphertext block pairs available becomes insufficient to derive an old session key by cryptanalysis during the (limited) lifetime of the long-term key K_{bs}, this protocol would be less vulnerable to replay attacks. (Note that the Kerberos protocol, which was derived from that of Needham and Schroeder's, includes confounders and a mechanism to limit the lifetime of session keys.)

The use of a logic in protocol analysis may inadvertently introduce assumptions about the underlying system. For example, the analysis of the Kerberos and Needham-Schroeder protocols above removes the first message from both protocols during the idealization step [2]. An assumption is made that a separate means of establishing the freshness of the second message exists [8]. For Kerberos, this assumption implies that (1) an *authenticated* source of time is available to all parties *prior* to running the authentication protocol; and (2) party A has a replay detection cache, and therefore has non-volatile memory, to verify the freshness of T_s. Assumption (1) is unrealistic in most systems, and (2) is unnecessarily restrictive (e.g., Kerberos is used with diskless workstations). For the Needham-Schroeder protocol, the assumption that the second message is fresh implies that party A remembers all the nonces sent in message 1 and not received in message 2, during the lifetime of the shared long-term key K_{as}. Again, this is both unrealistic and unnecessarily restrictive (e.g., use of non-volatile memory becomes necessary).

4.3 Andrew Secure RPC Handshake

In this example, we illustrate a protocol that allows an attacker to force the reuse of an expired session key, which is unknown to the attacker. The purpose of the attack is that of obtaining plaintext-ciphertext pairs, which can lead to the eventual discovery and replay of that key.

The simplified protocol description [2] for the Andrew Secure RPC handshake is shown below.

Protocol Description

  1. $A \to B : A, \{N_a\}_{K_{ab}}$

  2. $B \to A : \{N_a + 1, N_b\}_{K_{ab}}$

  3. $A \to B : \{N_b + 1\}_{K_{ab}}$

  4. $B \to A : \{K'_{ab}, N'_b\}_{K_{ab}}$

At the end of the protocol run, party A has no evidence to believe that K'_{ab} has indeed originated from B, since A does not use any handshake messages to verify whether B actually knows K'_{ab} or not. Hence, an intruder may replay message 4 consisting of some session key which is expired, and possibly make party A encrypt some message with $K'_{ab(expired)}$.

Unlike the Needham-Schroeder protocol, this protocol is vulnerable to replays irrespective of whether the expired key has been compromised or not. In addition, in the Needham-Schroeder protocol, an intruder could not obtain known plaintext-ciphertext block pairs for a session key beyond a preset limit, if this limit is specified and enforced within the underlying system (e.g., within the cryptographic facility [15]). In contrast, the Andrew Secure RPC Protocol permits an intruder to obtain known plaintext-ciphertext pairs even beyond the limits set by the underlying system. For example, the fourth message can be repeatedly inserted by party B possibly causing an unsuspecting party A to encrypt text with an expired session key K'_{ab}. Thus, unlike the Needham-Schroeder protocol, this protocol may, in fact, aid the discovery and forced reuse of the expired session key K'_{ab} by B.

Analyses using logics cannot distinguish between the properties of Needham-Schroeder and

Andrew-Secure RPC protocols. The first protocol is shown to be weak because it does not result in the belief B believes $A \xleftrightarrow{K_{ab}} B$, and the second protocol is shown to be weak because it does not yield A believes $A \xleftrightarrow{K_{ab}} B$. These examples, and the Kerberos IV example, show that the nalyses cannot take into account the dependency (or lack thereof) of the protocol properties on specific cryptographic properties (e.g., limited key lifetime) that may differ from protocol to protocol.

5   Dependency Graph

We have identified several property dependencies in cryptographic protocols. Figure 4 shows a graph which has properties (or assumptions) as its vertices and arcs showing dependencies. Cyclic dependencies are not illustrated because their elimination is reasonably well-understood [6]. The dependencies shown here are of two types - (policy) goal-specific and protocol-specific. The first type of dependencies (shown with straight arcs in Figure 4) appears in the analysis of all protocols that aim to establish specific (policy) goals (shown with dark ovals). The second type of dependencies (shown with dashed arcs) appears in the analysis of individual protocols, and take into account specific protocol attributes (e.g., types of encryption keys used, message freshness criteria, and so on). We briefly discuss both *goal-specific* and *protocol-specific* dependencies shown in the graph of Figure 4. In the dependency graph, the leaf nodes should be properties which can be related to actual design parameters (e.g., [13]).

We begin at the root of the dependency graph and proceed towards the leaves. The property *identity authentication using distributed session key or ticket* is dependent on

1. *Freshness of Session Key distributed*: Without this property, an old (possibly compromised) session key could be used for encryption, and the authenticity of encrypted messages cannot be ensured.

2. *Message sender recognition*: This is the property by virtue of which the sender of a message can be held accountable for the contents of the message. Recognizing the sender's identity is necessary in authenticating specific messages (and their contents), since not all parties are authorized to send all messages[2].

3. *Possession of Session Key only by trusted parties*: If this property is not supported, the source of an encrypted message cannot be identified with certainty.

The property *Freshness of Distributed Session Key* is established using

1. *Nonce Verification Rule* [2], and

---

[2]For instance, a message which has a secret key or ticket may originate only from a ticket granting server.

2. *Jurisdiction Rule* [2].

Hence, the *freshness of distributed session keys* is dependent on the validity of these two rules. In some protocols (e.g., [16]), freshness of session keys is also dependent on the fact that *old session keys are unbreakable*.

In protocols which fail to establish the freshness of session keys and do not allow session members to detect lack of key freshness using handshake messages (e.g., Andrew Secure RPC Protocol), the freshness of session key would depend on the lifetime restriction on the session key.

Possession of session keys only by trusted parties (i.e., confidentiality) can be ensured only if

1. the party which generates the session key has the authority to do so. Otherwise, the key may be leaked to intruders by the entity (process or machine) which generates the key.

2. encrypted message contents are secret among session members. Since session keys are often encrypted with key encrypting keys, if secrecy of encrypted message contents cannot be ensured, the confidentiality of the session key cannot be established.

3. session members are honest (i.e., they have only trusted code). Without this property, confidentiality cannot be established.

In protocols which transmit plaintext messages, confidentiality is dependent on whether responses to plaintext messages are checked to see whether the plaintext was modified (e.g., [12]).

The validity of the *nonce verification rule* in analyzing protocol message exchanges depends on whether message freshness can be the basis for transforming beliefs of the form X **believes** Y **said** $Message$ to X **believes** Y **believes** $Message$, if X **believes** that $Message$ has some contents which are fresh. The validity of this rule is also dependent on the validity of the *message meaning rule* in establishing the statement X **believes** Y **said** $Message$.

This rule has an inherent dependency on the method used to establish the freshness of message contents. Freshness of entire messages can be inferred from the freshness of specific message contents only if message integrity is preserved. Hence, freshness of message contents is dependent on *message integrity*. Message freshness may be established using (and hence, is dependent on) one or more of

• Synchronous nature of clocks,

• Randomness of strings (nonces), or

• Monotonicity of timestamps.

Depending on the types of replay threats assumed, the establishment of message freshness may have additional property dependencies [9]. In addition, in systems which permit the simultaneous generation of the same nonce in two or more sessions, the freshness of message contents would depend on the fact that principals can identify the source which generates the nonce (e.g., an interleaved attack on the ISO protocol in [1]).

The validity of *jurisdiction rule* [2] in analyzing protocol messages is dependent on the

1. competence of the key generation source in generating random keys.

2. competence of the key generation source in preserving the secrecy of the key, and

3. the *message meaning rule*, since the former uses the latter to establish the message source.

Message meaning rule [2] can be applied for the analysis of protocol messages only if the sender of the message can be identified. Without message sender recognition, statements of the form "X **said** $Message$" cannot be derived. Both *nonce verification rule* and *jurisdiction rule* are use *message meaning rule*, and hence, are dependent on it.

Message sender recognition is valid only if message integrity is preserved. Without message integrity, messages or parts thereof can be inserted or modified by intruders. In addition, depending on the types of keys that are used by the protocol for encryption, message sender recognition is also dependent on the fact that

• Long-term keys are unbreakable or

• Current session-keys are unbreakable.

Secrecy of encrypted material is a property which can hold only if the following properties are supported:

1. *Message integrity*: If message integrity protection mechanisms guarantee detection of message integrity compromises, then secrecy of messages which lack integrity may not be a concern. Unfortunately, this may not always be the case. For instance, in cipher-block chaining, message portions of known plaintext-ciphertext pairs can be spliced together to form new messages which pass checksum tests with high probability [17]. The contents of such spliced messages are not secret. Hence, secrecy is a function of whether integrity is preserved or not.

2. *Non-derivability of plaintext*: from the knowledge of ciphertext, without knowing the encryption key.

3. *Honesty of session members*: This property can be enforced by using only trusted software on trusted hardware platforms.

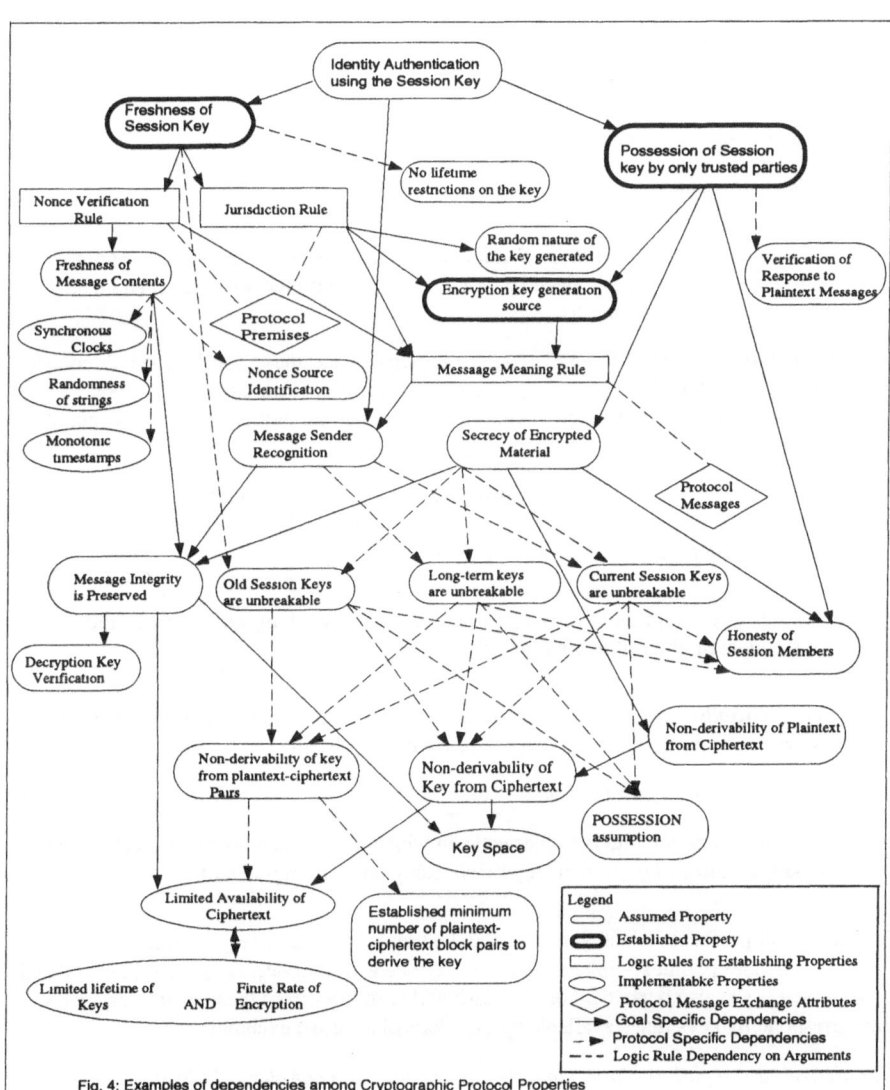

Fig. 4: Examples of dependencies among Cryptographic Protocol Properties

Further, based on the types of keys used for encrypting messages, *secrecy of encrypted material* is dependent on one or more of the following properties:

• Long-term keys are unbreakable.

• Session keys are unbreakable.

• Old session keys are unbreakable.

Message integrity depends on

1. *verification of decryption keys*: Without decryption key verification, one cannot distinguish between a message which has been decrypted with the wrong key, and a message whose integrity is not preserved.

2. *limited availability of plaintext-ciphertext pairs*, since these can be used to compose (splice) known ciphertext message portions to form new messages. The likelihood of success in this attack is a non-decreasing function of the number of available plaintext-ciphertext pairs.

3. *Size of encryption key space*, in protocols which use encryption for secrecy. If the encryption key space is small, an exhaustive search may have high rate of success, and may comprise a threat to the integrity of messages.

Unbreakability of encryption keys (i.e., current session keys, long-term keys or old session keys) is dependent on the following properties:

1. *Non-derivability of keys from ciphertext*: This property is dependent on the size of the encryption key space and the strength of the cryptosystem.

2. *Honesty of session members*, and

3. *Possession Assumption*: All other ways in which the key can be compromised are very unlikely to succeed (e.g., accidentally chancing upon the right key.)

In addition, in protocols which generate plaintext-ciphertext pairs (e.g., Needham-Schroeder Protocol generates plaintext-ciphertext pairs encrypted with long-term keys [16]), unbreakability of these keys is also dependent on non-derivability of keys from these pairs encrypted with the corresponding keys made available by the protocol message exchanges.

Non-derivability of plaintext from ciphertext is dependent on *non-derivability of the encryption key from ciphertext*. The latter property is dependent on the size of the encryption key. If the key size is small, this property cannot be supported, since an exhaustive search can be used to deduce the right key (Note that this attack need not be dependent on the availability of plaintext-ciphertext pairs. If the decryption algorithm is used on a ciphertext block with all possible keys, the ciphertext will decrypt properly only if the right key is used) with a high probability of success.

Non-derivability of encryption key from plaintext-ciphertext pairs made available by the protocol messages is dependent on the limited availability of such pairs, and on the established minimum number of plaintext-ciphertext block pairs that are required to derive the key (e.g., [18]). *Limited availability of ciphertext* is dependent on the fact that the key lifetime is limited, and the rate of encryption is limited.

6   Conclusions

To demonstrate the security effectiveness of cryptographic protocols, protocol properties must be specified in terms of the threats that need to be countered and the properties of lower level mechanisms that are assumed. The protocol properties derived using logics may then be validated by analyzing whether the underlying properties are supported. To perform this analysis, property dependencies must be identified. Ignoring these dependencies can lead to erroneous (false positive or false negative) conclusions. Once the properties established and assumed are all validated, then their threat resistance features can be analyzed.

In this paper, we illustrate the security effectiveness of cryptographic protocols with respect to some general cryptographic threats; e.g., the availability of plaintext-ciphertext pairs to an attacker. Deriving the dependency graph such as the one presented in this paper, is a necessary step in analyzing the security effectiveness of cryptographic protocols. The dependency graph may also be used to derive quantitative measures of security effectiveness based on the evaluation levels of the leaf properties.

Security effectiveness analysis is intended to complement existing analysis methods to make them useful in practice. The approach of evaluating the effectiveness of functions by property abstraction, dependency analysis, and goal analysis (in this case, threat resistance) is extensible to the analysis of other system functions.

Acknowledgements

The first two authors are grateful to Tom Tamburo and Marty Simmons of IBM Federal Systems Company for their continued support and encouragement.

References

[1] R. Bird, I. Gopal, A. Herzberg, P. Janson, S. Kutten, R. Molva, and M. Jung. Systematic design of two-party authentication protocols. *IEEE Journal on Selected Areas of Communications*, 1993.

[2] M. Burrows, M. Abadi, and R.M. Needham. A logic of authentication. Technical Report 39, DEC Systems Research Center, 1989.

[3] P.-C. Cheng and V.D. Gligor. On the formal specification and verification of a multiparty session protocol. *IEEE Symposium on Research in Security and Privacy*, 1990.

[4] D. Davis and R. Swick. Network security via private-key certificates. *ACM Operating Systems Review*, 24(4), October 1990.

[5] D.E. Denning and G.M. Sacco. Timestamps in key distribution protocols. *Communications of the ACM*, 1981.

[6] Federal criteria for information technology security. Vol. 1, Protection Profile Development, version 1.0, National Institute of Standards and Technology and National Security Agency, 1992.

[7] Des modes of operation. Federal Information Processing Standard, National Bureau of Standards, 1980.

[8] V.D. Gligor, R. Kailar, S.G. Stubblebine, and L. Gong. Logics for cryptographic protocols - virtues and limitations. *IEEE Computer Security Foundations Workshop*, 1991.

[9] L. Gong. Variations on the themes of message freshness and replay. *IEEE Computer Security Foundations Workshop*, June 1993.

[10] L. Gong, R. Needham, and R. Yahalom. Reasoning about beliefs in cryptographic protocols. *IEEE Computer Society Symposium on Research in Security and Privacy*, 1990.

[11] R. Kailar and V.D. Gligor. On belief evolution in authentication protocols. *IEEE Computer Security Foundations Workshop*, 1991.

[12] R. Kailar, V.D. Gligor, and L. Gong. On the security effectiveness of cryptographic protocols. Technical Report 93066, Electrical Engineering Department, University of Maryland, College park, MD 20742, December 1993.

[13] R. Kailar, V.D. Gligor, and S.G. Stubblebine. Reasoning about message integrity. *IFIP Conference on Dependable Computing for Critical Applications*, January 1994.

[14] J. Kohl and B.C. Neuman. The kerberos network authentication service, draft (revision 5). MIT Project Athena.

[15] S.M. Matyas. Key handling with control vectors. *IBM Systems Journal*, 30(2), 1991.

[16] R.M. Needham and M.D. Schroeder. Using encryption for authentication in large networks of computers. *Communications of the ACM*, 21(12), December 1978.

[17] S.G. Stubblebine and V.D. Gligor. On message integrity in cryptographic protocols. *IEEE Symposium on Research and Privacy*, 1992.

[18] M. Weiner. Cryptanalysis of short rsa secret exponents. *IEEE Transactions on Information Theory*, 36(3), May 1990.

Assessment of Dependability

Assessing the Dependability of Embedded Software Systems Using the Dynamic Flowgraph Methodology

Chris Garrett, Michael Yau, Sergio Guarro and George Apostolakis
School of Engineering and Applied Sciences
University of California
Los Angeles, California 90024-1597
U.S.A.

Abstract

The Dynamic Flowgraph Methodology (DFM) is an integrated methodological approach to modeling and analyzing the behavior of software-driven embedded systems for the purpose of dependability assessment and verification. The methodology has two fundamental goals: 1) to identify how events can occur in a system; and 2) to identify an appropriate testing strategy based on an analysis of system functional behavior. To achieve these goals, the methodology employs a modeling framework in which models expressing the logic of the system being analyzed are developed in terms of causal relationships between physical variables and temporal characteristics of the execution of software modules. These models are then analyzed to determined how a certain state (desirable or undesirable) can be reached. This is done by developing timed fault trees which take the form of logical combinations of static trees relating the system parameters at different points in time. The resulting information concerning the hardware and software states that can lead to certain events of interest can then be used to increase confidence in the system, eliminate unsafe execution paths, and identify testing criteria for safety critical software functions.

Keywords: embedded systems, software dependability, software safety, fault tree analysis, Dynamic Flowgraph Methodology

1 Introduction

Embedded systems are systems in which the functions of mechanical and physical devices are controlled and managed by dedicated digital processors and computers. These latter devices, in turn, execute software routines (often of considerable complexity) to implement specific control functions and strategies. Embedded systems have gained a pervasive presence in all types of applications, from the defense and aerospace to the medical, manufacturing, and energy fields. The advantage of using embedded systems is that very sophisticated and complex logic can be executed by relatively inexpensive microprocessors loaded with the appropriate software instructions. The originally implemented logic can also be modified at any point in the life of the system it is designed to control by uploading new software instructions. Due to this flexibility, embedded systems are being used increasingly in a number of safety critical applications.

While the cost-effectiveness and flexibility of embedded systems are almost universally recognized and accepted, it is also increasingly recognized that the task of providing high assurance of the dependability and safety of embedded system software is becoming quite difficult to accomplish, due precisely to its very complex and flexible nature. Software, unlike hardware, is unique in that its only failure modes are the result of design flaws as opposed to any kind of physical mechanisms such as aging[1-3]. As a result, traditional safety assessment techniques, which have tended to focus upon physical component failures rather than system design faults, have been unable to close the widening gap between the extraordinarily powerful capabilities of modern software systems and the levels of reliability which we are capable of exacting from them.

Although the recognition is growing that it would be very desirable, for reliability and safety assurance purposes, to integrate in one process the modeling and analysis of the hardware and software components of a digitally controlled system[2], the current state of the art does not offer practically implementable blueprints for such an approach. The approaches that have been proposed and/or developed in the past generally follow the philosophy of separating out the hardware and software portions of the assurance analysis. The hardware reliability and safety analysts evaluate the hardware portion of the problem under the artificial assumption of perfect software behavior. The software analysts, on the other hand, usually attempt to verify or test the correctness of the logic implemented and executed by the software against a given set of design specifications, but do not have any means to verify the adequacy of these specifications against unusual circumstances developing on the hardware side of the overall system, including hardware fault scenarios and conditions not explicitly envisioned by the software designer.

Currently, embedded system software assurance is not treated much differently from that of any other type of software for real-time applications (such as communications software). Three principal types of software assurance philosophies can be recognized in the published literature, which are assurance by testing, formal verification, and discrete state simulation.

Assurance by testing is the most common approach. Testing is often performed by feeding random inputs into the software and observing the produced output to discover incorrect behavior. Because of the extremely complex nature of today's modern computer systems, however, these techniques often result in the generation of an enormous number of test cases. Indeed, Ontario Hydro's validation testing of its Darlington Nuclear Generating Station's new computerized emergency reactor shutdown systems required a minimum of 7000 separate tests to demonstrate 99.99% reliability at 50% confidence[4].

Formal verification is another approach to software assurance which applies logic and mathematical theorems to prove that certain abstract representations of software, in the form of logic statements and assertions, are consistent with the specifications expressing the desired software behavior. Recent work has been directed at developing varieties of this type of technique specifically for the handling of timing and concurrency problems[5,6]. The abstract nature of the formalisms adopted in formal verification make this approach rather difficult to use properly by practitioners with non-specialized mathematical backgrounds. This practical difficulty is compounded by the growth in complexity and size of the process control software of the present generation. Finally, the issue of modeling and representation of hardware/software interaction, which is an important open issue in embedded system assurance analysis, does not appear to have surfaced as one of the current objectives of formal verification research.

The third type of approach to software assurance is one that analyzes the timing and logic characteristics of software executions by means of **discrete state simulation models**, such as queue networks and Petri-nets[7-10]. Simulated executions are analyzed to discover undesirable execution paths. Although this approach can be extended to model combined hardware/software behavior (since the hardware behavior can in principle be approximated in terms of transitions within a set of predefined discrete states), difficulties arise from the "march-forward" nature (in time and causality) of this type of analysis, which forces the analyst to assume knowledge of the initial conditions from which a system simulation can be started. In large systems, many combinations of initial states may exist and the solution space may become unmanageable. A different approach, which reverses the search logic by using fault trees to trace backward from undesirable outcomes to possible cause conditions, offers an interesting solution to this problem, but encounters difficulties due to limitations in its ability to represent dynamic effects, and to the fact that a separate model needs to be constructed for each software state whose initiating causes are to be identified[3,11].

All the methods discussed above have merit, but none direct a special effort toward the philosophy of developing a "systems approach" to tackling the central issue of integrated hardware-software analysis in embedded system assurance. The Dynamic Flowgraph Methodology (DFM) solves this problem by providing a deductive (i.e., reverse causality backtracking) analysis capability, while at the same time also providing the ability to keep track of the complex dynamic effects associated with sequential and time dependent software executions and embedded system behavior.

DFM is based on the Logic Flowgraph Methodology (LFM)[12-15], which is a concept for analyzing systems with limited dynamic features. LFM was originally developed as a method for analyzing/diagnosing plant processes with feedback and feedforward control loops. Research conducted over the past decade has demonstrated LFM's usefulness as a tool for computer-automated failure and diagnostic analysis which shows broader potential applicability and efficiency than most other approaches that have been proposed for such objectives[12,13]. As part of the LFM research effort, models of nuclear power plants and space-systems[14,16] have been derived; in addition, procedures to be applied in an expert system capable of assisting an analyst in the construction of LFM models have been identified[17]. Recently, the use of LFM and other logic modeling approaches for the safety and reliability analysis of aerospace embedded systems has also been investigated[18,19]. The system under consideration in LFM is represented as a logic network relating process parameters at steady state. This LFM model takes the form of a directed graph, with relations of causality and conditional switching actions represented by "causality edges" and "conditional edges" that connect network nodes and special operators. Of these, causality edges represent important process variables and parameters, and conditional edges represent the different types of possible causal or sequential interactions among them. The LFM models provide, with certain limitations, a complete representation of the way a system of interconnected and interacting components and parameters is supposed to work and how this working order can be compromised by failures and/or abnormal conditions and interactions.

DFM aims at extending the LFM concept to analyzing time-dependent systems. Certain features and rules are added to the LFM to address issues relevant in embedded system analysis not covered by LFM. These issues are:

1)   The need for a framework to represent time transitions. Discrete time transitions are almost always present in embedded system software, and often present even in embedded system hardware (eg., as a result of relay actions).

2) The need to identify and represent in a distinguishable fashion the continuous/functional relations and the discontinuous/discrete logic influences that are present in embedded systems.

The Dynamic Flowgraph Methodology (DFM) is a methodology for analyzing and testing embedded system software dependability[20]. It specifically addresses the fact that embedded system dependability analyses cannot be performed effectively if the software and hardware portions of the analysis and qualification process are carried out in relative isolation without a well integrated understanding and representation of the overall system functions and interfaces. Because software testing is expected to remain an important pillar of any dependability verification procedure, this methodology is designed to provide assistance to the software testing process, by helping to identify the most effective test criteria and reduce the amount of "brute force" testing required for dependability verification. Thus, the ultimate objective is to show how, by using effective analysis techniques, the amount of resources to be committed to the embedded system software verification process could be significantly reduced with respect to what would be required if verification were to be conducted by testing only.

The methodology has two fundamental goals: 1) to identify how certain postulated events may occur in a system; and 2) to identify an appropriate testing strategy based on an analysis of system functional behavior. System models which express the logic of the system in terms of causal relationships between physical variables and temporal characteristics of the execution of software modules are analyzed to determine how a certain state (desirable or undesirable) can be reached. This is done by developing timed fault (or success) trees for a given top event (translated in terms of the state(s) of one or more system variables) by backtracking through the model in a systematic manner. These timed trees simply take the form of logical combinations of static trees relating the system parameters at different points in time. The resulting information concerning the hardware and software states that can lead to certain events of interest can then be used to identify those software execution paths and associated hardware and environmental conditions which are potentially capable of leading to serious failures in the implemented system. The verification effort can then be focused specifically on these safety critical features of the system to ensure that such failures cannot be realized. This paper illustrates DFM by applying it to the problem of modeling and analyzing the Titan II Space Launch Vehicle (SLV) Digital Flight Control System (DFCS).

2 Basic Features of DFM

The application of DFM is a two-step process, as follows:

Step 1: Build a model of the embedded system being analyzed.

Step 2: Analyze the model to produce fault (or success) trees which relate the events, in both the physical system and the software, which can combine to cause system failures, including the time sequences in which they occur, and identify system failure (success) modes in the software and associated hardware as prime implicants (multi-state analogue of minimal cut sets) of the timed fault (success) trees.

The first step consists of building a model that expresses the logical and dynamic behavior of the system in terms of its physical and software variables. The second step uses the model developed in the first step to build timed fault (or success) trees that identify logical combinations of hardware and software conditions that cause certain specific system states of interest, and the time sequences in which these conditions come about. These system states can be desirable or undesirable, depending on the objective of the analysis. This is accomplished by backtracking through the DFM model of the embedded system in a systematic, specified manner. The information contained in the fault trees concerning the hardware and software conditions that can lead to system states of interest can be used to uncover undesirable or unanticipated software/hardware interactions, thereby allowing improvement of the system design by eliminating unsafe software execution paths, and to guide functional testing to focus on a particular domain of inputs and system conditions.

DFM models take the form of directed graphs, with relations of causality and conditional switching actions represented by arcs that connect network nodes and special operators. The nodes represent important process variables and parameters, while the operators represent the different types of possible causal or sequential interactions among them. DFM's usefulness as a tool for the analysis of embedded systems derives from its direct and intuitive applicability to the modeling of causality driven processes. DFM models provide, with certain limitations, a complete representation of the way a system of interconnected and interacting components and parameters is supposed to work and how this working order can be compromised by failures and/or abnormal conditions and interactions.

The application of DFM to a simple hardware system is illustrated in Fig. 1, where a valve is used to control a downstream flowrate. In Fig. 1(a), a piping and instrumentation diagram (P&ID) is drawn to describe the functional layout of the system, its components, and other elements of basic engineering data regarding the process. Other important attributes, most notably the ones linked to operational logic and control modes as well as the analyst's own understanding of the system, while not directly contained nor implicitly expressed in the P&ID, are nevertheless represented in the DFM model of the system (Fig. 1(b)). The DFM model is built with physical parameters UP, F, FM, and VX as continuous variable nodes, where:

UP	=	Upstream pressure,
F	=	Flow rate,
FM	=	Flow rate measure,
VX	=	Valve position at the present time,

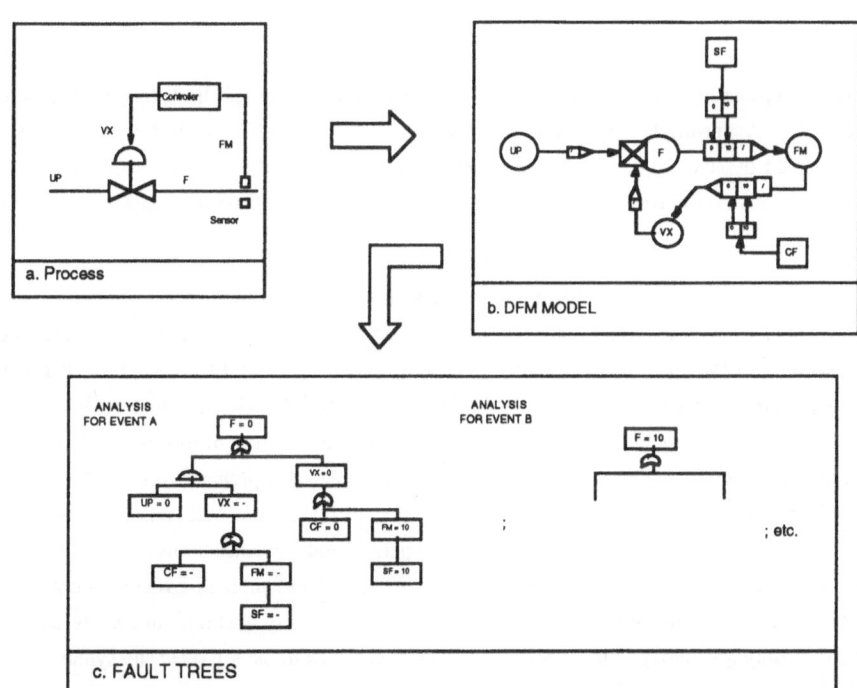

Figure 1 : Example of DFM Model and Fault Analysis

and SF and CF as discrete variable nodes, where:

SF   =   Sensor state,

CF   =   Control Function.

The relationships between parameters are represented by gains in transfer boxes, which may be different for different conditions. Edges connect nodes through transfer boxes. An example of how relationships are represented in the DFM model is the direct proportionality relationship between nodes UP and F. This is represented by a "/" in the transfer box between the nodes. The two nodes are connected through the transfer box using directed edges (Fig. 1(b)). The flowrate sensor, SF, has two possible degraded states. The sensor may fail high (state 10) or fail low (state 0). Accordingly, the relationship between the nodes F and FM may change, as is represented by states 0 and 10 in the transfer box. Failure of the valve control, CF, is handled similarly. State 0 signifies that the valve is failed shut, and state 10 signifies that the valve is failed open.

It should be noted that the results of a DFM analysis are obtained in the form of fault trees, which show how the investigated system/process states may occur. DFM thus shares, in the form of the results it provides, many of the features of fault tree analysis. The difference, however, is that it provides a documented model of the system's behavior and interactions, which fault tree analysis does not provide nor document directly. The most important feature of this methodology is that it is possible to produce, from a single DFM system model, a comprehensive set of fault trees that may be of interest for a given system. This is a most useful feature since, once a DFM model has been developed, it is not necessary to perform separate model constructions for each system state of interest (as is the case in fault tree analysis). Because DFM modeling focuses on parameters, rather than components, it also offers greater modeling flexibility than fault tree analysis, although this flexibility goes along with proportionately more complex modeling rules and syntax.

In Fig. 1(c), the fault tree for the top event, "flow rate is zero," is derived. By working backward through the DFM model starting from the flow rate node F, we can determine that the state, "flow rate F is zero," is caused by either "the valve is shut," OR "upstream pressure UP is zero" AND "the valve may be in any position" (represented by the "don't care" state (-)). This information is implicitly contained in the DFM input operator before the node F. Underlying this input operator is a decision table constructed by determining the states of F from the combinations of the states of UP and VX. Thus, given a particular state of F, the information organized in the decision table can be used in reverse to determine the combinations of UP and VX which cause this particular state of F. This information is explicitly denoted in the resulting fault tree by connecting events with logical AND and OR gates. Backtracking then continues until basic events are reached. To the extent to which it would be applied for systems which do not depend significantly on software, DFM is functionally equivalent to the Logic Flowgraph Methodology (LFM), a technique for computer-automated failure and diagnostic analysis. Details and examples concerning the application of LFM can be found in[12,18].

3.1 Modeling Dynamic Embedded System Behavior

An important issue that needs to be addressed in the modeling and analysis of embedded systems is the dynamic nature of their behavior. For the purpose of modeling such behavior and representing its effects in the system fault trees, the execution of software is modeled as a series of discrete state transitions within the DFM model. The usefulness of this approach is that it provides the ability to represent changes in the system logic at discrete points in time. This allows the development of fault trees which reflect static relationships between the variables at different points in time.

Embedded system software is generally real-time control software which receives real-time data from the physical part of the system and performs appropriate control actions on the basis of this data. The first step in modeling this type of software is to identify those portions of it which explicitly involve time dependencies such as interrupts and synchronization routines. In most cases, the majority of time dependencies are present in the controlling, or "main," module, however other modules may contain time dependent elements and one must make sure that they are isolated. The remainder of the code, which has no real time dependence other than its own execution time, is divided into modules which represent physically meaningful actions. An examination of the specifications, data dictionary, structure chart or even the code itself will usually suggest an appropriate division of modules. The control flow among the above defined software "components" is then represented by a state transition network in which the execution of each particular software component is represented by a transition between states. Associated with each transition is a time which represents the execution time of the software routine to which it corresponds. The modeling of the data flow between the software components is completely analogous to the modeling of causality flow. Associated with each transition in the state transition network is a DFM transfer box which, instead of describing a one-to-one causal relationship between physical variables, describes, in general, several many-to-one mappings between the corresponding software component's inputs and outputs and any relevant global variables.

The state transition network is then incorporated into the DFM model of the hardware system. The principal step in the integration of the software and hardware portions of the model is the identification of all data that is exchanged between the software and the physical world, i.e., the identification of all of the software "images" of hardware parameters and physical variables. Corresponding parameters are then connected through a transfer box and an edge which indicates the direction of information flow. In almost all cases, an interface point is constituted either by a sensor that measures a physical parameter to be input to the software or an actuator which translates a software command into a physical movement. Thus, one has to include in the hardware portion of the model nodes representing both the measured physical parameter and the associated measurement, or the software command and the actuator movement, respectively. The sensors and actuators themselves will be modeled as transfer functions normally representing direct proportionality but also containing conditional faulted mappings which represent different degraded states. The software portion of the model will begin with the measured value nodes which represent the images of physical parameters which are used as input to the software and end with the software commands which are output to the actuators. It is also important to note that, in addition to having a direct causal effect on the physical system via sensor measurements and actuator commands, software actions will, in general, also have an indirect effect on the physical system by conditionally changing the gains between physical variable nodes. This will occur whenever software controls the functions of physical devices such as pumps, valves, etc., and is modeled by edges connecting the relevant software variable nodes with the transfer boxes in question.

Fault trees are developed by an automated backtracking procedure which takes into account the timing effects introduced by the state transition network, resulting in trees which are time dependent. After a top event has been identified and translated in terms of the state(s) of one or more nodes of the system, backtracking proceeds until a transfer box is reached which is associated with a time transition (software module). Development of that branch of the tree is then suspended and the remaining branches are developed until they reach either a basic event or another time transition. The result is a fault tree for the system at time t_n. Then, from the state transition network, the transition is found which maps the system state from time t_n to time t_{n-1}, i.e., the last transition to have occurred. From that transition's transfer box, the input values associated with that transition at time t_{n-1} are found, and backtracking continues as before until, again, all branches end with either basic events or the outputs of other time transitions. This is the fault tree for time t_{n-1}. This process continues backward through successive transitions in the state transition network. The result is a series of linked, static fault trees which represent the state of the system at different times and which are linked through the inputs and outputs of time transitions (software module executions). The nature of the particular system dynamics being studied will provide heuristic rules for determining how far back in time one should go before halting the procedure.

4 The Titan II SLV Digital Flight Control System (DFCS)

The function of the Titan II SLV DFCS[21-22] is to stabilize the vehicle from launch through payload separation. Vehicle attitude control is accomplished via thrust vector control from liftoff through Stage II shutdown, and attitude control thrusters from Stage II shutdown to payload separation. The system also establishes the flight path of the vehicle by implementing all steering commands issued by the guidance system. The Digital Flight Control System consists of:

-        The Missile Guidance Computer and the Flight Control Software

-        The Attitude Rate Sensing System

-        The Inertial Measurement Unit (IMU)

-        Hydraulic actuators

The Inertial Measurement Unit measures the current vehicle orientation and acceleration, while the Attitude Rate Sensing System determines the pitch rate and yaw rate. The measurements from these sensors are then used by the flight control software to determine the appropriate engine nozzle deflection commands to the hydraulic actuators.

5 Modeling the Titan II SLV DFCS with DFM

A DFM model is developed for the Titan II SLV DFCS for Stage I flight. The essential parameters for capturing the behavior of this embedded system are listed in Table I.

The DFM model of the embedded system is shown in Figure 2. Due to the limitation in space, the representation of the flight control software is expanded in all the detail in Figure 3. In a DFM model, the circles are nodes representing hardware or software parameters, the boxes are DFM operators describing the relationships between the nodes, the solid arrows are causality edges indicating functional relationships, and the dotted arrows are conditioning edges representing conditional relationships. The principal step in integrating the software and hardware portions of the model is to identify all data that is exchanged between the software and the physical world. In the Titan II system, the IMU measures the gimbal angles , , , and $_R$. These measurements are represented in the software as the variables H7CH8, H7CH9, H7CH10, H7CH11, H7CH12, H7CH13, and H7CH16. Similarly, the measurements a_{um}, a_{vm}, and a_{wm} are read in as accelerometer counts and represented as D5NUC, D5NVC, and D5NWC in the software. Finally, the pitch rate and yaw rate are represented as W7P1IL, and W7Y1IL in the software. On the other hand, the D/A outputs W2DA11, W2DA21, W2DA31 in the software are converted into the thrust chamber deflection angles p, R, and Y.

In Figure 2, transfer boxes A and F model the IMU, where B_i and B_f are compared to generate the gimbal angle measurements , , , and R, and the accelerations a_{um}, a_{vm}, and a_{wm} are measured. Transfer box D represents the gyros which measure the pitch rate PR and yaw rate YR. These sensor inputs are used by the flight control software to calculate the thruster deflections p, R, and Y. Finally, transfer box C represents the rocket itself in which the body axes and the current acceleration depends on F, M, I, CM, p, R, and Y.

Figure 3 is the time-transition network constructed based on the control flow in the flight control software. The transition boxes represent software modules, and the nodes represent the essential input and output parameters of the software modules. This figure shows how the software calculates the commands to the thrust chambers from the sensor readings.

Note that most of the process parameter nodes are linked to transfer/transition boxes via causality edges. A conditioning edge links the node N8L to the transition box BLOCK 4. N8L is the time kept by the software, and this variable dictates the maneuvers to be executed. The execution of different maneuvers causes a discrete jump in the software in the form of using different equations to calculate the desirable body axes.

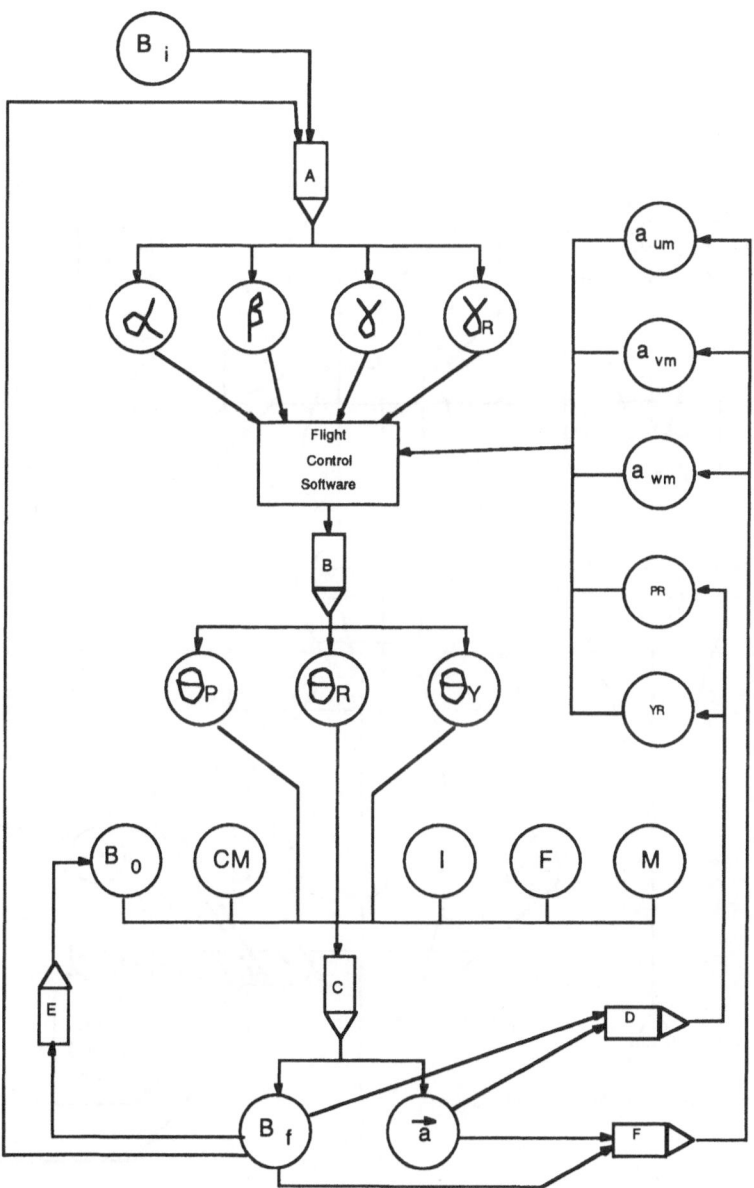

Figure 2 : DFM Model of the Titan II Flight Control System

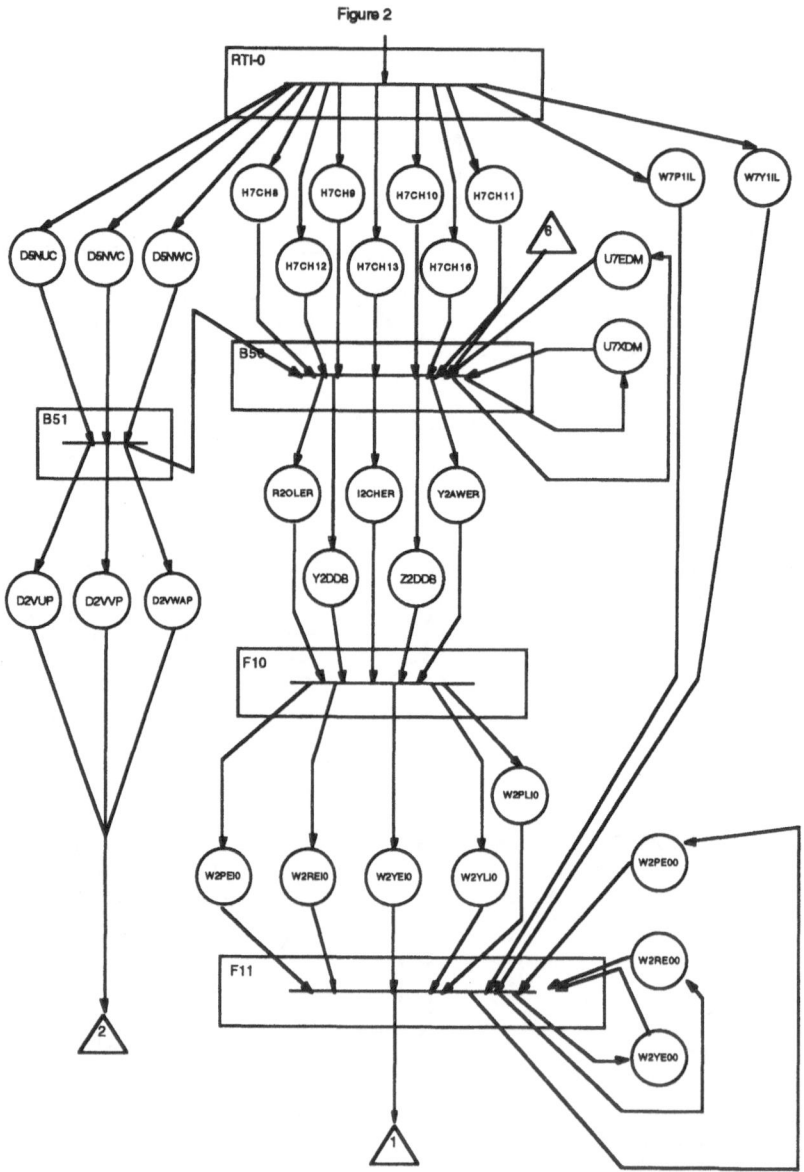

Figure 3 : DFM Model of the Flight Control Software (1/3)

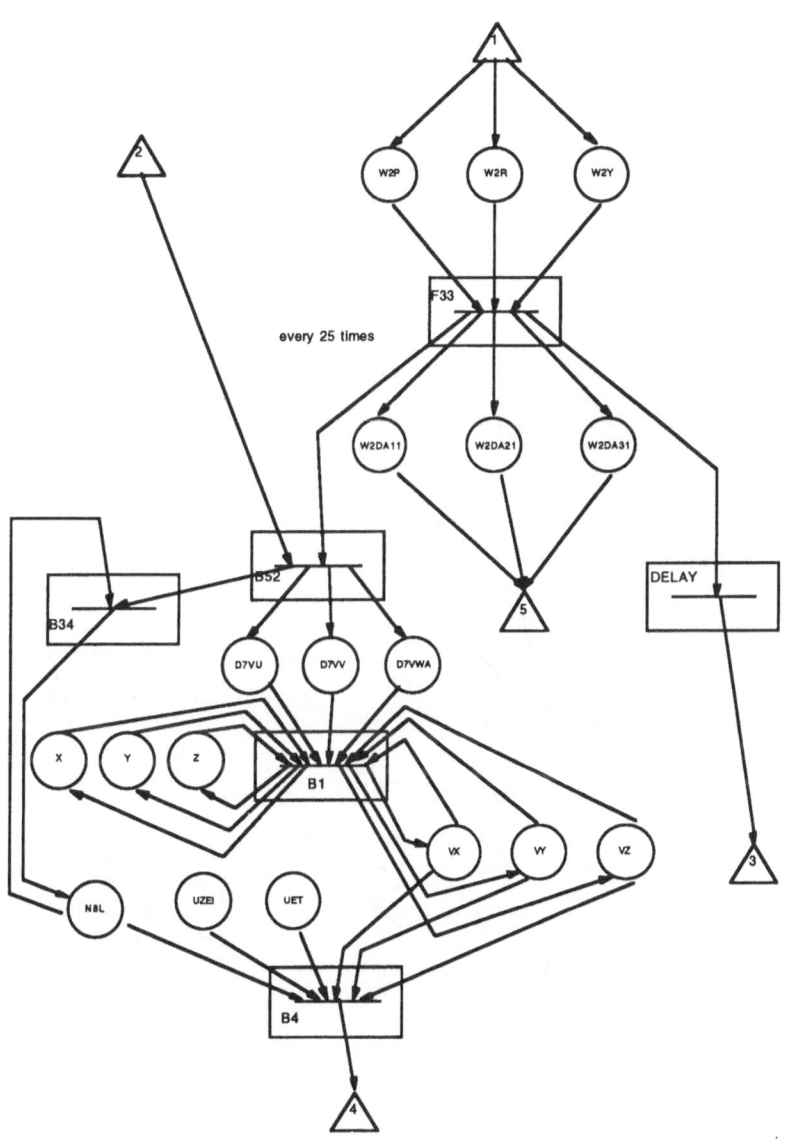

Figure 3 : DFM Model of the Flight Control Software (2/3)

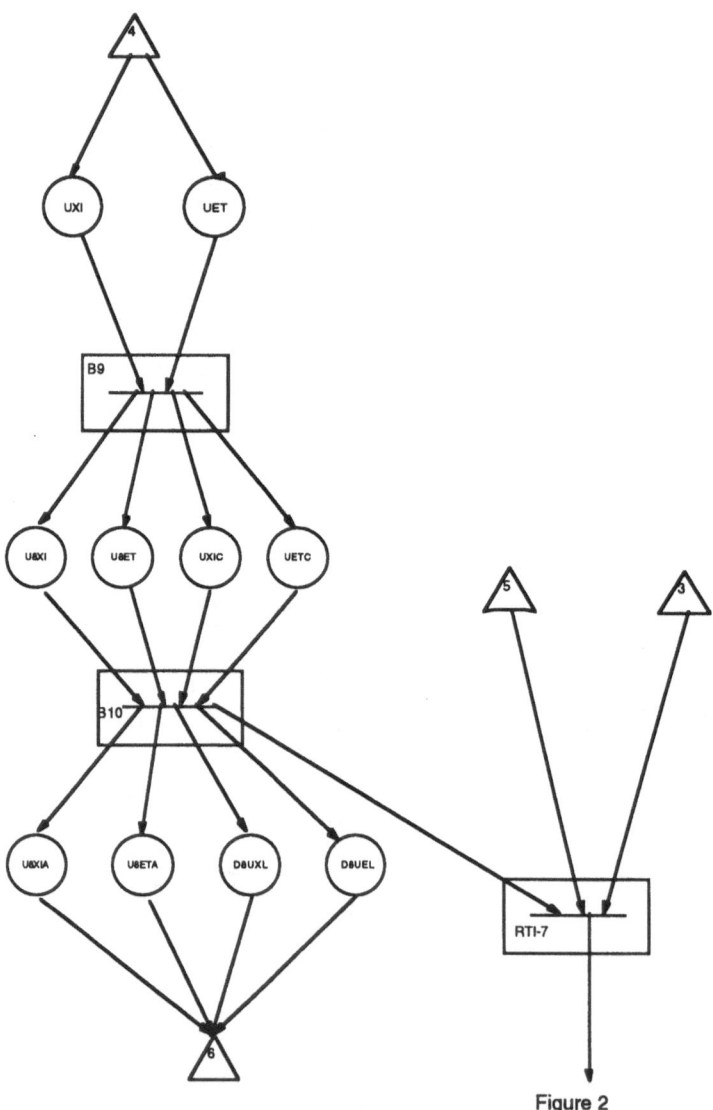

Figure 2

Figure 3 : DFM Model of the Flight Control Software (3/3)

Variable	Description
a_{um}	Acceleration along the accelerometer um-axis
a_{vm}	Acceleration along the accelerometer vm-axis
a_{wm}	Acceleration along the accelerometer wm-axis
a	Vector acceleration of the vehicle
B_0	Body axes of the vehicle at the beginning of a 40 msec cycle
B_f	Body axes of the vehicle at the end of a 40 msec cycle
B_i	Body axes of the vehicle at go inertial
CM	Center of mass of the vehicle
F	Thrust
I	Moment of inertia about the center of mass
M	Mass of the vehicle
PR	Pitch Rate
YR	Yaw Rate
α	Platform gimbal angle
β	Middle gimbal angle
γ_R	Outer redundant gimbal angle
γ_R	Gamma gimbal angle
θ_P	Angle of deflection of the engines for pitch correction
θ_R	Angle of deflection of the engines for roll correction
θ_Y	Angle of deflection of the engines for yaw correction
D5NUC	Number of Accelerometer Counts for the uc Accelerometer

D5NVC	Number of Accelerometer Counts for the vc Accelerometer
D5NWC	Number of Accelerometer Counts for the wc Accelerometer
H7CH8	K sin(-120°)
H7CH9	K sin(-60°)

Table I : Parameters of the Titan II DFM Model (1/2)

Variable	Description
H7CH10	K sin(β-120°)
H7CH11	K sin(β -60°)
H7CH12	K sin(γ_R-120°)
H7CH13	K sin(γ_R-60°)
H7CH16	Inner Gamma Resolver Input
W2DA11	D/A Output
W2DA21	D/A Output
W2DA31	D/A Output
W7P1IL	Pitch Rate 1 Gyro Input
W7Y1IL	Yaw Rate 1 Gyro Input

Table I : Parameters of the Titan II DFM Model (2/2)

After linking up the parameters by the transfer boxes and transition boxes, the next step is to construct decision tables to represent the relationships between the parameters. Combinatorial explosion is encountered in the construction of decision tables for the software modules. One of the modules will be used to illustrate the problem. In the module BLOCK 1, the variables X, Y, Z, VX, VY, VZ, D7VU, D7VV, and D7VWA are used to calculate new values for X, Y, Z, VX, VY, and VZ for the next second. The variables (X,Y,Z) represent the location of the rocket, (VX,VY,VZ) represent the velocity of the rocket, and (D7VU,D7VV,D7VWA) represents the acceleration of the rocket due to thrust alone. These nine variables are each discretized into 5 states, representing a large negative deviation, a moderate deviation, a value close to 0, a moderate positive deviation, and a large positive deviation.

To complete the BLOCK 1 decision table, we need to sample combinations of these 9 input variables. In our model, all these 9 input variables are each discretized into 5 different states. This means that we have to sample at least 5^9 times. In addition, the decision table produced will be huge, consisting of 5^9 rows. It will be very time consuming to look up many tables of this size during the model analysis step. A possible solution to this problem is suggested in the next section.

6 A Solution to the Problem of Combinatorial Explosion

The approach to solving the combinatorial explosion problem in the decision table construction is based on the intention to bypass the construction step. In the analysis, the DFM model is backtracked to identify causes for certain top-events. The decision tables provide intermediate information on the backtracking step, namely in finding the parameter states in-between the top events and their causes. If we are able to find these intermediate causes without looking up the decision tables, we can avoid the combinatorial explosion problem altogether.

As the Titan II flight control software subroutines implement equations with distinct physical meaning such as equations of motion or control laws, we can solve the equations implemented in the subroutines in reverse. For instance, a particular subroutine is encountered in backtracking the DFM model and certain outputs from this subroutine are found to eventually produce the top event. The next step is to find the combinations of inputs that produce those outputs via this particular subroutine. Instead of looking up the decision tables constructed previously for this subroutine in Step 1, we can try to solve the equations implemented in the subroutine.

For example, the analysts are interested in finding out why the thrust chambers have deflected 2°, 2.5^o, and 1.5^o respectively for roll, pitch, and yaw corrections. This condition is represented as the top event in the fault tree shown in Figure 4. In the DFM model in Figure 2 and Figure 3, we find that the condition $(p, R, Y) = (2^o, 2.5^o, 1.5^o)$ is backtracked through the transfer box B.

This condition is found to be caused by W2DA11 = 52, W2DA21 = 65, and W2DA31 = 39. Next, we backtrack the delay time transition which waits for RTI-7 to execute. The parameters W2DA11, W2DA21, and W2DA31 retain their values. Continuing the backtracking process, the DFM model shows that W2DA11, W2DA21, and W2DA31 are calculated by the subroutine FIG 33 with the input variables W2P, W2R, and W2Y. We need to find what values of these input variables produce the output values of interest in order to enter the gate in the next level in the fault tree. The equations implemented in this software module FIG 33 are solved, and the input values are found to be W2P = 51.7, W2R = -12.8, and W2Y = 51.8. This information is, then, entered into the fault tree. The backtracking process is continued and the equation solving procedure is repeated, if necessary, for the next subroutine. The approach for solving the equations is based on the Newton-Raphson method for solving a system of non-linear equations.

7 Utilization of Results

Once the fault tree has been fully developed, the prime implicants obtained from the fault tree analysis represent a complete set of the possible catastrophic failure modes of the system. This information can then be used to go back to the design of the system and remove those failure modes or add new design features to mitigate their consequences. The fault tree prime implicants can also be used to identify criteria for testing which focus on the safety-critical software functions and associated hardware and environmental conditions. The application of DFM to a system implementation prior to testing can sharply reduce the inductive effort required on the part of the analyst to select test cases. DFM relieves the analyst of the burden of identifying all of the test cases necessary to verify the dependability of the system. Instead, the analyst must only identify the physical variables, environmental conditions, and software functions which are necessary to characterize the logical and dynamic behavior of the system.

8 Conclusion

The Dynamic Flowgraph Methodology provides a relatively simple and effective method for modeling and analyzing embedded systems. DFM models represent cause-and-effect as well as timing relationships between individual software functions and interfacing hardware and process parameters associated with the physical system. The analysis of a DFM model results in the development of "timed" fault trees which reflect static relationships between the system variables at discrete points in time. The development of dynamic, multi-valued (i.e., non-binary) fault trees represents a significant technical advancement in the modeling and analysis of embedded systems.

These fault trees are capable of explicitly identifying the causes of system events due to both cause-and-effect and timing relationships. This information can then be used during design to identify and eliminate system faults resulting from unanticipated combinations of software logic errors, hardware failures and adverse environmental conditions, and to direct testing activity to more efficiently eliminate implementation errors by focusing on the neighborhood of potential failure modes arising from these combinations of system conditions.

Work is also currently in progress to identify which of the DFM analytical procedures are amenable to software implementation, and to develop a prototype package of software tools. Envisioned as fundamental components of this software package are a graphical model editor to assist in building system models, featuring a

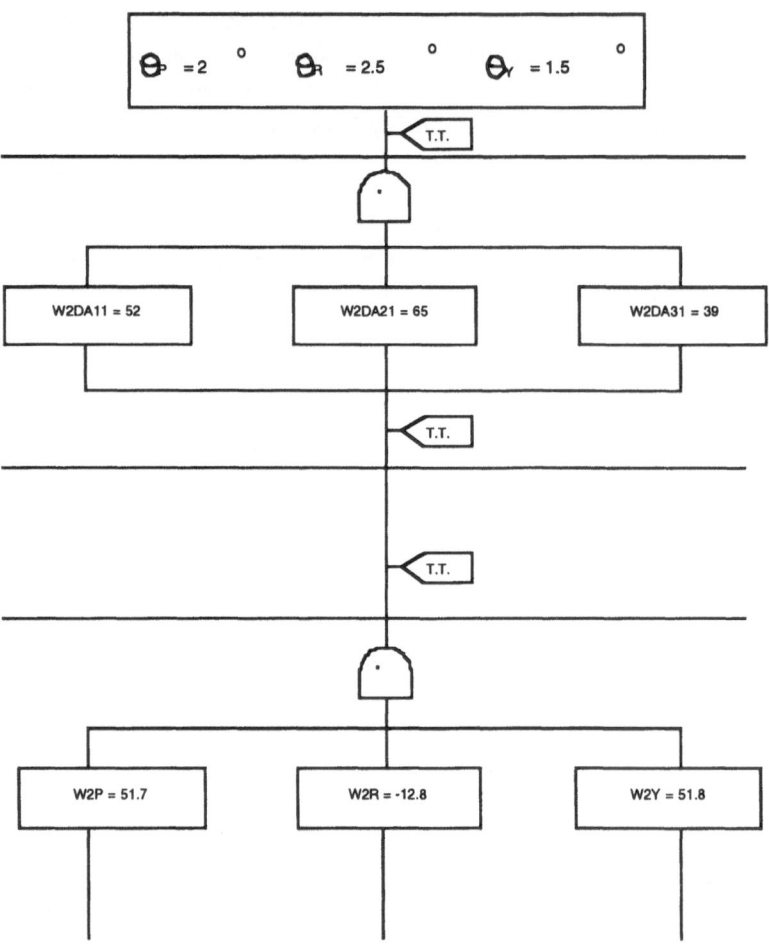

Figure 4. Fault Tree for (P, R, Y) = (2⁰, 2.5⁰, 1.5⁰)

relational database supporting a library of predefined LFM-like constructs for common system components (e.g., valves, flow sensors, etc.), and a tool which will automatically backtrack through system models to build fault trees. The rationale for this objective derives from the realization that the practical acceptance and utility of a new framework for embedded system reliability and safety analysis will depend heavily on the development and demonstration of computer based tools capable of facilitating its use. The development of the analytical approach which is at the center of the DFM project is based in part on the LFM methodology concept, which focuses on the derivation of hardware systems fault trees, starting from a digraph system model which expresses cause and effect relationships among system parameters. Even though the application of the LFM approach to the safety or reliability analysis of a hardware system presents a considerably lesser degree of complexity than an equivalent application to an embedded system of the approach that we are in the process of developing, LFM itself relies for its execution on the use of a software tool that implements and automates its procedures for fault tree derivation. Thus, it would be unrealistic to expect that this new method being developed could find practical applicability without the aid of an equivalent set of software tools.

References

1. J.A. McDermid, Issues in developing Software for Safety Critical Systems, *Reliability Engineering and System Safety*, Vol. 32, 1991.

2. S.B. Guarro, J.S. Wu, et al., Findings of a Workshop on Embedded System Software Reliability and Safety, UCLA-ENG 90-25, June 1990.

3. N.G. Leveson, P.R. Harvey, Analyzing Software Safety, *IEEE Transactions on Software Engineering*, SE-9, pp.569-579, 1983.

4. S. Petrella, P. Michael, et al., Random Testing of Reactor Shutdown System Software, Proc. of the Intnl. Conf. on Probabilistic Safety Assessment and Management, Beverly Hills, CA, Feb. 4-7, 1991.

5. K.T. Narayana, A.A. Aaby, Specification of Real-Time Systems in Real-Time Temporal Interval Logic, Proc. of the 1988 Conference on Real-Time Systems, IEEE Press, 1988.

6. R.R. Razouk, M.M. Gorlick, A Real-Time Interval Logic for Reasoning about Executions of Real-Time Programs, Proc.of the ACM SIGSOFT '89, ACM Press Software Engineering Notes Vol. 14 No. 8, Dec. 1989.

7. International Workshop on Timed Petri Nets. Torino, Italy, July 1-3,1985. IEEE Computer Society Order # 674

8. E.T. Morgan, R.R. Razouk, Interactive State-Space Analysis of Concurrent Systems, *IEEE Transactions on Software Engineering*, SE-13, No. 10, Oct. 1987.

9. T. Murata, Petri Nets: Properties, Analysis and Applications, Proc. of the IEEE, Vol. 77, No. 4, April 1989.

10. N.G. Leveson, J.L. Stolzy, Safety Analysis Using Petri Nets, *IEEE Transactions on Software Engineering*, SE-13, No. 3, March 1987.

11. S.S. Cha, N.G. Leveson, et al., Safety Verification in Murphy Using Fault Tree Analysis, Proc. of the International Conference on Software Engineering. Singapore, 1988, IEEE Press, 1988.

12. S.B. Guarro, D. Okrent. The Logic Flowgraph: A New Approach to Process Failure Modeling and Diagnosis for Disturbance Analysis Applications, *Nuclear Technology*, Vol. 67, 1984.

13. S.B. Guarro, D. Okrent, The Logic Flowgraph: A New Approach to Process Failure Modeling and Diagnosis for Disturbance Analysis Applications, UCLA-ENG 8507, Dec. 1985.

14. S.B. Guarro, A Logic Flowgraph Based Concept for Decision Support and Management of Nuclear Plant Operation, *Reliability Engineering & System Safety*, Vol. 22, 1988.

15. S.B. Guarro, Diagnostic Models for Engineering Process Management: A Critical Review of Objectives, Constraints and Applicable Tools, *Reliability Engineering and System Safety*, Vol. 30, pp.21-50, 1990.

16. Y.T.D. Ting, Space Nuclear Reactor System Diagnosis: A Knowledge Based Approach, Ph. D. Dissertation, UCLA, 1990.

17. S.B. Guarro, PROLGRAF-B: A Knowledge Based System for the Automated Construction of Nuclear Plant Diagnostic Models, International Topical Meeting on Artificial Intelligence and Other Innovative Computer Applications in the Nuclear Industry, Snowbird, UT, Aug 31-Sep 2, 1987.

18. S.B. Guarro, J.S. Wu, et al., Embedded System Reliability and Safety Analysis in the UCLA ESSAE Project, Proc. of the Intnl. Conf. on Probabilistic Safety Assessment and Management (PSAM), Beverly Hills, CA, Feb. 4-7, 1991

19. C.T. Muthukumar, S.B. Guarro, G.E. Apostolakis, Logic Flowgraph Methodology: A Tool for Modeling Embedded Systems, IEEE/AIAA Digital Avionics Systems Conference, Los Angeles, CA Oct. 14-17, 1991.

20. C.J. Garrett, S. B. Guarro, G. E. Apostolakis, Development of a Methodology for Assessing the Safety of Embedded Systems, 2nd Annual AIAA/USRA/AHS/ASEE/ISPA Aerospace Design Conference, Irvine CA, February 16-19, 1993.

21. Martin Marietta Astronautics, "Guidance, Control, and Ground Equations for Flight Plan XX Volume I: Guidance Equations XX-U001-I-05," February 18, 1991.

22. Martin Marietta Astronautics, "Guidance, Control, and Ground Equations for Flight Plan XX Volume II: Flight Control Equations XX-T001-II-08," June 24, 1988.

On Managing Fault-Tolerant Design Risks

Danforth Ball* and Amir Abouelnaga†
* The MITRE Corporation
Center for Advanced Aviation System Development
McLean, Virginia 22102
ball@mitre.org
† TRW Inc.
Systems Integration Group
Fairfax, Virginia 22033
U.S.A.

Abstract

The traditional process for tracking the design and implementation of large distributed fault-tolerant real-time processing systems consists of a set of formal reviews and tests to assess the design's correctness and compliance with requirements. Typically, there are no established formal procedures for monitoring the progress of the design during the implementation phase, the interval between the review and approval of the "paper" design and the formal testing of the completed design. This process is particularly inadequate for managing the fault-tolerant design, because critical technical risk areas, such as failure coverage and real-time performance, are not manifested until the implementation phase. The Fault Tolerance Scorecard (FTS) process augments the traditional review process by providing for frequent monitoring of specific critical attributes of the evolving fault-tolerant design. It uses metrics designed to monitor core features of the fault-tolerant design and their interaction with other system features.

Index terms: Fault-Tolerant Design, Fault Tolerance Scorecard, Real-Time Systems, Distributed Systems, Design Process.

1 Introduction

In this paper, we discuss an approach to managing the implementation risks attributable to the fault-tolerant design of large real-time distributed safety-critical computing systems (e.g., Air Traffic Control [CDD90]). The fault-tolerant design of such systems incorporates systematic development and design refinement processes such as those described in [2, 11] based on the strategies of using protective redundancy and nested defense. Key attributes of such systems include extensive use of software in a large number of networked autonomous computing stations that cooperate in solving complex algorithms in a hard real-time environment, providing a set of safety-critical services. The execution model is typically client/server and fault tolerance architecture is hierarchical, with protective redundancy at each level, allowing for recovery at one or more computing stations. Typically, most of the fault tolerance protocols are synchronous, to ensure hard real-time requirements satisfaction, and are implemented in the system's infrastructure software (i.e., the operating system or its extensions). The human operator is considered the ultimate detection/recovery entity. Finally, such systems tend to encapsulate both environmental parameters and system parameters in look-up tables. This is to permit the system to be easily adapted to changes in its operating environment, allow its parameters to be tuned for variations in processing profile, and optimize the match between computing resources and applications.

The Traditional Development Process. Current development procedures for large systems reflect a life cycle that consists of three consecutive major phases: Design, Implementation, and Operation. Monitoring procedures employed during the acquisition of these systems are based on a series of technical reviews and audits that are conducted according to formal standards as a way to mitigate the risk of non-compliance with requirements, as well as to provide discrete incremental project approval milestones. For example, U. S. Government projects use the Department of Defense standard MIL-STD-1521 [10], which requires the system's requirements and design to be reviewed during the System Requirements Review (SRR), the System Design Review (SDR), the Preliminary Design Review (PDR), and, finally, the Critical Design Review (CDR), thus providing "snap shots" of the state of the design. The standard also states that, "The Critical Design Review is normally accomplished immediately prior to releasing for coding the Computer Program Configuration Item (CPCI)," thus beginning the implementation phase. After CDR, and during the implementation phase, no additional CPCI formal technical reviews are required. However, the phase generally includes a wide range of activities such as software detailed design, coding and unit testing, development and integration testing. The formal process resumes at the Functional and Physical Configuration Audits (FCA and PCA) and formal acceptance testing at the conclusion of the implementation phase. Finally, the operation phase terminates when the system is retired.

The success of this process is based on the assumption that once the design has been thoroughly reviewed and approved, its implementation proceeds without major difficulty. However, the effectiveness of this design review process in the development of today's complex data processing systems, and especially in hard real-time systems, has become suspect. For example, the Defense Science Board Task Force on "Transition from Development to Production" [8] states:

> *"Design reviews have frequently become a time-consuming exercise contributing little to the assurance of design maturity. The Government contributes to this problem in permitting irrelevant discussion by unqualified [review] participants, which fosters the perception that a design review is conducted for the purpose of familiarizing people with an overview of the hardware design."*

Despite the demonstrated limitations of paper design reviews, there are few, if any, prescribed guidelines and processes for monitoring the progress of the design implementation. Boehm has shown that approximately 84% of the total development cycle costs are incurred during the period following the design approval [4]. Boehm also presents data illustrating that the cost to fix or change software escalates rapidly during this period because this change process becomes much more formalized and there are many more documents that must be changed. Design problems that are not discovered until formal testing has started can cost three to six times as much to fix as problems discovered during development testing [4].

The Conventional Risk Management Process. In general, the process of assessing the risk that the system may not meet its design or operational requirements starts very early in the design phase and continues through the implementation phase. System performance and RMA (Reliability, Maintainability, Availability) are usually identified as risk categories during the initial risk assessment activities; the categories evolve with the design, as new risks are identified and existing risks are mitigated. Fault tolerance risk is usually considered a component of RMA risk because of the contribution of fault tolerance to the enhancement of system availability. Risk categories are typically characterized in terms of quantitative parameters called Technical Performance Measurements (TPM)s. A TPM is "the continuing prediction and demonstration of the degree of anticipated or actual achievement of selected technical objectives" [9]. For example, a typical TPM covering system reliability would be its Mean Time Between Failure (MTBF). When the observed values of TPMs vary significantly from their expected values, a need for corrective actions to mitigate the heightened risk is indicated.

In section II we discuss why the limitations of conventional risk management processes are particularly severe with respect to the fault tolerance aspects of the design. In section III we present the Fault Tolerance Scorecard (FTS) process by which the progress of the full scale development of the fault tolerance design for a large system can be more closely monitored. In section IV we describe how the FTS process was used in managing the fault tolerance implementation risk for the Federal Aviation Administration's Advanced Automation System (AAS). In section V. we address the FTS effectiveness in managing risks.

2 Limitations of Conventional Risk Management Processes

Definition of Fault Tolerance Risks. Certain challenges face the developers of large distributed systems with respect to performing credible and productive fault tolerance risk identification, assessment and mitigation. As stated above, fault tolerance design risk is considered to be a contributor to system availability risk. Consequently, TPMs that address system RMA characteristics, such as Mean Time Between Failure (MTBF), Mean Time To Repair (MTTR), Mean Up Time (MUT), and other measures of availability and reliability, are often used to assess and track system fault tolerance risk. Although this approach has been typical, it tends to beg the question by stating, in effect, that the fault tolerance design risk is that the design may not be fault-tolerant. This obscures the real fault tolerance design issues and prevents assessment of the fault-tolerant design until implementation is nearly complete and the RMA TPMs begin to be collected.

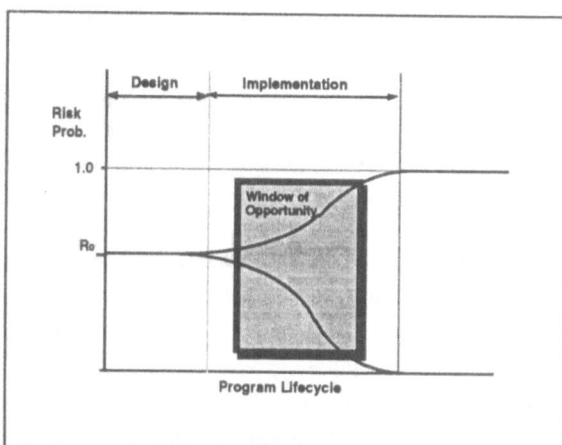

Figure 1: Risk Management "Window of Opportunity"

Windows of Opportunity. It is critical, then, to plan the TPM assessment processes correctly during the design and/or implementation phases. Figure 1 illustrates that there are "windows of opportunity" during which effective risk management can be performed. Before the "window of opportunity" opens, there is insufficient information available to warrant modification of the initial risk probability, R_0. As the implementation of the fault-tolerant design nears completion, the "window of opportunity" for effective risk management closes because the risk has either not materialized (R=0), or has become a reality (R=1.0) that must be dealt with on the program's critical path It is during the "window of opportunity" that developing design and implementation problems can be identified and corrected early enough to be accommodated within the overall program cost and schedule constraints.

System Availability and Fault Tolerance Risk Tracking Incompatibility. Attempting to use RMA TPMs for fault tolerance risk assessment is ineffective because of the mismatch between the windows of opportunity for assessing system availability and fault tolerance. This suggests that the conventional practice of incorporating fault-tolerant design risk with system availability risk is unacceptable for effective fault tolerance risk mitigation.

System availability addresses the probability of obtaining the correct system service when the service is required. In order to assess system availability, it is implied that the design would have to reach a level of implementation maturity sufficient to demonstrate system-level behavior. This usually occurs towards the end of the implementation phase. On the other hand, system fault tolerance addresses the capability of the design to "cover" (or to correctly detect and recover from) a failure. Fault tolerance features are (or should be) typically incorporated very early in the implementation phase because of their strong binding with the system's infrastructure software and hardware. Accordingly, fault tolerance implementation risk assessment window of opportunity should begin early in this phase to identify resource, performance, or coverage problems due to integrating fault tolerance with other infrastructure features. Moreover, relying on the measurement of RMA characteristics to determine the effectiveness of the fault-tolerant design increases the chance of costly design updates because of lack of focus on infrastructure integration problems--the focus is usually on the infrastructure's ability to support execution environments without failures during the early stages of implementation.

The situation is further aggravated if the design depends on adaptation data to define timing-related features such as deadlines and/or execution periods of system application tasks. The desire to achieve design functionality often overshadows the need to assess the viability of the fully-integrated infrastructure, one that includes all specialty engineering features such as fault tolerance, early in the implementation phase, leading to the use of "dummy" adaptation values for the sake of achieving demonstrable functionality. Lack of such assessment discipline would delay the discovery of fundamental design problems such as the inability of the hardware platform to support the integrated infrastructure or the incompatibility of fault tolerance protocols with other protocols hosted by the infrastructure.

The use of adaptation tables to encapsulate key system parameters can ultimately provide significant long-term benefits by facilitating the software change and maintenance process. However, it can also present a powerful temptation to defer the definition of some of the most critical aspects of the fault-tolerant design until late in the implementation phase. The risk in postponing the definitions is that the developed hardware and software infrastructure may not be capable of supporting the design performance assumptions, especially those related to the fault tolerance protocols. These parameter values determine such key factors as the fault tolerance mechanism's use of system resources, fault latency time, and service interruption time. In a safety-critical system with stringent real-time requirements, these factors are at the heart of the fault tolerance risk.

Finally, although performance and availability models are useful for assessing the basic characteristics of the fault tolerance architecture and infrastructure during the design phase, these models are of limited usefulness for risk assessment during the implementation phase because the model results are most sensitive to the same features of the design that present the greatest risk, namely performance faults, software design faults, and fault coverage.

Role of Models and TPMs. The above discussion should not be construed as an argument against the use of modeling or TPMs in assessing system RMA. We assert that modeling is an important function during the design phase because it identifies important design drivers and risk areas. We also believe that the availability growth modeling activities often associated with the system Failure Reporting And Corrective Action System (FRACAS) program during the test phase of the system is an important component in the system's RMA program. The successful implementation of the fault-tolerant design will need to be followed by an aggressive reliability growth program to achieve the high levels of reliability and availability required for safety-critical systems. However, we are skeptical that system RMA modeling or FRACAS TPMs are effective for fault-tolerant design risk tracking during the implementation phase.

3 The Fault Tolerance Scorecard Process

Definition of the Scorecard Concept. In its abstract form, a scorecard is a risk management tool. The scorecard itself is a set of templates that display selected information and data related to the fault tolerance design to help the decision maker identify specific risk trends in the system under observation. It has an associated process that specifies the steps needed to derive the templates, and collect, process, analyze, and react to information about the system in order to mitigate the risk. The templates are developed by decomposing generic fault tolerance risks into design-specific constituent risks that are measurable early in the implementation phase. The typical challenges in developing the scorecard are the identification of the risks to be tracked and the definition of the specific information and data needed to support the risk tracking and assessment. Challenges faced in implementing the scorecard process include assuring that the fault tolerance design allocations are sufficiently mature to permit quantification of the attributes to be measured, structuring the development schedule to permit early identification of necessary design refinements, establishing a schedule for the timely collection and reduction of the data, minimizing the resources, human and otherwise, required to support the scorecard monitoring process, and integrating this risk management activity with other similar activities. In other words, the quality of the scorecard depends not only on its information content, but the resources required to support it and its use in the overall risk management process.

Application of the Scorecard Concept to Fault Tolerance Risk Management. We propose using scorecard techniques to manage fault tolerance risk. The Fault Tolerance Scorecard (FTS), along with its associated process, is used to assess the evolution of the fault-tolerant design during the implementation phase. The FTS and its process satisfy three critical risk management objectives: (1) Focusing on design aspects that reflect the system's fault-tolerant operations, (2) establishing a discipline to collect data about the implementation, and (3) identifying unique fault-tolerant design/implementation risks and their effect on other system design risks.

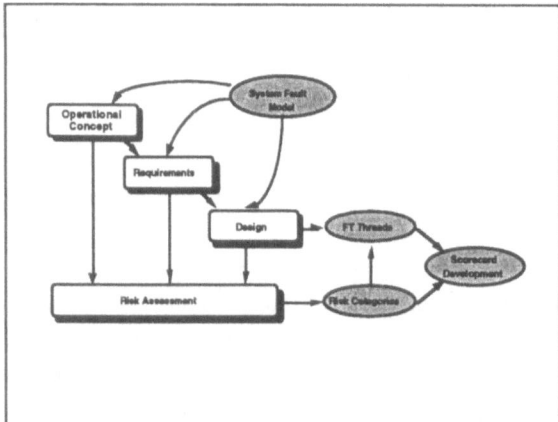

Figure 2. Interrelationship Between FTS and Program Development Process

Figure 2 highlights the interrelationship between the FTS and the traditional program development and risk assessment processes. The FTS is derived from an analysis of the specific fault-tolerant design, and an assessment of its associated design risks. The FTS process differs from more conventional RMA and software quality metrics in the degree to which it is tailored to the particular features of the design being monitored. In contrast, conventional metrics tend to treat the system as a "black box." These metrics may be useful in monitoring programmatic aspects of the implementation, but they offer little help in technical risk management during the implementation of the fault-tolerant design. The "front-end" analysis of the design and its associated risks used in the development of the scorecard is what enables the FTS to provide early visibility into the critical internal areas of the design, as well as identifying schedule risks associated with the late development of critical components.

Figure 2 also shows the activities related to identifying the risks to be tracked, thus determining the need and focus of the scorecard. The (implicit or explicit) existence of a fault model is the basis for the fault-tolerant design in typical safety-critical systems [2]. The model is used to contribute to generating correct, precise, and consistent system dependability requirements that, in turn, are used to identify risk categories. Another activity in risk identification is the definition of execution threads that would expose the particular fault-tolerant design feature under investigation (e.g., covering a specific fault type). These fault tolerance threads are execution threads that would be chosen to demonstrate fault tolerance features of a system function or service. Thread definition is typically an iterative process that begins with simple threads, possibly to test simple fault tolerance mechanisms, and continues to produce more complex threads, possibly to test a system service, as the implementation matures. This incremental process allows for the early validation of system fault tolerance features and is often consistent with the implementation process for large systems, where every implementation increment provides increased functionality. It is also consistent with implementations that use concurrent engineering processes because of its responsiveness to the amount of functionality available in each of the concurrent activities and its utility in mitigating the risk of interface incompatibility.

As a risk mitigation support process, the FTS process must specify the schedule and techniques governing the collection of system data to populate the FTS templates, consistent with the system's implementation and integration plans, as well as the administrative activity needed to plan and execute risk mitigation activity.

Fault-tolerant Design Risk Categories. It is clear that the fault-tolerant design of a safety-critical system exacts a toll on practically all aspects of the system's design. This is primarily due to the proliferation of fault tolerance or system availability management functionality at all levels of the distributed data processing hierarchy, from the node hardware to the application level. Generally speaking, fault-tolerant design risk is the risk that fault tolerance and system availability management functions do not provide the level of fault tolerance needed to support overall system service(s) availability. This risk can be decomposed into the risk categories discussed below.

Each of the risk categories forms the basis of one or more FTS templates. The templates are interdependent and all must be considered together to assess the state of the design. Our analysis indicates that the identity and number of fault tolerance risk categories, and hence FTS templates, generally depends upon the important attributes of the design. However, we have identified the following three generic risk categories as the most frequent bases of any set of templates.

1. <u>System Performance</u>: The operation of the fault tolerance mechanisms in large real-time systems is governed by a wide variety of time-based system parameters (i.e., timing

parameters) that control such factors as the frequency of software task health status reporting, timeout intervals, error thresholds, retry counts, etc. Fixed and parameterized timing values are usually established during the design phase as a result of performance modeling and requirements analysis. These "baseline" values not only affect the operations of fault tolerance mechanisms themselves, but other factors such as the additional processing required by the mechanisms to perform their tasks. The values are selected to optimize system performance while minimizing false alarms. It is critical that these baseline values be established prior to the implementation phase because they reflect key design assumptions that may be challenged during system implementation; the absence of these definitions would seriously jeopardize the successful system implementation because these parameter values often represent key drivers in the real-time performance characteristics of the system. Actual measurements of the timing values should be taken under both a failure-free and a faulty execution environment. The fault-free environment measurements serve as the control sample for all the faulty environments executed. The baseline values are used as the basis for comparison with measured values. The FTS is used to record and highlight deviations between comparable baseline and measured values that form the basis of the assessment information presented in the FTS templates. These deviations are indications of design problems and require immediate attention. Unless this is accomplished, it would be very difficult to predict either fault detection and recovery times or the overall burden on the system caused by fault tolerance.

2. <u>System Resource Usage</u>: The operation of fault tolerance mechanisms is also governed by the amount of system resources needed by the fault-tolerant design. Budgets and measurements of system resources include areas such as: (1) The processing and communications bandwidth used by the fault tolerance protocols to manage the failure detection functions, such as processing and exchanging health status messages across the system and coordinating system recovery and service restoration; (2) the amount of storage used by checkpoint data to support system state recovery; (3) the checkpoint frequency needed to support recovery to most recent correct state; and (4) the amount of memory used by the fault tolerance mechanisms to detect and recover from failures. These budgets are also usually established during the design phase and should be treated with the same level of care as the performance allocations, measurements, and analysis activities.

3. Systems Failure Coverage: The quality of the fault-tolerant design depends upon its effectiveness in detecting and recovering from (i.e., covering) system failures. This template provides visibility into the process of selecting fault tolerance threads and the extent of their selection to cover the system fault tolerance features. Information presented herein indicates the consistency between the system fault model and its design, the overall coverage of the testing, the success rate, false alarm rate, and a rough measure of the general stability of the software under test and the quality of the restored service.

Decomposing the Risk Categories into Constituent Risks. Each of the generic risk categories discussed above must be further decomposed to define a set or sets of *design-specific* constituent risks, where each member of the set contributes to the risk identified by the category. This step is a critical element of the FTS process. Tailoring the generic risks to the specific characteristics of the design being monitored is what provides the basis for identifiable, measurable, and trackable design features, thus providing the basis for an FTS template. In addition, the risk decomposition process serves to focus both technical and management attention on the specific risks associated with the chosen design approach.

The FTS template format must contain important information elements to ease the project management and technical staff's ability to assess each constituent risk. All the necessary information must be displayed in a manner that accentuates deviations from expected system behavior that may be due to many factors, including improper design assumptions, incomplete specification, erroneous implementation, or incorrect operations.

We discovered that, as a minimum, the following information elements are necessary for each constituent risk, i.e. FTS template:

1. Execution Thread Specification: Describes the execution environment and thread to enable the collection of correct system behavior data. This includes specifications of the operational environment and the hardware and software functions to be exercised.

2. Design Allocation Information: Displays information developed during the design phase about the specific risk category. This includes information such as modeling results, resource allocation, and performance budget data.

3. <u>Measurement Information</u>: Displays information developed about the system due to measurement activities. This includes filtered raw and computed data that highlight the specific attribute in the execution thread that is relevant to the risk category under investigation.

4. <u>Prediction Information</u>: Reflects additional modeling or analysis activity based on a mix of actual and design information. Initially the predictions are based solely on modeling and analysis results; however as the implementation proceeds, measurements are used to update and refine the initial modeling projections.

5. <u>Assessment Information</u>: States the technical specialist's interpretation of the findings. This includes information about the quality of the data, risk trend, analysis of problems encountered during data collection and/or analysis, a proposed rationale for observed deviations, and so on.

6. <u>Risk Criticality Level</u>: States the criticality level of the risk to help the decision maker plan necessary mitigation activity priorities. A three-level structure is typical, such as high, medium, or low.

The interdependency of these scorecard templates is illustrated in Figure 3.

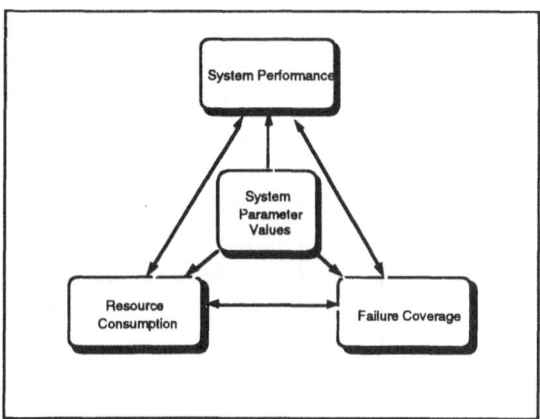

Figure 3. Scorecard Template Interdependency

System design and performance budgets are a synthesis of several factors such as the modeling activities performed during the design phase and the measurements collected during the execution of the FT execution threads. As the implementation progresses, actual measurements of system loads and performance would be used to evaluate the accuracy of the baseline values. Moreover, predictions may be made about the values of some parameters based on measured values of other related parameters. Figure 4 illustrates the closed-loop monitoring of the design and implementation progress that is achieved through the FTS process.

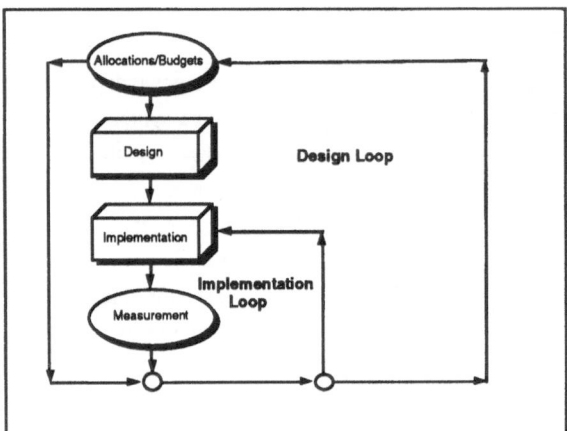

Figure 4. Closed Loop Monitoring of Design and Implementation with FTS

If the measured or predicted values are not consistent with the baseline values, then the assumptions used to develop the baseline values must be re-assessed. Corrective action is necessary to bring these interrelated factors back into agreement. It is possible for the corrective action to be limited to a simple "tuning" of the values. It is also possible that the necessary corrective action may be extensive, requiring fundamental changes in the software or hardware. It may also be necessary to relax the fault tolerance design requirements to optimize system costs and performance.

Need for Good Metrics. Metrics needed to assess the constituent risks must be selected to maximize ease of data collection, analysis, and relevance to risk assessment; and minimize measurement or prediction corruption due to irrelevant data. A constituent risk must be associated with at least one metric. A metric may be added or removed based on the state of the design and its quality in providing useful information. Once again, our analysis revealed that some metrics would remain in each of the sets for the same reasons stated about the constituent risk set.

The constituent risk metrics are selected to provide mechanisms for monitoring progress in the implementation of critical areas of the fault-tolerant design and not to provide a complete characterization of all design aspects. The metrics form the basis of a periodic sample of only those selected fault-tolerant design aspects that present a significant design risk. They are chosen carefully to provide the necessary insight into the implementation without excessive overhead (e.g., data collection and analysis) to facilitate and encourage the use of the FTS monitoring process.

Defining the Fault Tolerance Execution Threads. Identifying the constituent risks facilitates the specification of fault tolerance threads. The threads are designed to exercise the suspect design features. In complex systems, this specification is challenging because of trying to satisfy the conflicting requirements of simplifying the test run and capturing as much high fidelity data as possible to ease risk assessment. Our analysis revealed that threads can be categorized in a multitude of ways. Function-based threads address the algorithms used in fault tolerance and availability management (i.e., parity check algorithm) and protocols (i.e., heartbeat protocol correctness). Performance-based threads address behavioral and resource usage aspects of the design (i.e., timeout interval in the heartbeat protocol). It is evident that the two thread types are related. The key issue is to focus the test run on specific data items related to the risk to be addressed. This is to reduce the number of instrumentation probes in order to minimize interference with the system's real-time behavior and to keep the volume of data collected to manageable levels for data reduction and analysis. Moreover, it is best to identify a small subset of critical data items to collect when running either thread type to serve as reference measurement points.

Perhaps the most important task in developing execution threads is to ensure their viability under conditions when the system is running free from faults (i.e., the normal or steady state) and when the system is running with faults (i.e., the recovery state). Thread specifications (i.e., execution scenarios) must include provisions to address not only the normal and recovery states, but also the critical transitions between them, thus emulating fielded system operations. A fault-free version of the thread is used to simulate steady-state operation and to serve as the "control" sample for the faulty versions of the same thread. A fault injection facility must be available to support this activity [11]. The facility must have the capability to specify the fault injection and behavior data collection points in the system.

4 Development of the Fault Tolerance Scorecard

The Advanced Automation System (AAS). The FTS concept evolved as part of the risk management program for the U. S. Federal Aviation Administration's AAS program to modernize the nations's air traffic control system. The system's RMA requirements are demanding, requiring the development of a fault-tolerant design [3, 5]. Two major levels of the designs's hierarchy are the applications and the computing platform. The applications contain error detection mechanisms including those implementing performance failure semantics [6], called "heartbeat" protocols, and mechanisms to "reconstitute" a recovered application with the current processing state of air traffic control. The computing platform performs applications redundancy management services. It uses synchronous protocols that maintain information consistency among a network of autonomous processors, called a group, to support automatic recovery. The protocols facilitate agreement on group state among group members. This approach has the advantage of enabling the group, with a maximum cardinality of four processors, to run as long as one correctly functioning processor remains, while allowing for bounded recovery times. The protocols critically depend on the reliability and timeliness properties of the underlying communications services used for the exchange of group state data among the processors. These properties depend on both recovery time requirements as well as the system's load and performance requirements [6].

Evolution of the Fault Tolerance Scorecard for AAS. The FTS concept evolved throughout the AAS development. The AAS program placed special emphasis on a series of incremental demonstrations and supporting analyses that were intended to identify and eliminate design weaknesses [1]. The FTS was developed to generalize our AAS experience as technical advisors to the FAA into a fault tolerance risk management process. We believe that the FTS approach is most effective when it includes the participation of both the customer's and the contractor's program management teams. In the rest of this section, we discuss the general approach used to derive the FTS templates for the three risk categories and identify some of the constituent risks associated with the performance risk category.

Identifying the Constituent Risks. Generally, a constituent risk is uniquely related to a specific physical component or function supporting fault-tolerant operation, as well as metric(s) to track its evolution. Our analysis indicates that membership in the set of constituent risks for a risk category is design-dependent and that each constituent risk has a lifetime consistent with the discussion presented earlier in this paper.

The *performance risk* category was initially decomposed into three constituent risks: End-to End Communications Delay for Group State Update Messages, Clock Synchronization Accuracy, and Recovery Time. The first two risks are a consequence of using synchronous protocols in managing a group. A fundamental assumption governing the use of these protocols is that the deviation between local processor clocks and the interprocessor communication delay for group state messages must have an absolute upper bound [6]. Recovery Time risk was identified to assess the ability of the redundancy management mechanisms to react within the required recovery time under various workload and failure conditions.

Clock synchronization accuracy deviation is a function of the clock drift rates, how often clock synchronization messages are sent, and the difference between the maximum and minimum round-trip delay of the synchronization messages. The clock drift rates are a given characteristic of the hardware clock, and the frequency of clock synchronization messages was determined as part of the overall resource consumption budgeting and allocation process. Thus, the primary clock synchronization risk was that the communications network would not satisfy the maximum round-trip synchronization message delay allocation.

As the implementation progressed, we identified additional constituent risks and had to decompose some of the existing ones. For example, a new constituent risk, called Event Processing Time was added to the performance risk category. Communication Delay risk was further decomposed into Message Processing/Communication Delay and Event Scheduling (Timer) Latency.

The Event Processing Time risk was recognized because of the hard real-time aspects of the design; certain air traffic control services have to meet response time deadlines. The design employs "heartbeat" protocols to ensure meeting the deadlines. However, wide variations in the time it takes an application to process a certain event can introduce "false alarms" and system instability because of executing unnecessary recoveries resulting from missed heartbeats. The ability to contain event processing time within the allocated heartbeat intervals was our objective in identifying this risk.

We also discovered that in addition to the network message delay, a major contributor to the interprocessor end-to-end communications delays was the design's tasking and event scheduling algorithms. Consequently, the Communication Delay risk was decomposed as described above. Once again, the new constituent risks helped us focus on design-specific and unique features to facilitate rapid risk assessment and tracking.

Decomposition of the *resource consumption risk* category was performed as part of the performance modeling and simulation activity. Each constituent risk addressed use of a specific resource (e.g. memory and CPU utilization) by the fault tolerance mechanisms under various workloads and failure conditions. Resource consumption allocations were derived from the same timing design allocations used in assessing performance risks, as well as the code length, and database sizes of the fault tolerance mechanisms. Metrics were established to assess the level of resource consumption from the operation of the fault tolerance mechanisms and establish allowable thresholds.

Decomposition of the *coverage risk* category continues to evolve. However, this risk category may be decomposed into constituent risks identified through the Failure Modes and Effects Analysis (FMEA) process. Each constituent risk addresses the system's ability to recover from a specific failure mode. Metrics were devised to assess the design's automatic recovery capability. Note that the condition of each risk is binary; if the design recovers from a specific failure mode, then the risk is removed from the constituent risk set, otherwise it remains a member of the set.

Figure 5 shows an example of the FTS templates related to the above discussion. The general risk categories, their constituent risk sets, and a member of the risk set with its associated template information are shown.

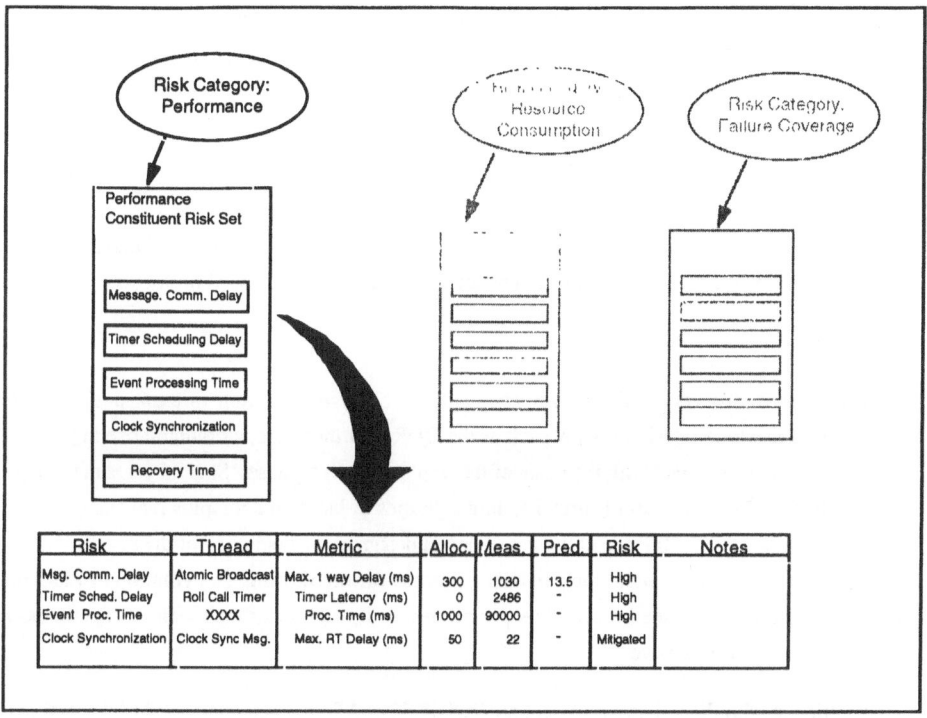

Risk	Thread	Metric	Alloc.	Meas.	Pred.	Risk	Notes
Msg. Comm. Delay	Atomic Broadcast	Max. 1 way Delay (ms)	300	1030	13.5	High	
Timer Sched. Delay	Roll Call Timer	Timer Latency (ms)	0	2486	-	High	
Event Proc. Time	XXXX	Proc. Time (ms)	1000	90000	-	High	
Clock Synchronization	Clock Sync Msg.	Max. RT Delay (ms)	50	22	-	Mitigated	

Figure 5. Example of Constituent Risk Set Members

5 Conclusion and Further Discussion

In addition to general questions about the effectiveness of the design review process, the design of distributed fault-tolerant computing systems poses some unique challenges:

* The extreme sensitivity of reliability models to minimal changes in the system's fault coverage and the inability to obtain credible estimates of this coverage for a yet to be implemented design makes it virtually impossible to predict the system's reliability and availability with any reasonable degree of accuracy or confidence during development.

* The successful operation of complex real-time fault-tolerant systems hinges upon the quality and efficiency of infrastructure software that executes needed protocols.

Thus, there is really very little known about the viability of the fault-tolerant design at the conclusion of a detailed design review such as the CDR. Furthermore, a dramatic rise in program expenditure rate is observed with the onset of the implementation phase. Several years may elapse between the CDR and the start of formal testing activities of large and complex real-time systems. Given the very limited confidence in the fault-tolerant design at the beginning of the implementation phase, prudent risk management dictates that a process for monitoring the evolution of the fault-tolerant design be established during this period of rapid program growth. The FTS process is designed to meet this objective.

Our experience with the risk management process on the AAS program convinced us that not only does the FTS provide a means of passively tracking the development, it can lend an organization and discipline to the overall fault-tolerant design implementation. For example, it is tempting to defer the definition of timing parameter values that are not needed to develop the software, assuming that the parameter values can be "tuned" after software development is completed. The need for allocations to support the FTS metrics serves to highlight critical risk areas of the design where system engineering is incomplete. The definition of execution threads also reveals the critical software components that should be developed early to provide adequate time to identify problems and make any necessary design refinements. With the cooperation and support of program management and the design and implementation activities, the FTS can help to establish the development priorities that are essential for effective risk management. Conversely, if the critical aspects of the fault-tolerant design are not scheduled until the end of the development cycle, there is little that can be done to manage the risk effectively.

The AAS risk management process revealed several significant problems in the performance risk category. Although the nominal time delays were well within tolerances, occasional "spikes" in the data revealed delays several orders of magnitude greater than the assumed upper bound of allowable delay times. These spikes were very infrequent. Nevertheless, given the anticipated message rates and number of processors in the system, even these relatively infrequent spikes would have caused unacceptable levels of instability. Several sources of the excessive delays were found, including a problem with the Ada run time environment, paging, tasking priorities. The solutions to these problems were non-trivial and included redesign of the tasking structure [7] and recoding the time-critical parts of the group membership protocols as operating system kernel extensions. The important point is that the risk management process brought these problems to the surface early enough that the necessary design modifications could be made "off the critical path" without disrupting the overall program.

The FTS integrates data from a continuing process of fault injection, testing, and performance measurements on the evolving design with analytical and model projections of the fault tolerance performance characteristics. Figure 6 shows the FTS as the vehicle for monitoring and summarizing the results of the on-going analysis and testing activities.

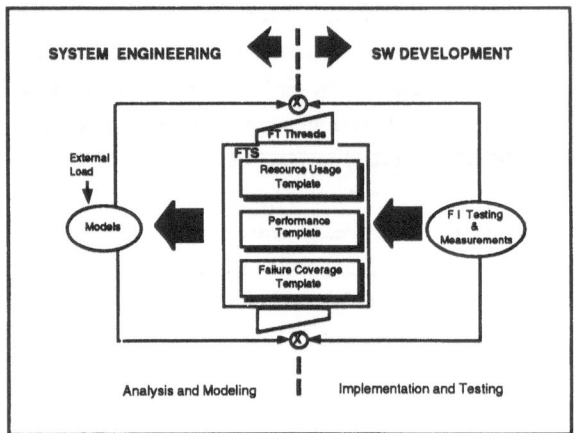

Figure 6.   Relationship of FTS to Analysis and

Testing Activities

One of the key benefits of the FTS is that it enforces convergence among the system engineering and software and hardware development organizations. The scorecard templates are designed to "close the loop" between the system's design and its implementation. Without the discipline of the FTS process, it is possible for the hardware or software development to proceed without being accountable for satisfying the stringent performance requirements implicit in a distributed real-time computing system. Strict rules of binding and coupling among the autonomous computing nodes makes this type of system unforgiving of performance delays. Performance delays that exceed the design assumptions cause far more serious side effects than just lengthening the response time; they result in the breakdown of fault tolerance management protocols. This, in turn, can cause false alarms and a general breakdown in the infrastructure that supports cooperation among the computing stations.

Over time, the FTS process will reveal the trend of the fault-tolerant design implementation. This will either build confidence that the development risk is being effectively managed, or provide early and increasingly urgent warnings that more drastic corrective action must be taken to avoid program cost and schedule growth.

Acknowledgements

The FTS concept evolved from tests and measurements performed by the IBM Federal Systems Company as part of the AAS program. The authors are grateful for the dedication and sustained hard work of the IBM technical staff who performed the fault tolerance risk reduction activities. The authors would also like to thank Flaviu Cristian and Algirdas Avizienis for their encouragement and helpful comments on the draft of this paper.

References

[1] A. Avizienis, *The Dependability Problem: Introduction and Verification of Fault Tolerance for a Very Complex System*, Proceedings of the 1987 Fall Joint Computer Conference, Dallas, TX , October 25-29, 1987, pp. 89-93.

[2] Algirdes Avizienis, *A Design Paradigm for Fault-Tolerant Systems*, AIAA Computers in Aerospace Conference, October 7-9, 1987, pp 52-57.

[3] A. Avizienis and D. Ball, *On the Achievement of a Highly Dependable and Fault-Tolerant Air Traffic Control System,* IEEE Computer, Vol 20, No. 2, February 1987, pp 84-90.

[4] Barry W. Boehm, Software Engineering Economics, Prentice-Hall, New Jersey. 1981, pages 40 and 65.

[5] F. Cristian, J. Dehn, B. Dancey, *Fault-Tolerance in the Advanced Automation System*, 20th Int. Symposium on Fault-Tolerant Computing, 1990.

[6] Flaviu Cristian, *Understanding Fault-Tolerant Distributed Systems*, Communications of the ACM, Vol 34, No. 2, February 1991.

[7] Jonathan Dehn, *The Ada Framework for Event Driven Programs Used in AAS*, Proceedings of the Second EUROSPACE Ada in AEROSPACE Symposium, November 1991.

[8] DoD 4245.7-M, Defense Science Board Task Force Report, *Transition from Development to Production*, 1988.

[9] MIL-STD-499A (USAF), Military Standard for Engineering Management, 1974.

[10] MIL-STD-1521A, Military Standard for Technical Reviews and Audits for Systems, Equipments, and Computer Programs, 1976.

[11] David A. Rennels, *The Evolution of a Fault-Tolerant Design: Contractor Milestones and Evaluation*, AIAA Computers in Aerospace Conference, October 7-9, 1987, pp 58-59.

Panel Session: Qualitative versus Quantitative Assessment of Security

Qualitative vs. Quantitative Assessment of Security:
A Panel Discussion

Teresa F. Lunt (moderator)
Computer Science Laboratory
SRI International
Menlo Park, California 94025
U.S.A.
lunt@csl.sri.com

The reliability community strives to prevent bad things from happening by accident, while the security community aims to prevent bad things from happening through malice. It has been argued that, since dependability is the trustworthiness of a computer system to justify the reliance that can be placed on the service it delivers, security is an attribute of dependability. This would lead us to conjecture that the methods for assessing dependability could be applied to security. The question is, can the quantitative assessment techniques used by the reliability community be applied to the assessment of a system's security, or is security a binary attribute: either you have it or you don't?

Security could be viewed as the measure of a system's defenses against attack, including the absence of malicious faults that have been intentionally inserted. It seems unlikely that measures could be devised for assessing the likelihood that intentional faults may have been introduced during system design or implementation. But how do we do we assess how secure an allegedly secure system is? Existing evaluation criteria generally provide only a qualitative assessment of security, for example, the so-called Orange Book, or *DoD Trusted Computer System Evaluation Criteria (TCSEC)* [3], and the emerging "harmonized" European criteria, *Information Technology Security Evaluation Criteria (ITSEC)* [4].

There are a few exceptions, however. The Orange Book has guidelines for covert channel bandwidths. The German criteria [2] include a qualitative estimation of "strength of mechanism." The concept of strength seems understood for mechanisms such as cryptographic algorithms, and is assessed by how hard is it to break. Other aspects of security mechanisms that seem amenable to such an assessment are length of cryptographic key or password; again, this would be assessed by how long it would take to do a brute force attack or guess. Thus, these assessments of strength all have to do with how much effort is required of the opponent to defeat the mechanism. How do you assess the strength of a mechanism such as an access control list? Should we look at how much effort is required to defeat the mechanism? There does not seem to be a way inherent to the mechanism itself (analogous to the length of password or strength of cryptographic algorithm, for example) to make such a determination. Should we then look at the strength of the implementation, look for weaknesses in the source code? Or should we decide on some other basis, for example, whether the access control rules allow permissions to propagate in such a way as to defeat the original intent of the creator of an access control list [1].

The established security criteria are for the most part qualitative, reflecting the degree of effort that has been made to make a system secure, rather than how secure the system really is. For example, it seems to be taken on faith that the use of certain software engineering practices during system development will result in a more secure system, and this is reflected in the various criteria. But can we put a measure on how *much* the use of such measures will increase the security of a particular system, or on which or how many of such methodologies is required in order to attain a given numerical security goal? And how would we express security quantitatively, if it were possible to do so: in number of break-ins per unit time, or time to first break-in, or probability of break-in under given circumstances or attack scenarios? Objections to such quantitative measures would include that once a given security flaw is known to an intruder or to the intruder community, the probability of its exploitation shoots up and the measure of security drops precipitously.

Nevertheless, some feel that analogies can be drawn between security and reliability assessment. Some have noted an analogy between failure cost and the reward or cost of a security breach. With reliability models, which deal with the stochastic process of failures in time (given a stated operational profile for the system), the longer the time, the greater the chance of failure. By analogy, stochastic security models could replace the operational profile with an attack profile and could reinterpret time as the effort expended in break-in attempts over time. Security would be expressed as a degree of protection given a certain expenditure of effort. This raises many questions, including how to measure effort and how to come up with the probabilities of attack. The latter is related to the specific threat and is likely to vary considerably from one installation to another of the same system. Even for a specific installation, the threat and certainly its probability distribution may be largely unknown. Even if we do without knowledge of the threat and specify the measure of security only in terms of effort, how do we account for the fact that a large number of people may know what some of the flaws are and how to exploit them (e.g., hackers' toolkits), so that little or no effort is needed for an attack?

These and other questions will be the topic of discussion for this panel.

References

[1] Matt Bishop. Theft of information in the take-grant protection model. In *Proceedings of the First Workshop on the Foundations of Computer Security*, Franconia, New Hampshire, June 1988.

[2] BSI. *IT-Security Criteria*, 1989.

[3] *Department of Defense Trusted Computer System Evaluation Criteria, DOD 5200.28-STD*. Department of Defense, December 1985.

[4] European Communities Commission. *Information Technology Security Evaluation Criteria (ITSEC), Provisional Harmonised Criteria (of France, Germany, the Netherlands, and the United Kingdom)*, June 1991. Version 1.2, ISBN 92-826-3004-8.

A Fault Forecasting Approach for Operational Security Monitoring

Marc Dacier
LAAS-CNRS
Toulouse 31077
France
dacier@laas.fr

So far, the security of computing systems has been mostly assessed from a qualitative point of view: a system is assigned a given security level with respect to the presence or absence of certain functional characteristics and the use of certain development techniques. This approach seems to be quite satisfactory for systems where security is a main concern. However, we think that such a static perception is not suitable for other application contexts where system ease of use and information sharing are among the prime objectives to achieve.

The inadequacy is all the more pronounced when low levels of security are considered. For example, in the *Orange Book*, if you decide not to implement a mandatory access policy, which is one of the minimal requirements to reach the B1 level, your system will be assigned a class C2 rating, at best. Thus, a system that contains every feature required at the B1 level, except mandatory access policy, is not evaluated as more secure than another system containing only the required features of the C2 level.

It could be argued that refinement of the various levels would solve the problem. This is partially true and has been, and still is, investigated within the various security criteria defined around the world. However, the approach still suffers from the lack of ability to deal with operational security. So far, neither the environment, nor the users, nor the way the system is managed and configured take part in the evaluation process. Yet, experience shows their importance. This is especially true for systems where the freedom of action left to the users is substantial (typically, systems at, or below, C2 level).

When dealing with the development of dependable computing systems, the combined utilization of a set of four methods - fault prevention, tolerance, removal and forecasting - is usually considered. Now, security is an attribute of dependability and it is worthwhile noting that the pre-defined criteria for a given security level can be seen as a fault prevention method. Indeed, the use of a prescribed development method prevents the introduction of design faults; also, the implementation of certain functional characteristics prevents operational faults. Searching, in the security area, for the equivalent of the other methods leads to some interesting remarks.

Although fault removal and fault tolerance have been considered in the literature, fault forecasting, to our knowledge, has rarely been investigated. This could result from the difficulty of envisaging the existence of features exploitable as security flaws without immediately trying to remove them. Common sense tends to say that if such a flaw exists then it is a fault and should be removed. However, if, for whatever reason, such unremovable faults exist, a fault forecasting approach would help us to assess the ability of the computing system to resist attacks performed by exploiting the identified flaws. It is our claim that there are systems where faults exist, are well-known, have an incidence on the overall security and, because they are useful features, should not be removed. In that particular case, their influence should be quantitatively assessed.

Due to the lack of space, we will merely give a trivial example of such a fault, taken in the context of *Unix*[1] operating systems. A feature exists in *Unix* that lets you give some other users access to your account and grant them all your privileges. This can be done by writing their user names in an ad hoc file called *".rhosts"*. Then, they can log into your system with your name without knowing your password. At that point, from the operating system point of view, they *"are"* you. This feature can be very useful when you have to share an account between different users you trust. This is effectively often used. The problem resides in the transitivity property of this feature that can be incompatible with the notion of trust that it is supposed to implement. If A trusts B and B trusts C, it does not imply that A trusts C. However, if B is in the *".rhosts"* file of A and C in the *".rhosts"* of B, then, from a practical point of view, C *is* in the *".rhosts"* of A. In that case, there is no easy solution to eliminate the problem other than negotiation between the users.

We stress the fact that this is just one example among many others. They all highlight a notion that we call *"transitive identity stealing"*, based on well-known, useful features that, if misused, can endanger the global security. The point is that they are not faulty by themselves. They are valuable features we want to keep in the system. However, an attacker can take advantage of their existence to stage an attack against the system. Therefore, it is our feeling that, in this case, the best way to assess the ability of the system to resist attacks is to conduct a white box approach: for a given system we must be able to automatically detect the existence of all these features, to define targets according to a security policy and to estimate the distribution of probability for an attacker to reach these targets with respect to his effort. Of course, theoretical and practical problems arise in building such a model and work is currently carried on at LAAS [1] to build a tool to automate the assessment process. Such a tool would also be able to discover faults that should be removed, implementing at the same time fault removal and fault forecasting.

---

[1] Unix is a trademark of X/Open

As already stated, this approach does not pretend to be the ultimate solution. It must be considered solely as a complement to other existing methods. Furthermore, it restricts itself to the domain of systems with a low-level of security, as explained before. Nevertheless, if successful, such modeling would offer a convenient way to follow up the operational security. Indeed, it could be used routinely by the administrators to monitor the evolution of the system security according to the modifications of system configuration such as installation of new software packages, addition or deletion of user accounts, etc. Finally, it would offer concrete data to the administrator for negotiating with a user whose configuration endangers the overall system security.

References

[1] M. Dacier, M. Kaâniche, Y. Deswarte. *Predictably Dependable Computing Systems (PDCS-2)*. First Year Report, ESPRIT Project 6362, September 1993, pp. 561-578.

Measurement of Operational Security

Bev Littlewood
Centre for Software Reliability
City University
London EC1V 0HB, UK
b.littlewood@cse.city.ac.uk

For all aspects of dependability, users want to know what to expect of *actual system behaviour in operation*. In the case of reliability, it is now possible to obtain such operational measures for systems even in the presence of design faults (e.g., software faults). Similarly, a measure of the *security* of a system should capture quantitatively the intuitive notion of 'the ability of the system to resist attack'. That is, it should be *operational*, reflecting the degree to which the system can be expected to remain free of security breaches under particular conditions of operation (including attack). So-called 'quality indicators' (e.g., properties of development process, structural properties of the product, etc.) do not provide such operational measures. In particular, current security *levels* at best merely reflect the extensiveness of safeguards introduced during the design and development of a system. Whilst we might expect a system developed to a higher level than another to exhibit 'more secure behaviour' in operation, this cannot be guaranteed; more particularly, we cannot infer from knowledge of such a level what the actual security behaviour will be.

The probabilistic approach to reliability is now well established, and it is generally accepted that it is sensible to expect there to be a 'natural' uncertainty about the failure behaviour of a system in the presence of design faults. It must be admitted that, in the early days of software reliability modelling, this was not an easy idea to get across to practitioners. This is reflected in the use of the term 'systematic' to describe design faults, which is used to convey the fact that such a fault will *always* cause a failure whenever the appropriate triggering circumstances are present. However, the use of the word *systematic* here refers to the fault mechanism, i.e., the mechanism whereby a fault reveals itself as a failure, and not to the failure *process*. We are uncertain about when the next failure will occur because we are uncertain about which inputs will be received in the future, and we do not know which, of all the possible inputs, do cause failure. This means that we cannot predict with certainty when an input will be received that will trigger a failure. The uncertainty here, then, lies in our imperfect knowledge of the operational environment.

In trying to provide a probabilistic theory for security, we need first to be convinced that there is similar uncertainty for which probability is the appropriate mechanism of representation - as in reliability, there is a tendency to view things deterministically. The uncertainty here arises in a similar way to that in reliability modelling: it is a result of the incompleteness of our knowledge. Thus to a system attacker, who has incomplete knowledge of the system she is attacking, there is uncertainty as to the effects of her attacks; to the system owner, there is uncertainty as to the nature of the attacks that will be received, and even as to whether particular attacks would result in security breach. In recent work [Littlewood, Brocklehurst et. al. (to appear)] we discuss in some detail similarities and differences between reliability and security with the intention of working towards probabilistic measures of 'operational security' of systems, similar to those that we have for reliability. Very informally, these measures could involve expressions such as the rate of occurrence of security breaches (cf. rate of occurrence of failures in reliability), or the probability that a specified 'mission' can be accomplished without a security breach (cf. reliability function).

In reliability we usually arrive at our measures via a model of a *stochastic process of failures* in *time*. In trying to imagine a similar stochastic process in the security context, security breach seems a natural analogy to system failure. Time, on the other hand, presents more of a problem. In reliability we can usually define a time variable that captures the notion of 'exposure': the greater the exposure the greater the chance of failure. In security, time alone does not seem adequate. If we restrict ourselves to the case of intentional security breaches, a system cannot be breached if not attacked (it can deliver 'ordinary functionality' in time); certain attacks, such as bribery, are like shocks and it is the magnitude of the shock that determines the chance of a breach; in some attacks, such as Trojan horses, attacks and their consequences are not synchronised. In security, then, we believe that a more general *effort* variable needs to be identified upon which to build the stochastic process of breaches. This variable representing the effort expended by an attacker could involve resources, money, and even ability.

In reliability, we are often interested in the consequences of failures as well as merely the number of them. Similarly in security, there is a reward (to the attacker) or loss (to the system owner) associated with a security breach. In addition there is the possibility of continuous accrual of reward - even 'failed' attacks can give useful information to an attacker - which seems to have no parallel in reliability.

Another area where security seems to pose special problems is in the importance of the *viewpoint*. Thus the measures of security represented by probabilities (e.g., probability of breach for a certain expenditure of effort) may differ from attacker to attacker and from attacker to system owner. This may mean that subjective, Bayesian probabilities are inevitable, rather than 'objective' frequentist ones. There will be similarly different subjective views of rewards and losses. Thus the *nature* of the reward to an attacker from a breach will usually differ from that of the loss to the system owner: the one may be simply the pleasure of intellectual achievement, the other monetary loss. Even when rewards and losses are of the same nature for different viewers, it is unlikely that they will perceive the *amounts* similarly. The attacker reward and the system owner loss from a breach will not be the same: it is not a zero-sum game.

From the above brief account of this new approach to security evaluation it should be clear that many questions need answering before it will be possible to build a workable theory. For example, we need to know how to construct a suitable effort variable to play the role that time plays in reliability models. The author and some colleagues have started to see whether such questions are open to experimental investigation. The results of a pilot study, involving volunteer student attackers of a university system in normal operation, are quite promising. An alternative approach might be to try to collect data on real systems with the co-operation of their owners and, if possible, attackers. In the meantime we would welcome views on the feasibility of this whole approach from members of the security community.

Acknowledgements

The work reported here was supported by the ESPRIT Basic Research Project number 6362, PDCS2.

References

[1] B. Littlewood, S. Brocklehurst, N.E. Fenton, P. Mellor, S. Page, D. Wright, J.E. Dobson, J.A. McDermid and D. Gollmann, "Towards operational measures of computer security," *Journal of Computer Security*, (to appear).

Quantitative Measures of Security

John McLean
Center for High Assurance Computer Systems
Naval Research Laboratory
Washington D.C. 20375
U.S.A.
mclean@itd.nrl.navy.mil

1   Introduction

To the casual, yet interested, reader, one of the most striking properties of the *Trusted Computer System Evaluation Criteria* [1] and its international successors is that none of these documents contain any attempt to relate their evaluation levels to a measure of how much effort must be expended to break into a system.[2] As a consequence, it's impossible to evaluate rationally the marginal benefit of spending the extra money necessary to obtain a higher rating than a lower one.

One reason for this gap between evaluation levels and cost of system penetration is the difficulty of quantifying penetration costs. In this paper I hope to shed some light on the questions of what is needed, what we have, and what it would be useful to have in the future.

2   The Cost of Penetration

Penetration has a variety of meanings, depending on the context. A system may be easy to penetrate for the purpose of withholding service, but hard to penetrate for the purpose of obtaining or modifying confidential data. Even if we limit ourselves to one type of penetration, say confidentiality attacks, systems differ. For example, obtaining some types of data is useless if the penetration can be detected after the fact. For other types of data, this is not an issue.

However, even if we leave these subtleties aside, confidentiality has properties that make it less suitable than, say, reliability, for certain types of quantitative assessment. The primary difference between security and reliability is that the former must concern itself with *malicious agents* who may affect the system, not simply once its fielded, but even during its development. As a result, concepts, such as the *reliability function*, the *rate of occurrence of failures*, and the *mean time to next failure*, which makes sense in the reliability world, have no relevance in the security world. If a computer system contains a Trojan Horse or a communication system employs a flawed cryptographic algorithm or protocol, then the *rate of occurrence of failures* is simply the rate at which the agent who planted the Trojan Horse or an agent who has detected the flaw decides to exercise it. Although it may be possible to exclude such attacks and determine functions that reflect the mean time to a system's spontaneously broadcasting classified material on an unclassified network or succumbing to a random hacker, such measures are of little comfort in the security world. Further, although testing and experiments subjecting a system to amateur hackers may be useful in giving us information about defects of the latter kind, they tell us nothing about defects of the former kind. Malicious defects are designed to be undetectable and flaws in crypto systems are rarely detected by amateurs, no matter how much time they are given.

3   Quantifying Security

Despite the skeptical conclusion of the previous section, we should note that there are already in place some quantitative measure for security. For encryption, there is complexity theory which helps quantify the effort needed to break a code.[3] For confidentiality in computer systems, there are applications of information theory employed in quantitative security models [4, 5] and techniques for covert channel analysis [6, 7]. The difference between these information-theoretic measures of computer security and the sort of quantitative measures one would obtain by applying techniques from the reliability arena is that whereas the latter is concerned with determining the mean time between failures, the former is concerned with measuring the amount of damage that can be done by exploiting failures. This difference reflects a difference in orientation between the security community and the reliability community. The reliability community is interested in determining how long it will take for a system to fail. The security community assumes that a flawed system will immediately fail (in the sense that a penetrator will immediately take advantage of the flaw) and is more concerned with quantifying the damage such a penetrator can effect. Although covert channel capacity and similar measures are not perfect indicators of this damage, they do, at least, approximate it since they give information about the time required to exploit a channel.[8]

A useful analogy might be to consider the different ways of evaluating a bank safe. We may be interested in the chance of a safe spontaneously opening on its own or being opened by a curious child. However, we would be unlikely to buy a safe for any serious purpose for which this was a realistic concern. A more relevant measure is how long it would take for an expert to open a safe given the appropriate tools and what sort of tools would be required. The former measures are akin to reliability, the latter to channel analysis. If we are interested only in shielding some rather trivial data from casual observers, then the former approach has merit.[1] However, in most real security applications, we assume a casual amateur cannot enter a system and are concerned with how long it would take a determined expert.

This is not to say that information theory provides all the information we would like. With respect to confidentiality, we are interested, not only in the time it would take to gain data from a system, but also in the tools that are required, the risk of exposure during the attack, the detectability of attack after the fact, the type of data at risk, etc. However, reliability measures are similarly silent on these issues. Turning to availability or integrity, although it's conceivable that information theory will be fruitful in these areas by quantifying how much secure information can flow in a system under denial of service or integrity attacks (a sort of inverse capacity measure which reflects the worst possible information rate rather than the best), there is no supporting research to demonstrate this. Performing such research would be a worthwhile endeavor. Reliability-like measures, by contrast, are unlikely to succeed with respect to integrity and availability for the same reason they fail with respect to confidentiality: they do not incorporate malicious agents.

4   Conclusion

To evaluate the effectiveness of techniques used to build secure systems some sort of quantitative measure of penetration resistance is necessary. However, the reliability model is the wrong way to go. Certainly with respect to confidentiality, and possibly with respect to integrity and availability as well, information theoretic approaches, though not perfect, are more suitable.

References

[1] National Computer Security Center, *Trusted Computer System Evaluation Criteria*, CSC-STD-001-83, Ft. Meade, MD, 1983.

[2] J. McLean, "New Paradigms for High-Assurance Systems," *Proc. of the New Paradigms Workshop*, IEEE Press, forthcoming.

[3] D. Denning, *Cryptography and Data Security*, Addison-Wesley, Reading, MA, 1982.

[4] J. McLean, "Security Models and Information Flow," *Proc. 1990 IEEE CS Symposium on Research in Security and Privacy*, IEEE Press, 1990.

---

[1]For example, I believe Kryptonite bicycle locks were first tested by leaving locked bikes in various locations in New York City.

[5] J. Gray, "Toward a Mathematical Foundation of Information Flow Security," *Journal of Computer Security*, Vol. 1, no. 3-4.

[6] J. Millen, "Covert Channel Capacity," *Proc. 1987 IEEE CS Symposium on Research in Security and Privacy*, IEEE Press, 1987.

[7] I. Moskowitz and A. Miller, "The Channel Capacity of a Certain Noisy Timing Channel," *IEEE Transactions on Information Theory*, Vol. 38, no. 4 , 1992.

[8] J. McLean, "Models of Confidentiality: Past, Present, and Future," *Proc. Computer Security Foundations Workshop VI*, IEEE Press, 1993.

The Feasibility of Quantitative Assessment of Security

Catherine Meadows
Naval Research Laboratory
Code 5543
Washington, DC 20375
U.S.A.
meadows@itd.nrl.navy.mil

In general, quantitative models of security have not met with a very warm acceptance. This has been a result of a number of factors. First of all, research in security has largely concentrated on high-assurance systems for which quantitative measures may not be realistic. Secondly, research in security has focussed on confidentiality, which is, at least in theory, possible to guarantee to a high degree of assurance even when most of the system is assumed to be actively trying to subvert the security policy. Thirdly, research in security generally operated in isolation from consideration of other system requirements.

However, all this has been changing. More interest in security is being shown by the commercial sector, which is more interested in cost-effort tradeoffs that achieving the highest degree of security possible. Also, more interest is being shown in systems that have other critical requirements besides security, such as real-time requirements or fault-tolerance. These requirements may come into conflict with security requirements, and some means are needed to quantify the tradeoffs between the competing properties.

As a result there has been a growing body of work on quantititative assessment of different security properties. The work of Kailar, Stubblebine and Gligor develops metrics for the the security of certain kinds of cryptographic protocols [3]. Other recent work has focussed on measuring the capacities of covert channels so they can be traded off against the real-time requirements of a system [2, 4]. Still other work, such as Gong's [1], trades off various security goals of key distribution protocols against such performance metrics as number of messages and number of rounds of communication.

Although this kind of work is a promising start to developing a quantitative assessment of security, in general recent work has concentrated on quantitative measures of isolated security properties. What is lacking is some means of integrating these various approaches. By integration I mean, not only the integration of metrics for different security properties, but integration of the results of applying a single metric to different parts of a system. For example, a great deal of work has been done on measuring the capacity of isolated covert channels. But although there has been some work on measuring the capacity of composed or parallel channels [6, 5, 7], in general the problem of measuring the "leak rate" of an entire system has not received very widespread attention.

Given the lack of theoretical underpinnings at the most basic level, it is not surprising that more ambitious attempts to measure the likelyhood that the security of a computer system can be breached are not very convincing. What we need if more work from the ground up. We need to study closely the results of applying a a few metrics to the same component, and the result of applying a single metric to different components to the same system. Once we understand the results of integrating metrics on this small scale, we will be much better able to understand the feasibility of integrating them over an entire system.

References

[1] L. Gong. Lower Bounds on Messages and Rounds for Network Authentication Protocols. In *Proceedings of the 1st ACM Conference on Computer and Communications Security*, pages 26–37. ACM Press, November 1993.

[2] J. W. Gray III. On Introducing Noise into the Bus Contention Channel. In *Proceedings of the 1993 IEEE Symposium on Security and Privacy*, pages 90–98. IEEE Computer Society Press, May 1993.

[3] R. Kailar and V. Gligor. On the Security Effectiveness of Cryptographic Protocols. In *Proceedings of DCCA4*. 1994.

[4] I. S. Moskowitz and A. R. Miller. The Influence of Delay Upon an Idealized Channel's Bandwidth. In *Proceedings of the 1992 IEEE Symposium on Security and Privacy*, pages 62–67. IEEE Computer Society Press, May 1992.

[5] J. Trostle. The Serial Product of Controlled Signalling Systems: A Preliminary Report. presented at Computer Security Foundations Workshop III, June 1990.

[6] C.-R. Tsai and V. D. Gligor. A Bandwidth Computation Model for Covert Storage Channels and its Applications. In *Proceedings of the 1988 IEEE Symposium on Security and Privacy*, pages 108–121. IEEE Computer Society Press, May 1988.

[7] T. Wittbold. Networks of Covert Channels. presented at Computer Security Foundations Workshop III, June 1990.

Quantitative Measures vs. Countermeasures

Jonathan K. Millen
The MITRE Corporation
Bedford, Massachusetts 01730
U.S.A.

The appeal of quantitative models of security is that, like mathematical models in general, they can be analyzed to yield insights that are not immediately obvious. Below is a simple example that illustrates the reward of setting up and analyzing a quantitative model. However, the conclusion drawn from this particular model is not to be trusted, because the model is not realistic.

Suppose that there are some multilevel secure systems which are not connected to one another. Each supports two levels of data: let us call them "S" and "U" data (for "sensitive" and "unimportant"). Each system has some users cleared to access S as well as U data, and other uncleared users who can access only U data.

Suppose also that an enemy wishes to compromise S data with the help of agents acting as users on several systems, as many as it can afford. There will be at most one agent per system. Some cost is involved to retain each agent for an indefinite period, as long as needed. We assume that it costs more to install or bribe an agent cleared for S data than it does to do the same for an uncleared agent.

Computer security measures in the multilevel systems will not prevent a cleared agent from accessing S data, so the value of a cleared agent is clear. Depending on the strength of the computer security measures, even an uncleared agent has some chance of compromising S data by finding and exploiting flaws or covert channels.

The enemy is faced with an optimization problem: to compromise the information on the greatest number of systems with the least total cost. Its cost and yield are affected by the allocation of cleared vs. uncleared agents. To investigate this problem, let us quantify the relevant parameters. Let u be the number of uncleared agents, and s the number of cleared agents. Let a be the cost of installing an uncleared agent, and let $b > a$ be the cost of installing a cleared agent. The amount by which b exceeds a acts as a measure of the sensitivity of S data.

Let p be the probability that an uncleared agent will eventually succeed in defeating the protection on a system, and thereby gain access to data presently and subsequently stored on that system. The probability p measures (inversely) the strength of the security protection.

The enemy wishes to minimize its average cost per compromised system. This average cost is the ratio of the total cost of all installed agents to the expected number of compromised systems. The expected number of compromised systems includes all systems with cleared agents, plus a proportion p of those with uncleared agents. Thus,

$$r = \frac{au + bs}{pu + s}.$$

This expression is easier to deal with if we define $x = s/u$, the ratio of cleared to uncleared agents, giving:

$$r = \frac{a + bx}{p + x}.$$

The enemy should choose $x \geq 0$ so as to minimize r. By plotting r against x, or inspecting the first derivative, we find that there are three cases:

- if $p = a/b$, the curve for r is flat, and any x will do.

- if $p < a/b$, the curve for r has positive slope, so $x = 0$, and $s = 0$.

- if $p > a/b$, the curve for r has negative slope, so $x = \infty$, and $u = 0$.

Thus, the enemy will do best by choosing between two pure strategies: all cleared agents, or all uncleared agents, depending on how the strength of security in the attacked systems compares with the clearance cost ratio a/b.

Now consider the results of this analysis from the point of view of an organization designing or acquiring the multilevel secure systems. We have found that once the system is strong enough to make $p < a/b$, the enemy will not install any uncleared agents, which means that the computer security protection mechanisms will not be challenged! In other words, a system is as secure as it needs to be when p is only as small as a/b; it is unnecessary to aim for perfect security, $p = 0$.

This pleasant insight illustrates the potential benefit of pursuing quantitative models of security, though this particular conclusion cannot be applied as it stands to the real world. The model is based on assumptions that we cannot expect to hold, and the results assume that the relevant parameters can be measured exactly. Nevertheless, there is room for hope that this kind of analysis, performed with more care and subtlety, would be rewarding.

Basic Problems in
Distributed Fault-Tolerant Systems

Continual On-Line Diagnosis of Hybrid Faults*

C. J. Walter, N. Suri and M. M. Hugue
AlliedSignal MTC
Columbia, Maryland 21045
U.S.A.

Abstract

An accurate system-state determination is essential in ensuring system dependability. An imprecise state assessment can lead to catastrophic failure through optimistic diagnosis, or underutilization of resources due to pessimistic diagnosis. Dependability is usually achieved through a fault detection, isolation and reconfiguration (FDIR) paradigm, of which the diagnosis procedure is a primary component. Fault resolution in on-line diagnosis is key to providing an accurate system-state assessment. Most diagnostic strategies are based on limited fault models that adopt either an optimistic (all faults s-a-X) or pessimistic (all faults Byzantine) bias. Our Hybrid Fault-Effects Model (HFM) handles a continuum of fault types that are distinguished by their impact on system operations. While this approach has been shown to enhance system functionality and dependability, *on-line* FDIR is required to make the HFM practical. In this paper, we develop a methodology for utilization of the system-state information to provide continual on-line diagnosis and reconfiguration as an integral part of the system operations. We present diagnosis algorithms applicable under the generalized HFM and introduce the notion of fault *decay time*. Our diagnosis approach is based primarily on monitoring the system's message traffic. Unlike existing approaches, no explicit test procedures are required.

Keywords: on-line diagnosis, fault-effects, Byzantine faults

1 Introduction

Strategies which incorporate fault identification with high resolution are becoming an important feature of fault tolerant systems. These systems are designed to provide maximal system

---

*Work supported in part by ONR # N00014-91-C-0014

dependability at the lowest possible cost to the user. The ability to accurately diagnose faults is an important aspect of achieving this goal. In general, diagnosis has two primary objectives: (a) to identify a faulty unit so as to restrict the fault-effect on the system operations and (b) to support the fault isolation and reconfiguration process with information concerning the state of the system resources. The better the diagnostic resolution to identify a fault at the lowest level, the more efficient is the FDIR procedure, with minimal and very specific resources targeted for repair. This translates into an efficient management of the system resources, and the enhanced resolution can result in lower repair costs (cost of spares, time latency, fault-management costs). Also, if FDIR can be done during system operation, there will be more resources available for sustained system operations.

A variety of approaches have been proposed for system diagnosis - see survey in [1], which have limited applicability for the following reasons. First, these methods are invariably off-line techniques due to the extensive overhead induced in collecting and interpreting the diagnostic information. Second, each approach is based on a particular system and fault model which influences the amount of diagnostic resolution obtainable and determines the achievable degree of system dependability. Third, diagnosis is treated as an isolated objective in the system and not as a continual and integral phase of the system operations.

The dependability objective requires an accurate assessment of the state of the operational system, with imprecise assessment potentially resulting in either underutilization of resources or system failure due to incorrect assessment of the system-state. Off-line diagnosis procedures support the first goal of identifying the faulty units in need of repair. The advantage of performing tests off-line is that interference with the system processing tasks does not occur. One major disadvantage in running the diagnosis off-line is that information is not available to the reconfiguration process which would make it more efficient. A second disadvantage is that transient or intermittent faults are not easily duplicated in an off-line environment. Actual stress placed on the system during active operation may be required to produce the fault, since it may be difficult to anticipate a test case or may be too expensive to simulate off-line. Also, these procedures impose a latency between the fault-occurrence and diagnosis phases, thus restricting the applicability of the diagnosis operations to actual system operations.

Although on-line diagnosis has some clear benefits, it also has its challenges. The first challenge is to implement a procedure with high resolution that does not severely degrade the performance of the system. Approaches, such as the PMC [4] model (and variations - centralized and distributed), which rely on units explicitly testing each other to obtain syndrome information, are too costly in this regard and are invariably off-line procedures. Other methods, such as comparison testing [6] also have their shortcomings with the reliance primarily on comparison of test results for fault detection.

1.1 Diagnostic Approach

In [5, 10], we showed that using a more rigorous fault model allows dependability to be more precisely and realistically evaluated for a given *fault set* instead of a single fault type. The stuck-at-X model usually results in overly optimistic evaluations as faults which exhibit less predictable manifestations are frequently present. The Byzantine model assumes worst-case behavior for each fault occurrence; this is a conservative model and results in a pessimistic assessment of dependability. The need is to find a more realistic and flexible fault model and to demonstrate its applicability in supporting fault-diagnosis.

The majority of diagnostic approaches assume permanent faults to prevent syndrome ambiguities due to arbitrarily test results from a node with a transient or intermittent fault. Thus, they are unable to handle Byzantine faulty nodes, which can send conflicting information to testing nodes. In fact, it is shown in [7] that identification of a node that is Byzantine faulty is impossible using single pass syndrome collection. [7] also provide an off-line algorithm to tolerate Byzantine faults. In [10], we have presented an on-line diagnosis and fault identification model that admits Byzantine faults.

We deviate from the conventional fixed severity (both time-domain and data-domain) fault models to develop a composite and cohesive fault classification basis. Our model handles a continuum of fault and error types, classified according to their impact on system operations. Although this approach is shown to provide better assessments of operational indices, its practical usage requires an on-line FDIR process. In this paper we present diagnosis protocols, developing on our previous work. In [10], we demonstrated that different error types could be detected, focusing on local detection primitives. We presented the basis for an on-line algorithm which handles mixed fault types. One important feature shown was that this approach could be implemented with low overhead. It was stated that in MAFT[9] the overhead was $\leq$ 6% even though the algorithm executes continually. This method allows diagnosis to be directed to identifying the cause of detected errors and their impact on system resources. Also, understanding the cause and effect relationship, allows the reconfiguration process to be properly guided.

Our previous classification of detectable faults as either symmetric or asymmetric was from a low-level perspective on information flow in the system. We will provide a higher level classification from the standpoint of how detected faults affect the global state of the system based on the HFM. This reconciling of local views to a higher-level global view is accomplished by first executing a distributed agreement algorithm to ensure consistent error syndrome information and penalty counts. This agreed upon information is used in the diagnosis phase by mapping low-level syndrome bits to high-level fault manifestation groups, such as benign, catastrophic, value, crash and Byzantine, which match the dependability model. Thus, we show that a more precise model for dependability can be constructed with significant benefits, supported by the requisite on-line algorithms. Our approach is not to treat diagnosis as a stand-alone function but to keep it as an integral and continually on-going phase of system operations.

Thus, the algorithm is not scheduled, but uses the system communication traffic as test stimuli. If necessary, specific tests can be run to handle episodic faults.

In section 1.2 we describe the HFM and system model. We present the progressively refined diagnosis algorithms in section 1.3. The formalism for the diagnosis algorithms are based on the Hybrid Oral Messages (HOM) algorithm, whose properties are also discussed. In section 1.4 we introduce the issue of the temporal behavior of faults through a notion of fault decay time.

2   System and Fault Model

We consider a distributed system framework, comprising *nodes* that communicate using a synchronous deterministic message passing protocol. Individual nodes make decisions and compute values based on information received in messages from other nodes. We consider the system communication model as:

A1: A direct path exists from each node to all other nodes. A node issues a single message which is broadcast to all connected nodes.

A2: A non-faulty node can identify the sender of an incoming message, and can detect the absence of an expected message.

A3: Node and link faults are considered indistinguishable (This assumption is later alleviated in Alg. HD).

Our goal is to identify faulty nodes and to prevent system failure in the presence of a specified number of faults. The system fails when consistent decisions or computations across the system are no longer possible. Unlike previous work, we distinguish faults by their behavior, and place no limitations on the duration of a fault. Under this framework, the sole indicator of a faulty node is an error in its transmitted message, as viewed by all receivers of the message. The receiver acquires a *local view* of the sending node's health by applying fault detection mechanisms, such as range and framing checks, to the incoming message. The exchange of local views among receiving nodes is then used to acquire a global perspective of the effects of a faulty unit.

2.1   Fault Model

Our interest is in considering faults with unrestricted behavior. In this paper, we adopt the Hybrid Fault-Effects Model (HFM), in which faults are classified according to the fault manifestations they present across the system. We provide a two-fold motivation for using HFM as linked to the diagnosis objective.

First, we are basing diagnosis on consensus algorithms. These algorithms are very robust but also expensive (time and space complexity) to implement. Designed to provide coverage to the worst case Byzantine faults, these assume all fault occurrences to be Byzantine faults and require a node-redundancy of $N > 3m$ to cover m faults. The HFM assumes perfect coverage to a limited number of arbitrary faults, but recognizes that weaker fault types are typically more common than the classical Byzantine faults. The algorithms under the HFM do not compromise the system's capability of tolerating Byzantine faults, however they provide additional and concurrent coverage to *fault sets* of weaker manifestations. This also facilitates a higher resolution of fault granularity compared to considering all faults of a single fault severity. The distinction here is that hybrid faults can be of any type, as long as they can be tolerated by the system implementation *without causing system failure*. If the system is destroyed, or a sufficiently large portion of the system is damaged, then the issue of diagnosis becomes moot.

Second, in HFM it is important to note that the fault classes are *disjoint*, preventing any ambiguities in discerning the fault behavior and effect and subsequent diagnostic ambiguities. Furthermore, as HFM considers fault classification based on the *effect* the fault causes to the system operation, it provides an uniform framework to handle both time-domain and data-domain faults.

Under the HFM, classification is based on (1) *Local-Classification* of fault-effects to the extent permitted by the fault-detection mechanisms built in at the node level, and (2) *Global-Classification* based on nodes exchanging their local-classification with other system nodes to develop a global opinion on the fault-effect.

HFM classifies fault-effects into two types: *benign* and *value* (or *malicious*) faults. *Benign or non-malicious* are detectable by all non-faulty nodes within the scope of the fault detection mechanisms implemented *locally* in each node. Examples of *benign* faults consist of locally discernible effects such as bit garbling, framing errors, range violations, message omission or early/late message delivery. Interestingly, the *benign* fault set encompasses a number of fault classes defined previously as timing, omission, crash and stopping faults. These faults can still be classified individually under the HFM. Any detectable fault in a message will result in the sender being classified as *locally benign faulty*-(**b**) by the receiver.

Value or malicious faults are the locally undetectable class of faults. Messages which pass all data validity and range deviance checks, but provide a valid but *incorrect* value constitute *value* faults. Essentially, none of the constraints on good values or messages are violated. For example, if the valid range for an expected datum value is [0 - 100] and the expected *correct* value is 50, but instead the value 75 arrives, the node, locally, cannot discern any discrepancy.

A further classification for both *benign* and *value* faults is of symmetry. Suppose at least one non-faulty node cannot detect a node fault, and the same fault is detected by every other good node. The sending node is thus asymmetric value faulty, but is identified as *locally benign* by the nodes that can detect it, showing that the local view is not accurate. Thus, exchange and accumulation of global information across several time periods is often necessary to diagnose such faults. Value faults are also partitioned into *symmetric-*(s) and *asymmetric-*(a) value faults, according to their presentation throughout the fault scope. It should be noted that diagnosis of nodes as symmetric or asymmetric faulty requires a global perspective, or several rounds of message exchange to emulate one. Byzantine agreement [2] algorithms are often implemented to ensure that distributed nodes or processes arrive at the same decisions and computational results in the presence of arbitrary faults. However, with simple modification of existing fault detection and masking techniques, correct computations can be also be guaranteed under the HFM. Unlike the $N > 3m$ redundancy requirement of the classical Byzantine models, N nodes will tolerate $> 3a + 2s + b$ composite *(a, b and s)* faults under the HFM. We extend the classical Byzantine fault model to address mixed fault types, while ensuring correct system operation in the presence of faults. Instead of assuming that all faults are *arbitrary*, we adopt the hybrid fault taxonomy and handle faults according to the errors they produce.

3   HOM - Agreement under HFM

Our diagnosis protocols are based on the Exact Agreement or Consensus paradigm. Prior to developing the diagnostic procedure, we first need to demonstrate that exact agreement is indeed possible under the HFM. Further, it is also required to demonstrate the stability and correctness of the diagnosis procedure in the presence of faults. This is unlike the off-line techniques which assume the sanity of diagnosis in a fault-free scenario. Both of these facets are handled by the HOM Agreement algorithm.

3.1   HOM(a) Algorithm

The *Oral Messages Algorithm* (OM) [2] demonstrated that Agreement can be guaranteed if $n > 3t$ and $r \geq t$ where t is the number of Byzantine faults, n is the number of nodes, and r is the number of rebroadcast rounds. However, it makes the pessimistic assumption that all faults are asymmetric value faulty. This implies that a 4, 5, or 6 node system can tolerate only a single arbitrary fault, and that 7 nodes will be required to tolerate two faults. Note that the model covers only 2 faults, not necessarily 2 *arbitrary* faults. In contrast, the HFM covers both Byzantine and non-Byzantine faults under a single framework, without increasing the complexity of the underlying algorithms.

The *Hybrid Oral Messages* (HOM(a)) -Table 1- is a r-rebroadcast round protocol based on the OM[2] algorithm, requiring $n > 2a + 2s + b + r$ nodes to mask $(a + s + b)$ faults, where b nodes are benign faulty, s nodes are symmetric value faulty, a nodes are asymmetric value faulty, and

Table 1: HOM(a) Algorithm

S1: The Transmitter sends its personal value, v, to all receivers.

S2: $\forall i$, let v_i denote the value that Receiver i gets from the Transmitter.

If $a = 0$, and a benign-faulty(**b**) value is received, Receiver i adopts $\mathcal{E}$. Otherwise, Receiver i adopts v_i. The algorithm then terminates.

If $a > 0$, each Receiver adopts $R(\mathcal{E})$, if a benign-faulty(**b**) value is received, and $R(v_i)$ otherwise. Each receiver then acts as the Transmitter in Algorithm HOM($a - 1$) sending its personal value to the other $N - 2$ nodes.

S3: $\forall i, j$, with $i \neq j$, let v_j denote the value Receiver i gets from sender j in Step 2 of HOM($a - 1$). If no message is received or v_j is obviously incorrect, Receiver i adopts $\mathcal{E}$ for v_j; otherwise, v_j is used.

Since all Receivers act as senders in HOM($a - 1$), each Receiver will have a vector containing (N-1) values at the end of HOM($a - 1$). Receiver i adopts $v = HOM - maj(v_1, v_2, \ldots v_{N-1})$ as the Transmitter's value.

$a \leq r$. As $n > 2a + 2s + b + r$ indicates, several composite (a,b,s) fault scenarios are covered by HOM(a), not just a fixed number of t arbitrary faults.

The algorithm $Z(r)$ of [8] tried to address the issue of consensus and composite fault types. However, as it was found to be incorrect [3], we present a hybrid algorithm that achieves Agreement under the assumption of hybrid faults, satisfying the Hybrid Agreement conditions HOM1 and HOM2:

HOM1 (Validity): If the Transmitter is *non-faulty*, then all non-faulty Receivers select the sender's original value. If the Transmitter is *benign faulty*, then all non-faulty Receivers will adopt a default value, $\mathcal{E}$. If the Transmitter is *symmetric faulty*, then all non-faulty Receivers will adopt z, the sender's original value.

HOM2 (Agreement): All non-faulty Receivers agree on the value of the Transmitter.

The HOM(a) uses a family of error values, $\{\mathcal{E}, R(\mathcal{E}), \ldots, R^r(\mathcal{E})\}$, where r is the number of rebroadcast rounds, to incorporate the HFM. If the value $R^k(\mathcal{E})$ is received, where $k \geq r - l$ in S2 of HOM($a - l$), then that too is recognized as an error, and $\mathcal{E}$ should be adopted. The function HOM-*maj*, used by HOM(a), is as follows. Given a set V of k values, $v_i, \ldots, v_k$, HOM-*maj*(V) is given by

$$\text{HOM} - maj((V)) = \begin{cases} \mathcal{E}, & \text{if all of the } v_i \text{ satisfy } v_i = \mathcal{E}. \\ R^{-1}(v_\mathcal{E}), & \text{if } v_\mathcal{E} = maj(exclude(V, \mathcal{E})) \text{ exists.} \\ v_0, & \text{otherwise.} \end{cases}$$

The provision which adopts $\mathcal{E}$ if all the v_i are $\mathcal{E}$ cannot occur on a good node, but is included to provide a fail safe default value should that case occur on a partially faulty node.

3.2   Algorithm HOM(a) - Properties

We defer detailing the proof of the algorithm [3] and limit our discussion to stating the relevant Lemmas and properties of interest which help formalize the diagnosis objective.

Lemma 1 *Algorithm HOM(a) achieves Validity (**HOM1**) for any integers $a, b, s, N, r \geq 0$, such that $N > 2a + 2s + b + r$, and $a \leq r$.*

Lemma 2 *If $N > 2a + 2s + b + r$, for any $N, a, s, b \geq 0$, $r > 0$ and $a \leq r$, then HOM(a) satisfies Agreement (**HOM2**).*

Taken together, Lemmas 1 and 2 prove the following theorem.

Theorem 1 *Algorithm HOM(a) achieves Byzantine Agreement when $N > 2a + 2s + b + r$, for any $r \geq 0$, any $a \leq r$, any $s \geq 0$, and any $b \geq 0$, where a is the number of asymmetric value faults, s is the number of symmetric value faults, and b is the number of benign faults in the system during execution of HOM.*

The HOM algorithm assured agreement on a specified information value. The extension to obtaining exact agreement on the values of all constituent nodes leads to the Hybrid Interactive Consistency (HIC) problem. The HOM forms the proof basis for the HIC algorithms too. We briefly mention the requirement for HIC.

Algorithm 1 *(HIC(a)) Let S be the set of nodes holding values upon which HIC is desired, with $|S| = N$. Each node sends its private value to all other nodes in S, acting as the transmitter in HOM(a), with the value of r identical for all nodes.*

At the conclusion, each good node in S will hold a final vector which satisfies the following conditions.

HIC1 (Validity): Each element of the final vector that corresponds to a non-faulty node is the private value of that node. Each element of the final vector that corresponds to a benign faulty node is $\mathcal{E}$.

HIC2 (Agreement): All non-faulty nodes compute the same vector of values.

Theorem 2 *Algorithm HIC(a) achieves Agreement when $N > 2a + 2s + b + r$, for any $r \geq 0$, any $a \leq r$, any $s \geq 0$, and any $b \geq 0$, where a is the number of asymmetric value faults, s is the number of symmetric value faults, and b is the number of benign faults in the system during execution of HOM.*

Proof: By Theorem 1, at the conclusion of HOM(a) with node i as the transmitter, HOM1 and HOM2 will be satisfied. Thus, each good node will have a consistent view of node i's personal value. All nodes then execute HOM(a). By the proof of agreement and validity for HOM, if $N > 2a + 2s + b + r$, and all nodes act as the Transmitter in HOM(a), with $a \leq r$, and all nodes using the same value of r, then conditions **HIC1** and **HIC2** will be satisfied. Since i is arbitrary, this guarantees that all good nodes will hold the same vector of values, either a good node's original value or an agreed upon default value for a faulty node's original value.   □

4   On-Line Diagnosis

In distributed diagnosis, each non-faulty node in the system is expected to cooperate with other non-faulty system nodes in detecting, identifying and removing faulty nodes from the system. In a typical diagnostic procedure, system components are tested and the results, or *syndromes,* are collected and analyzed by all active components. At a minimum, the status of a node (faulty or good) should be agreed upon by all non-faulty nodes. Once a faulty node has been identified, the decision to add or to exclude a node should also be consistent in all non-faulty components. Essentially, we desire a complementary usage of consensus and diagnosis facets to provide a 2-phase approach: *Phase(a):* local diagnosis, followed by *Phase(b):* global diagnosis based on consensus principles.

Our primary goal in on-line diagnosis is to identify faulty system nodes while maintaining correct system operation. A node's analysis of messages received from another node, as well as messages expected but not received, constitutes implicit testing of the sending node. We define diagnosis intervals, in which the following primitives are executed: *local detection and diagnosis, global information collection* and *global diagnosis* – on a concurrent, on-line and continual basis. The information collected locally by each node during diagnosis interval k, $\mathcal{D}(k)$, is broadcast to all other nodes, which then collect and analyze the information during $D(k + 1)$, to formulate a global perspective on the fault behavior. The length of the diagnosis interval is bounded by the assumed frequency of asymmetric value faults in the system.

We present the diagnosis algorithms on a progressive basis. Algorithm **DD** (*Distributed Diagnosis*) represents the basic two-phase, on-line diagnostic approach linking the diagnosis and consensus procedures. In Algorithm **HD** (*Hybrid Diagnosis*), we extend on **DD** to provide the capability of discriminating between node and link faults, where possible; to assess the severity of the node fault from temporal perspectives, and to incorporate the facets of node recovery and re-integration.

Table 2: Algorithm DD for Node i : Basic Consensus Syndrome

DD0. Initialize s_i^k to be the zero health vector.

DD1. *local detection*: Node i monitors message traffic from all other nodes throughout $\mathcal{D}(k)$. If a benign-faulty message is received from from node j during diagnosis interval k then set $\sigma_{ij}^k = (\sigma_{ij}^k$ OR $1)$, thus forming the local k^{th} round health vector s_i^k.

DD2. *Global assimilation*: Node i collects health vectors s_i^{k-1} computed during diagnosis interval $\mathcal{D}(k-1)$ into the $n \times n$ global syndrome matrix $S_i^{(k-1)}$. Row l of $S_i^{(k-1)}$ is the syndrome vector $s_i^{(k-1)}$ received from node l.

DD3. *Syndrome Matrix/Diagnosis*: Combine the values in column j of $S_i^{(k-1)}$, corresponding to other nodes' views of the health of node j, by using a hybrid voting function to generate a consistent health value, faulty or non-faulty, for node j. Node exclusion/inclusion as per consensus based heath value.

DD4. At the end of diagnosis interval k, node i sends its health vector, $s_i^k =< \sigma_{i1}^k, \sigma_{i2}^k, \ldots \sigma_{in}^k >$, to all other nodes.

4.1   Distributed Diagnosis: Node Health Monitoring

The goal of algorithm **DD**, shown in Table 2, is for all non-faulty nodes to acquire a consistent view of the health of all other nodes in the system. This is done by an exchange of local health assessments and subsequent computation of a consistent global health vector, using the HOM algorithm. It is pertinent to mention that the diagnosis algorithm running during diagnosis interval $\mathcal{D}(k)$ utilizes information collected across the system over the previous round $\mathcal{D}(k-1)$. All operations can be considered to be on-going in a pipelined manner.

Inter-node messages are considered as the sole indicators of the health of a node. Based on this, step **DD1**, the local detection phase, examines all received messages for errors. Since the detection of an error in a message by its receiver implies that the sender is *locally benign faulty*, local detection utilizes the parity checks, checksums, message framing checks, range checks, sanity checks, and comparison techniques. The failure to receive an expected message from a node or an early/delayed message is also logged as an error for that node.

During $\mathcal{D}(k)$, each node i locally formulates a health vector, $s_i^k =< \sigma_{i1}^k, \sigma_{i2}^k, \ldots, \sigma_{in}^k >$, containing an entry, σ_{ij}^k, corresponding to the perceived status of each system node, j. If any error is detected from a given node, j, its entry σ_{ij}^k is set to 1; otherwise it remains at the fault-free value of 0. This local diagnosis step is equivalent to the identity mapping, as no further local diagnosis occurs following detection.

To achieve consistency of this local diagnosis, and to build the global perspective to handle the "locally undetectable" class of *value* faults, information dispersal across the system is necessitated. At the end of each detection interval, in step **DD4**, the local health vector s_i^k of node i is sent to all other nodes. Thus, each node compiles (and analyzes) a global syndrome matrix during $\mathcal{D}(k+1)$ that contains the local health assessments of every node by other nodes over $\mathcal{D}(k)$.

During $\mathcal{D}(k)$, global diagnosis of each node's health during $\mathcal{D}(k-1)$ is performed in step **DD3**. The local health vectors computed during $\mathcal{D}(k-1)$ form the rows of the global health matrix $S_i^{(k-1)}$. If no health vector or an obviously erroneous vector is received from node l, then an error indicator value, $\mathcal{E}$, is adopted for each $\sigma_{ij}^{(k-1)}$ in $s_l^{(k-1)}$, and node l is assessed for an error by updating σ_{ij}^k. The global health vector held by each non-faulty node i is denoted by $h_i^{(k-1)}$, with entries $\eta_{ij}^{(k-1)}$ giving the global status of node j during $\mathcal{D}(k-1)$. The global health vector for $\mathcal{D}(k-1)$ is computed during $\mathcal{D}(k)$ by applying a hybrid majority voting function, described below, to each column of $S_i^{(k-1)}$.

First, all elements of column j equal to $\mathcal{E}$ are excluded, along with node j's opinion of itself $(\sigma_{jj}^{(k-1)})$. The final value, $\eta_{ij}^{(k-1)}$, is the majority of the remaining values. If no majority exists, the value 0 should be adopted to ensure that a good node is not identified as faulty. At the conclusion of $\mathcal{D}(k)$, each good node will contain a global health vector $h_i^{(k-1)} =< \eta_{i1}^{(k-1)}, \eta_{i2}^{(k-1)}, \ldots, \eta_{in}^{(k-1)} >$, where $\eta_{ij}^{(k-1)} = 1$ means that node i has diagnosed node j as being faulty during $\mathcal{D}(k-1)$. It needs to be stated that this diagnosis is consistently achieved by all non-faulty nodes, as this is the Agreement condition HIC1.

4.2 Hybrid Diagnosis: Fault Identification and Severity

In **DD**, an error in a single message from a node during a single diagnosis interval is sufficient to cause that node to be removed from the system. The local detection mechanism described previously for DD1 treats a node that sends a single erroneous message as if all messages from the node were indeed faulty. Essentially, a transient and a permanent fault will have an identical fault effect here. This is not an efficient strategy, and could lead to rapid depletion of system resources. Further, **DD** provides only the fault detection and isolation facets of FDIR. Recovery of a faulty node or node re-integration require refined temporal considerations, which the **DD** does not fully support. These issues, and increasing the diagnostic resolution, are dealt with in **HD** (Table 4), building on the basic framework of **DD**. We also add temporal fault detection to the local diagnosis primitive, and replace the simple good–bad local diagnosis with a preliminary assessment of the node and/or link fault symmetry.

4.2.1 Local Primitives in HD

The scalar status value, σ_{ij}^k, used in step DD1 of the previous algorithm, corresponding to node i's local assessment of node j's health during $\mathcal{D}(k)$, is replaced in step HD1 by a local diagnosis vector, e_{ij}^k. This local diagnosis vector is used to indicate the type of detection mechanism

Table 3: Fault Classification under HD

Type		Recorded in:
$\mathcal{MM}$	Missing Message	ϵ_{ij1}^k
$\mathcal{IFM}$	Improperly Formatted Message	ϵ_{ij2}^k
$\mathcal{ILM}$	Improper Logical Message	ϵ_{ij3}^k
$\mathcal{CVM}$	Failing Comparison to Voted Value	ϵ_{ij4}^k

which found errors in messages from node j, providing a preliminary diagnosis of the fault type. While the entries in $e_{ij}^k =< \epsilon_{ij1}^k, \epsilon_{ij2}^k, \ldots, \epsilon_{ijm}^k >$ can have a one-to-one correspondence with the m fault detection mechanisms implemented in the system, as in MAFT [9, 10], we consider four potential diagnoses shown in Table 3.

We assign a penalty weight to each error type, commensurate with its assumed severity in the system implementation, and accumulate the weights for each node over $\mathcal{D}(k)$. By definition, these detected errors result from *benign* faulty nodes. However, discerning the potential symmetry of the errors is useful in discriminating between a crash faulty node and a less severe faulty communications link. The relationships among ω_{MM}, ω_{ILM}, ω_{IFM}, ω_{CVM} forms the basis of inferences on fault-type. Of course, more or fewer weights could be used. Also, an additional correlated weight can be used if a faulty node exhibits several of these behaviors during a single diagnosis interval. The final extended health vectors and accumulated penalty weights from $\mathcal{D}(k)$ are sent to all nodes at the end of $\mathcal{D}(k)$.

4.2.2 HD: Global Diagnosis/Properties

During $\mathcal{D}(k)$, the extended health vectors and penalty weights from all nodes during $\mathcal{D}(k-1)$ are analyzed in a fashion similar to that in DD. Steps HD3 and HD4 ensure a consistent global perspective on the cumulative penalty values associated with each and every system node following global information collection in HD2. The overall relative value of a fault type e.g., $\epsilon_{ij1} > \epsilon_{ij2}$'s, etc are useful in attempting to identify the type of fault. The penalty weight for each node under diagnosis is its initial value in HOM(1). The asymmetric fault coverage is limited to one fault by the single re-broadcast round.

Since behavior of a faulty node is unrestricted, and a faulty node can send different corrupt health vectors (or none) to other nodes, good nodes may receive different health matrices. So, we must prove that the final health vectors h_i computed by all good nodes i during each diagnosis interval are consistent. Furthermore, we must assess the correctness and completeness of diagnosis. Global diagnosis is *correct* for $\mathcal{D}(k-1)$ if each node identified as faulty by a good node in step **HD3** of HD (during $\mathcal{D}(k)$) is indeed faulty. Similarly, global diagnosis is *complete* for $\mathcal{D}(k-1)$ if all nodes that were faulty during $\mathcal{D}(k-1)$ are identified as such in step HD3 in that diagnosis interval.

While the statement of algorithm *HD* does not explicitly invoke any fault tolerance algorithm,

Table 4: Algorithm HD for node i : Refined Syndrome

HD0. The expanded health vector, $s_i^k = < e_{i1}^k, e_{i2}^k, \ldots, e_{in}^k >$, is initialized to zero at the beginning of $\mathcal{D}(k)$; restore penalty weight vector $p_i^k = < \rho_{i1}^k, \rho_{i2}^k, \ldots, \rho_{in}^k >$ from D(k-1).

HD1 *local diagnosis:* Node i monitors message traffic from all other nodes. The following steps for each received message from each node j, and for each expected message not received from each node j.

    HD1.1 If no message ($\mathcal{MM}$) is received from node j, then set $\epsilon_{ij1}^k = \max(\epsilon_{ij}^k, 1)$ and $\rho_{ij}^k = \rho_{ij}^k + \omega_{MM}$.

    HD1.2 If an incorrectly formatted message ($\mathcal{IFM}$) is received from j, then set $\epsilon_{ij2}^k = \max(\epsilon_{ij2}^k, 1)$ and $\rho_{ij}^k = \rho_{ij}^k + \omega_{ILM}$.

    HD1.3 If an incorrect logical message ($\mathcal{ILM}$) is received from node j, then set $\epsilon_{ij3}^k = \max(\epsilon_{ij3}^k, 1)$ and $\rho_{ij}^k = \rho_{ij}^k + \omega_{IFM}$.

    HD1.4 If an incorrect value ($\mathcal{CVM}$) is detected in a message from node j by comparison to a voted value, j, then set $\epsilon_{ij4}^k = \max(\epsilon_{ij}^k, 1)$, and $\rho_{ij}^k = \rho_{ij}^k + \omega_{CVM}$.

HD2. *Information assimilation:* Node i collects the health vectors computed during $\mathcal{D}(k-1)$ into the $n \times n$ syndrome matrix $S_i^{(k-1)}$, where row l of $S_i^{(k-1)}$ is the syndrome vector $s_i^{(k-1)}$ received from node l. Similarly, the penalty counts computed during $\mathcal{D}(k-1)$ are collected into the matrix $P_i^{(k-1)}$, where column j of $P_i^{(k-1)}$ contains the weights received from all nodes regarding node j.

HD3. *Diagnosis: Consensus:* Combine the values in column j of $S_i^{(k-1)}$ to generate a health value, faulty or good, for node j by first OR-*ing* the entries in $e_{ij}^{(k-1)}$, and then performing a hybrid majority vote down the column. Combine the values in column j of $P_i^{(k-1)}$ to generate a consistent incremental penalty count for node j during $\mathcal{D}(k-1)$ using a hybrid voting function. *Threshold Matching: Inclusion/Exclusion*

HD4 At the end of diagnosis interval k, node i sends its health vector, $s_i^k = < \sigma_{i1}^k, \sigma_{i2}^k, \ldots \sigma_{in}^k >$, and its penalty count vector, $p_i^k = < \rho_{i1}^k, \rho_{i2}^k, \ldots, \rho_{in}^k >$, to all other nodes.

---

it is implicit in the definition of HD and in the fault masking used in the the **HIC(a)** algorithm. The local health of node j as viewed by node i during $\mathcal{D}(k-1)$, assessed as either 0 or 1, is equivalent to node j holding a personal value of either 0 or 1 and transmitting it to node i during $\mathcal{D}(k-1)$. This corresponds to the initial round of the **HOM(a)**, with $r = 1$. The sending by node i of its local assessment of node j's health to other nodes at the end of $\mathcal{D}(k-1)$ represents the rebroadcast round of HOM(1), with the hybrid majority column vote of $S_i^{(k)}$ during interval $\mathcal{D}(k)$ equivalent to the final value for j as computed by node i in HOM(1).

Since *HD* is executed on all nodes, and each node i monitors all other nodes, completion of *HD* is equivalent to all good nodes achieving interactive consistency during $\mathcal{D}(k)$ on the health of all other nodes during $\mathcal{D}(k-1)$. Theorems 1 and 2 provide conditions under which the global health vectors h_i^{k-1} are guaranteed to be consistent on all good nodes. These theorems also permit us to prove the following correctness and completeness results.

Theorem 3 *(Correctness) If* $n > 2t - b + 1$, *where* $t = a + b + s$ *faulty nodes are present during* both $\mathcal{D}(k-1)$ *and* $\mathcal{D}(k)$, *then diagnosis under* Algorithm HD *during* $\mathcal{D}(k)$ *is guaranteed to be correct for* $\mathcal{D}(k-1)$.

Proof: By Theorem 1, with the default majority value set to 0, all good nodes will have $h_{il}^{(k-1)} = 0$ for the values of all good nodes l. Thus, any node j for which $h_{ij}^{(k-1)} = 1$ must be faulty.   □

Theorem 4 *(Completeness) If* $n > 2t - b + 1$ *as defined above, and node* j *was faulty during* $\mathcal{D}(k-1)$, *then under* Algorithm HD, *node* j *will be diagnosed during* $\mathcal{D}(k)$ *as having been faulty during* $\mathcal{D}(k-1)$.

Proof: By definition, a node that is benign faulty during $\mathcal{D}(k-1)$ will be detected locally on each good node. Thus, the sender's initial value for these nodes is 1, with all good nodes adopting 1 (faulty) as the local diagnosis during $\mathcal{D}(k-1)$, and agreeing on 1 at the conclusion of $\mathcal{D}(k)$, by Thm 1.   □

This covers the cases for data and node faults. Following the formalism of the HD algorithm, we now motivate the fault resolution and the temporal aspect of aggregating the penalty weights in steps HD1 and HD3.

4.3   HD: Temporal Perspectives

Algorithm HD improves the judgment of fault severity at any interval in time so that units with less severe fault indications are left operational. Additional processing is required in HD as we are interested in handling fault-effects over a longer period of time than the diagnostic interval. We can build on the results obtained by instantaneous diagnosis(DD) by placing them into a temporal framework(HD).

Errors can be viewed as the manifestation of faults which exist in the system. *Duration* is defined as the total time a fault and its effects are present in the system during actual operation. We can introduce the concept of *decay-time* to be the length of time an error would be present if the fault was instantaneously applied and removed. Thus, the error is the effect of an instantaneous fault injected at time t_0 which lasts for a time Δt_f. The concept of *decay-time* allows error information to be carried across multiple intervals so that instantaneous diagnosis information can be related over time.

It must be noted that the *decay-time* does not always correlate directly with the severity level of the error. If a function in the system hard core is impacted by the transient fault, it may be necessary to immediately deal with the effects rather than waiting for them to die out. This method can account for the possibility that errors may propagate due to lack of containment and cause other errors which have their own *decay-time* and severity.

Errors which have shorter *decay-times* will have less time to further impact system operation. For example, a lost bit on a communication link due to a transient fault should be considered as an error with a short *decay-time*. If a noise pulse affects the link, some time will need to pass before the energy is dissipated from the medium. During this time, the messages being sent may be corrupted, depending on the level of noise. Another example would be a memory module with scrubbing. When an an error occurs, there will be a time period where the error could propagate and induce further errors. Once the scrubbing mechanism detects the error and removes it, the immediate danger of error propagation will have lapsed (even though the faulty source may still be present and/or intermittent).

Decay rates can therefore be determined if there exist regular and predictable times where errors can be detected or removed. If the times of detection are not regular, it is prudent to assume a worst case scenario which can be arrived at in a number of ways. The first approach would be to assess a penalty so severe it causes exclusion immediately so that further reliance on fault detection is not needed. A second method would be to attempt to identify the worst case detection time by a higher level mechanism. This method may allow for some error propagation until it begins to affect a critical higher level function. A third alternative is to schedule more extensive FDIR tasks to attempt to collect more information while imposing greater overhead on the system.

Based on the concept of *decay-time*, we can assume that if the fault is applied and then goes away immediately, the fault-effects should only last for a certain period of time. Faults with their associated *decay-times* and handled as follows. First, a penalty weight (W_e) should be assessed against the faulty unit in the interval $\mathcal{D}(k)$. Second, during the following diagnostic intervals, for the node displaying sustained fault-free behavior, the penalty weight against it is reduced by a predetermined amount, referred to as the *decrement-count* (DC). The ratio, W_e/DC provides the *decay-time* for the given error.

This *decrement-count* is introduced so that temporary malfunctions do not result in permanent exclusion. Note, that if the fault persists, penalties will continue to accrue and the decrement amount will be offset by new increases in penalty weights. The duration aspect of faults is also handled in this model. For the fault being transient, the source of the fault is removed and the effects should disappear after the appropriate decay time has passed. For a permanent fault, the source of the fault will remain present and new penalties will continually accrue over each new diagnostic interval until an exclusion threshold is reached. In order to handle transient and intermittent faults which are severe enough to cause exclusion but allow re-admission, the exclusion and re-admission thresholds must be appropriately separated.

This approach alternately supports modeling of permanent faults by setting the decrement count to zero. If one wants to support graceful re-admission of system units with on-line repair, even permanent faults can have a relatively short *decay-time*. The decay time can also be based on the time of re-admission, since once the unit is repaired its count must be decremented for it not to appear faulty anymore.

5 Discussion

Most existing diagnosis strategies treat diagnosis as a stand-alone process in the system operations, and are primarily off-line techniques. In this paper, we have addressed the problem of performing on-line diagnosis as an integral phase of the system FDIR process. Unlike existing approaches, the strategy is based on monitoring the system message traffic rather than using explicit test procedures.

Extending beyond the fixed fault severity models (time-domain and data-domain, s-a-X, Byzantine faults), the HFM framework is developed, which permits handling a continuum of fault types as groups of faults of varying fault manifestations under a single algorithm. The HFM's applicability to diagnosis is formally shown through the development of a distributed agreement algorithm (HOM) and diagnosis algorithms DD and HD. The integration of HFM into the diagnosis domain facilitates increased diagnostic resolution which can be used for improved resource management strategies.

The paper has also introduced the concept of fault *decay-time* and its impact in handling transient, intermittent and permanent faults in conjunction with Byzantine faults. The usage of penalty counts is shown as a basis for graceful node exclusion and re-admission protocols. In future work a detailed penalty count model will be developed. Overall we have shown that a more precise dependability model can be constructed, supported by on-line diagnosis algorithms under a generalized hybrid fault model.

References

[1] M. Barborak et al. The consensus problem in fault-tolerant computing. *ACM Computing surveys*, 25(2):171-220, June 1993.

[2] L. Lamport et al. The byzantine generals problem. *ACM Trans. on Prog. Languages and Systems*, 4:382-401, July 1982.

[3] P. Lincoln and J. Rushby. A formally verified algorithm for interactive consistency under a hybrid fault model. *FTCS-23*, pages 402-411, 1993.

[4] F. Preparata, G. Metze, and R. T. Chien. On the connection assignment problem of diagnosable systems. *IEEE Trans. on Electronic Computing*, ec-16:848-854, Dec 1967.

[5] N. Suri et al. Reliability modeling of large fault-tolerant systems. *FTCS-22*, pages 212-220, 1992.

[6] A. Sengupta and A. Dahbura. On self-diagnosable multiprocessor systems: diagnosis by the comparison approach. *IEEE TOC*, 41(11):1386-1396, Nov 1992.

[7] K. G. Shin and P. Ramanathan. Diagnosis of processors with byzantine faults in a distributed computing system. *FTCS-17*, pages 55-60, June 1987.

[8] P. Thambidurai and Y. K. Park. Interactive consistency with multiple failure modes. *Proc. of RDS*, pages 93-100, 1988.

[9] C. J. Walter et al. MAFT: A multicomputer architecture for fault-tolerance in real-time control systems. *RTSS*, Dec 1985.

[10] C. J. Walter. Identifying the cause of detected errors. *FTCS-20*, June 1990.

The General Convergence Problem: A Unification of Synchronous and Asynchronous Systems

M.H. Azadmanesh* and R.M. Kieckhafer†
*Standard Manufacturing Company, Box 3844
Omaha, Nebraska
mazadman@cwis.unomaha.edu
† Department of Computer Science and Engineering
University of Nebraska, Lincoln
Lincoln, Nebraska
rogerk@cse.unl.edu
U.S.A.

Abstract

An important problem in fault-tolerant distributed systems is maintaining agreement between non-faulty processes in the presence of undiagnosed faults. To achieve agreement, processes must exchange their local "opinions" of a particular value, and then vote on the values received to arrive at a "consensus". Approximate Agreement defines a condition in which it is not necessary for consensus values to be identical. Rather, they need only agree to within a predefined tolerance.

Approximate Agreement can be achieved through a sequence of convergent voting rounds, in which the range of values held by non-faulty processes is reduced in each round. Convergent voting has been addressed in the context of two classes of systems, *Synchronous* and *Asynchronous* systems. In addition, recent studies have addressed both *Completely Connected* and *Partially Connected* systems. Together, the properties of synchrony and connectivity define the voting environment. In every study to date, each system environment was treated as a separate problem.

This paper presents a single unifying model of the convergence problem, and a set of *General Convergence Relations* applicable to any of the above voting environments. These relations permit the properties of a convergent voting algorithm to be derived in the simplest environment (synchronous, completely connected) then extrapolated to the other environments.

1 Introduction

An important issue in fault-tolerant distributed computing is the ability of non-faulty processes to reach agreement on data values in the presence of faulty processes. This issue arises whenever non-faulty processes legitimately form differing "opinions" regarding the correct value. They must then exchange and vote upon their local values to arrive at a single consensus value. The problem is significantly more complex if a faulty process is permitted to send conflicting values to different non-faulty processes. A faulty process with this property has been called malicious, two-faced, Byzantine, or asymmetric.

In many real world applications, such as sensor data management and fault-tolerant clock synchronization [5, 8, 10, 12], non-faulty processes need not achieve exact agreement. Rather, they need only agree on a value to within a specified tolerance. This criterion is known as *Approximate Agreement*. Given an arbitrarily small positive real value ϵ, Approximate Agreement is defined by two conditions [3, 4]:

A1: AGREEMENT – The voting algorithms executed by all non-faulty processes eventually halt with voted values that are within ϵ of each other.

A2: VALIDITY – The voted value held by each non-faulty process is within the range of the initial values held by the non-faulty processes.

Several algorithms for achieving Approximate Agreement have been published. Many employ multiple rounds of message exchange with convergent voting algorithms which guarantee that the range of values held by the non-faulty processes is reduced in each round [3, 4, 6, 7, 8]. This property, called single-step convergence, guarantees that the range of values will eventually be less than ϵ, given enough rounds.

In the context of convergent voting, systems are partitioned into two classes: *synchronous* and *asynchronous*. In a synchronous system there are finite bounds on processing and communication delays in non-faulty processes [3]. There is thus a point in time by which any process executing a voting algorithm will have received all data from all non-faulty processes. Any data arriving after that time must have come from a faulty process. By contrast, an *asynchronous* system imposes no finite bounds on process operation [3]. Thus, each non-faulty process must at some point decide to proceed with the voting algorithm without knowing whether all data has been received from all non-faulty processes. Not surprisingly, asynchronous systems are less fault tolerant than synchronous systems.

It is not uncommon to confuse synchrony in voting algorithms with clock synchronization between processors. This confusion is partly due to the fact that distributed clock synchronization is one application for convergent voting algorithms. In reality, the two forms of synchrony are independent of each other. A distributed system with synchronized clocks may still require an asynchronous voting algorithm, if the application software does not bound the time required to generate and broadcast an opinion (or if the bound is too long to be of practical use). Similarly, in a system executing a cyclic real-time workload, a bound on the elapsed time between the transmission of successive opinions can exist, even without clock synchronization. Then, if propagation delays are also bounded, a synchronous voting algorithm can be employed.

Recent research has addressed convergent voting in the simultaneous presence of multiple fault modes [1, 2, 6, 7]. Faults were partitioned into three modes: asymmetric (Byzantine), symmetric (single-valued) and benign (self-incriminating). Using this mixed-mode fault model, simple expressions were derived for the performance and fault-tolerance of a broad family of convergent voting algorithms called *Mean-Subsequence-Reduced* (MSR) algorithms. Results have been obtained for both synchronous systems and asynchronous systems. In addition, convergent voting has been analyzed in both completely connected and partially connected networks. In the case of partially connected systems, the convergence properties were analyzed without the benefit of message relays. This constraint allows convergent voting in large sparsely connected systems where communication overhead makes large scale message relaying impractical.

Whether a system is synchronous or asynchronous is independent of whether it is completely or partially connected. There are thus four separate system types in which convergent voting has been analyzed. Previous studies have treated each system type as a separate problem. This paper presents a single set of equations which define the fault-tolerance and convergence properties of any MSR algorithm in *all four* system types. In doing so, we demonstrate that the different system types are just special cases of a more general convergent voting problem.

Section 2 presents some background material necessary to understanding the convergent voting process. Section 3 describes the *General Convergence Problem* and presents a set of *General Convergence Relations* applicable to all four system types. These relations employ two *environment parameters* which allow them to be customized to each system type. In Section 4, the results of [1, 2, 6, 7] are examined for each individual system type. Simple expressions are then derived for the environment parameters in each system type. Section 5 summarizes the results and discusses ongoing research.

2   Background

2.1   Real–Valued Multisets

Approximate Agreement requires the manipulation of multisets of real values. A multiset is a collection of objects similar to a set. However, it differs from a set in that the elements of a multiset need not be distinct. For example, a set of real numbers contains no more than one occurrence of any given value, while a multiset of real numbers may contain multiple occurrences of the same value. The number of times a particular object (value) appears in a multiset is called the *Multiplicity* of that object. A finite multiset V of real values may be represented as a mapping $V : \Re \rightarrow \aleph$. For each real value r, $V(r)$ is defined as the multiplicity of r in V. The size of V is $V = |V| = \sum_{r \in \Re} V(r)$.

An alternative representation for a multiset of real numbers is a monotonically increasing sequence of the real values of its elements, i.e. $V = \langle v_1, \dots, v_V \rangle$ ordered such that: $v_i \leq v_{i+1} \; \forall \, i \in \{1, \dots, V-1\}$ [9]. Both representations of a multiset are equivalent, but for certain operations one form or the other is more convenient. To avoid confusion, we use upper-case symbols for multiplicities in the real-to-integer mapping form, e.g. $V(r)$. Similarly, we use angle-braces and lower-case symbols for elements in the sequence form, e.g. $V = \langle v_1, \dots, v_V \rangle = \langle v_i \rangle \; \forall \, i \in \{1, \dots, V\}$.

Real Parameters –   A multiset of real numbers has several useful parameters.

$\min(V) = \min(r \in \Re : V(r) > 0) = v_1$; the minimum value of the elements in V.

$\max(V) = \max(r \in \Re : V(r) > 0) = v_V$; the maximum value of the elements in V.

$\rho(V) \quad = [\min(V), \max(V)] \quad = \; [v_1, v_V]$; the real interval spanned by V. $\rho(V)$ is called the *range* of V.

$\delta(V) \quad = \max(V) - \min(V) \quad = \; v_V - v_1$; the difference between the maximum and minimum values of V. $\delta(V)$ is called the *diameter* of V.

Subsequences –   Given two sequences U and V, U is a *subsequence* of V if all elements of U are selected from the elements of V, and arranged in the same order as their relative order in V. While a subsequence is also a submultiset, it has the important property that the *index* of an element in V is the sole criterion for its inclusion in U. Thus, a *subsequence selection function* is a mapping from the *indices* of U to the *indices* of V.

Formally, let $I_V = \{1, \dots, V\}$ be the set of indices for multiset V, and let $I_U = \{1, \dots, U\}$ be the set of indices for multiset U. Then, U is a subsequence of V if there is a fixed one-to-one (injective) mapping function $k : I_U \longrightarrow I_V$ which preserves order. Thus, each index $i \in \{1, \dots, U\}$ corresponds to exactly one index $k(i) \in \{1, \dots, V\}$, where $k(i) \leq k(i+1)$. It follows that $u_i = v_{k(i)}$. Furthermore, since V is a monotonically increasing sequence of real numbers, $u_i \leq u_{(i+1)} \; \forall \, i \in \{1, \dots, U-1\}$.

2.2   Multiple Mode Fault Model

In real-world systems, truly Byzantine behavior occurs only under highly improbable conditions. Accordingly, Meyer and Pradhan [11] have partitioned the space of all possible faults into two distinct modes: *Benign* faults, defined as those which are self-incriminating, or immediately self-evident to *all* non-faulty processes, and *Malicious* faults, defined as all faults which do not qualify as benign. Thambidurai and Park [13] have further partitioned malicious faults into two sub-modes: *Symmetric* faults, whose behavior is perceived identically by all non-faulty processes, and *Asymmetric* faults, whose behavior may be perceived differently by different non-faulty processes. Given a asymmetric, s symmetric, and b benign faults, the total number of faults is $t = a + s + b$.

An asymmetric error can occur only if there is a *communication* fault which causes a message to be received differently by different receivers. This fact can be used to bound the probability of asymmetric fault occurrence. For example, each processor in the Multicomputer Architecture for Fault-Tolerance (MAFT) [5] has a hardware transmitter subsystem which broadcasts messages to all other processors. If the transmitter components (including the medium) are non-faulty, then any value output by the processor will be broadcast symmetrically to all other processors. Therefore, an asymmetric fault can originate only in the transmitter subsystem, not in the processor hardware or software. Since the transmitter consists of a few simple chips and a passive medium, its fault rate is significantly lower than that of the entire processing node. Hence the asymmetric fault rate is orders of magnitude less than the symmetric fault rate [5, 13].

By exploiting the relative rarity of asymmetric faults, the Thambidurai and Park fault model permits more accurate analysis of the performance and fault-tolerance of agreement algorithms. This fault model has been applied to Byzantine Agreement algorithms [13], and to both synchronous [6, 7] and asynchronous [1, 2] Approximate Agreement algorithms.

2.3   MSR Voting Algorithms

One large family of convergent voting algorithms has the general form [6]:

$$F(\mathbf{V}) = \text{mean} \left[Sel_\sigma \left(Red^\tau (\mathbf{V}) \right) \right].$$

The reduction function Red^τ removes the τ largest and τ smallest elements from multiset $\mathbf{V}$. The function Sel_σ selects a submultiset of σ elements from the reduced multiset $Red^\tau (\mathbf{V})$. The final voted value is the arithmetic mean of the selected multiset.

If Sel_σ produces a subsequence of $Red^\tau (\mathbf{V})$, then $F(\mathbf{V})$ is the *Mean* of a *Subsequence* of the *Reduced* multiset. The family of all voting algorithms with this property are called *Mean-Subsequence-Reduced* (MSR) algorithms [6]. Members of the MSR family differ from each other only in their definition of the selection function Sel_σ. Some examples of MSR algorithms are [3, 4, 6]: the Fault-Tolerant Midpoint, Fault-Tolerant Mean, Dolev's Optimal algorithm, Mixed-Mode Optimal algorithm, and Binary Mean.

3   The General Convergence Problem

A system may be synchronous or asynchronous independent of whether it is completely or partially connected. There are thus four separate system types in which the convergent voting problem must be addressed. Previous studies of convergent voting have treated each system type as a separate problem. In fact, several earlier studies treated each individual voting algorithm as a separate entity [3, 4, 8].

Recent research has produced simple expressions for the fault-tolerance and convergence properties of the entire MSR family of algorithms [1, 2, 6, 7]. However, each system type was still treated as a separate case. This section presents a unifying set of *General Convergence Relations* for the fault-tolerance and convergence properties of MSR voting algorithms applicable to all four system types.

Throughout this paper the following definitions are used frequently in the context of MSR algorithms. For simplicity, subscripts are omitted when there is no room for confusion.

$\mathbf{V}_i$ = The multiset of values used in a given voting round by non-faulty process i.

$\mathbf{M}_i$ = $Red^\tau(\mathbf{V}_i)$, the *Medial Multiset* of $\mathbf{V}_i$.

$\mathbf{S}_i$ = $Sel_\sigma(\mathbf{M}_i)$, the *Selected Multiset* generated by $F(\mathbf{V}_i)$. The number of selected elements σ is identical for all non-faulty processes.

$\mathbf{P}_i$ = The set of processes whose values are receivable by process i. $\mathbf{P}_j$ is similarly defined for process j.

$\mathbf{P}_{i\cap j}$ = $\mathbf{P}_i \cap \mathbf{P}_j$, the set of processes whose values are receivable by both processes i and j.

$\mathbf{P}_{i\cup j}$ = $\mathbf{P}_i \cup \mathbf{P}_j$, the set of processes whose values are receivable by either process i, process j, or both.

$\mathbf{U}_i$ = The submultiset of $\mathbf{V}_i$ containing all *correct* values received by process i. A correct value is defined as any value received from a non-faulty process. Similarly, an *error* is any value received from a faulty process, regardless of whether its actual value coincidentally falls within the range of correct values.

$\mathbf{U}_{i\cap j}$ = The multiset of all values generated by *non-faulty* processes in $\mathbf{P}_{i\cap j}$.

$\mathbf{U}_{i\cup j}$ = The multiset of all values generated by *non-faulty* processes in $\mathbf{P}_{i\cup j}$.

3.1   Single-Step Convergence

It has been shown that Approximate Agreement can be achieved if a voting algorithm is *single-step convergent*. The precise definition of single-step convergence depends on whether the system is completely connected or partially connected. However, in general, an algorithm is single-step convergent if it guarantees that both of the following conditions are true following every voting round:

VALIDITY – For each non-faulty process, the voted value is within the range of correct values it receives.

CONVERGENCE – For each pair of non-faulty processes, the difference between their voted values is strictly less than the diameter of the submultiset of correct values received.

3.1.1 Completely Connected Systems

In a completely connected system, $U_i \equiv U_{i \cap j} \equiv U_{i \cup j}$. Each non-faulty process i executes a convergent voting algorithm, producing voted value $F(V_i)$. A voting algorithm is convergent if both of the following conditions are true for every voting round:

[C1] For each non-faulty process i, $F(V_i) \in \rho(U_{i \cap j})$.

[C2] For each pair of non-faulty processes (i, j),
$|F(V_i) - F(V_j)| \leq C \, \delta(U_{i \cap j})$, where $0 \leq C < 1$.

As defined in Subsection 2.1, $\delta(U_{i \cap j})$ and $\rho(U_{i \cap j})$ are the diameter and range, respectively, of the correct values received by both i and j. Parameter C is the *Convergence Rate* of a voting algorithm, the primary measure of its effectiveness.

3.1.2 Local Convergence with Partial Connectivity

A partially connected system differs from a completely connected system in that a given process does not receive values from all non-faulty processes. Rather, it receives values only from a specific subset of processes. There are now two types of convergence to be considered: local convergence over a specified subgraph, and global convergence over the entire system graph. This paper addresses the process of local convergence without the benefit of message relays. Global convergence is the topic of ongoing research, and is beyond the scope of this paper. The following constraints are placed on the system. (1) the system topology is a non-hierarchical, regular, homogeneous, undirected graph of N processing nodes, each with degree d. (2) Messages received by a voting process may not be relayed to another process. Thus, the physical and logical connectivity are identical. Each voting process receives its own value as well as those of its immediate neighbors so that $V = d + 1$ for all non-faulty processes. (3) $N >> V$ so that "wrap-around" effects can not assist the local convergence process in a given voting round.

In a completely connected system, $\mathbf{U}_{i\cap j} \equiv \mathbf{U}_{i\cup j}$. However, in a partially connected system, $\mathbf{U}_{i\cap j} \subset \mathbf{U}_{i\cup j}$. We define two types of local convergence for partially connected systems.

Intersection Convergence: Given a voting algorithm $F(\mathbf{V})$, two processes i and j are single-step *Intersection Convergent* if the following conditions are both true:

[I1] $F(\mathbf{V}_i) \in \rho(\mathbf{U}_{i\cap j})$, and $F(\mathbf{V}_j) \in \rho(\mathbf{U}_{i\cap j})$,

[I2] $|F(\mathbf{V}_i) - F(\mathbf{V}_j)| \leq C\,\delta(\mathbf{U}_{i\cap j})$, where $0 \leq C < 1$.

Union Convergence: Given a voting algorithm $F(\mathbf{V})$, two processes i and j are single-step *Union Convergent* if the following conditions are both true:

[U1] $F(\mathbf{V}_i) \in \rho(\mathbf{U}_{i\cup j})$, and $F(\mathbf{V}_j) \in \rho(\mathbf{U}_{i\cup j})$,

[U2] $|F(\mathbf{V}_i) - F(\mathbf{V}_j)| \leq C\,\delta(\mathbf{U}_{i\cup j})$, where $0 \leq C < 1$.

3.2   General Convergence Relations

Figure 1 summarizes the relationships between the system environments described above. Each of the six leaf-nodes in Figure 1 represents a separate environment for the convergent voting problem. For an MSR voting algorithm executing in any system environment, there are two questions of primary interest: (1) under what conditions is the algorithm convergent, (2) if the algorithm is convergent, what is its convergence rate C? In this subsection we present general parameterized expressions which answer both of these questions in all six environments.

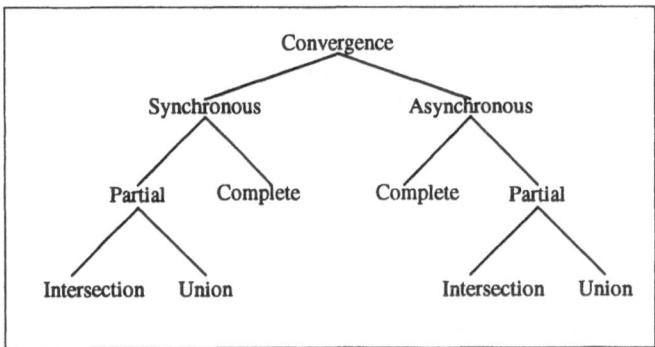

Figure 1: Taxonomy of Convergent Voting Environments

3.2.1 Fault-Tolerance

An MSR voting algorithm $F(\mathbf{V}) = \text{mean}\,[Sel_\sigma(Red^\tau(\mathbf{V}))]$ has three main parameters: (1) V, the size of the voting multiset, (2) τ, the number of elements removed from each extreme by the reduction function, (3) σ, the number of elements chosen by the selection function. We will show that an MSR voting algorithm can be convergent only if:

$$\tau \geq \tau_0, \tag{3.1}$$

$$\sigma \begin{cases} \geq 1 & : \alpha = 0 \\ \geq 2 & : \alpha > 0 \end{cases}, \tag{3.2}$$

$$V \geq 2\tau + \max(\alpha + 1, \sigma). \tag{3.3}$$

These relations employ two new *environment parameters* α and τ_0. Parameter α is defined as the *effective* number of asymmetric errors in the voting set $\mathbf{V}$, while τ_0 is the *effective* number of malicious errors in $\mathbf{V}$. In Section 4 it will be shown that simple expressions for α and τ_0 can be found for each system environment.

3.2.2 Convergence Rate

An important result of [1, 2, 6, 7] is the ease with which the convergence rate C can be determined for *any* MSR voting algorithm. If an MSR algorithm is convergent, then its convergence rate is given by:

$$C = \frac{\gamma_\alpha}{\sigma}, \tag{3.4}$$

where γ_α is a function of parameter α defined as below.

Recall that the *Medial Multiset* $\mathbf{M} = Red^\tau(\mathbf{V}) = \langle m_1, \ldots, m_M \rangle$, and the *Selected Multiset* $\mathbf{S} = Sel_\sigma(\mathbf{M}) = \langle s_1, \ldots, s_\sigma \rangle$. By the definition of an MSR algorithm, $\mathbf{S}$ is a subsequence of $\mathbf{M}$. Thus, if g is an index into $\mathbf{S}$, then for each integer $g \in \{1, \ldots, \sigma\}$ there exists exactly one integer $k(g) \in \{1, \ldots, M\}$ which guarantees that $s_g = m_{k(g)}$.

Given two indices $g, h \in \{1, \ldots, \sigma\}$ where $g \leq h$, we define $\Delta k(g, h)$ as the number of elements in $\mathbf{M}$ *spanned* by elements $\langle s_g, \ldots, s_h \rangle$ in $\mathbf{S}$, i.e.

$$\Delta k(g, h) = [k(h) - k(g)]. \tag{3.5}$$

If $g = h$, then $\Delta k(g, h) = 0$. However, if $g < h$, then $\Delta k(g, h)$ is the number of elements of $\mathbf{M}$ in the submultiset $\langle m_{k(g)+1}, \ldots, m_{k(h)} \rangle$.

For a given non-negative integer $\alpha < M$, it will be useful to know the minimum value of $(h - g)$ for which $\Delta k(g, h) \geq \alpha$. Thus, for each $g \in \{1, \ldots, \sigma\}$, we define the quantity $\Delta_\alpha(g)$ as follows:

IF: $\Delta k(g, \sigma) \geq \alpha$,
THEN: $\Delta_\alpha(g) = $ the minimum value of $(h - g)$ such that $\Delta k(g, h) \geq \alpha$,
ELSE: $\Delta_\alpha(g)$ does not exist for this value of g.

The ELSE clause is required because if $\Delta k(g, \sigma) < \alpha$, then there is no $h \in \{g + 1, \ldots, \sigma\}$ for which $\Delta k(g, h) \geq \alpha$. Thus, if $\Delta_\alpha(g)$ exists, then $\Delta k(g, g + \Delta_\alpha(g)) \geq \alpha$. Finally, we define the parameter γ_α as:

$$\gamma_\alpha = \max_{\forall\, g \in \{1, \ldots, \sigma\}} (\Delta_\alpha(g)). \qquad (3.6)$$

Thus, given any $g \in \{1, \ldots, \sigma - \gamma_\alpha\}$, it is assured that $\Delta k(g, g + \gamma_\alpha) \geq \alpha$. In other words, the submultiset $\langle s_g, \ldots, s_{g+\gamma_\alpha} \rangle$ is guaranteed to span α elements of $\mathbf{M}$, for all $g \in \{1, \ldots, \sigma - \gamma_\alpha\}$.

As a practical matter, obtaining the values of γ_α and σ is relatively simple, given constants M and α, and a specified selection function Sel_σ. Parameter σ is just the number of elements selected by Sel_σ (M). To determine γ_α, one simply determines $\Delta_\alpha(g)$ for each $g \in \{1, \ldots, \sigma\}$ by inspection and selects the maximum thereof. If there is no value of g for which $\Delta_\alpha(g)$ exists, then γ_α does not exist, and the algorithm is not convergent.

4   System Environment Parameters

This section reviews recent results [1, 2, 6, 7] on the performance and fault tolerance of synchronous and asynchronous MSR algorithms for completely and partially connected systems. Then, simple expressions for the environment parameters α and τ_0 will be derived for each case.

4.1   Synchronous Systems

In a synchronous system there is a finite bound on the processing and communication delays of each round. A convergent voting algorithm expects to receive data from all non-fauly processes within this bound. An arbitrary default value is chosen for any data not received within the bound. Thus, at the time the voting algorithm starts, all non-faulty processes have the same number of elements V in their voting multisets.

4.1.1   Convergence with Complete Connectivity

Convergent voting in a completely connected synchronous system containing a asymmetric faults, s symmetric faults and b benign faults was addressed in [6]. It was shown that it is advantageous to discard identified benign errors priori to voting. Thus, Given N processes, the size of the voting multiset is $V = N - b$. It was also shown that an MSR voting algorithm can be convergent only if [6]:

$$\tau \geq a + s \tag{4.7}$$

$$\sigma \begin{cases} \geq 1 & : a = 0 \\ \geq 2 & : a > 0 \end{cases} \tag{4.8}$$

$$V \geq 2\tau + \max(a + 1, \sigma). \tag{4.9}$$

Furthermore, the convergence rate of an algorithm is [6]:

$$C = \frac{\gamma_a}{\sigma}. \tag{4.10}$$

Equations (4.1) – (4.4) become (3.1) – (3.4), respectively, using:

$$\alpha = a \quad \text{and } \tau_0 = a + s. \tag{4.11}$$

4.1.2   Local Convergence with Partial Connectivity

A major difference between completely connected and partially connected systems is the strategy for handling benign errors. In a completely connected system, benign errors can be discarded because *all* processes can delete the benign errors from V and vote with a smaller sized multiset. However, in a partially connected system, *no* fault is self-evident to *all* processes. For example, a processor containing a benign fault may be adjacent to processor i, but not to processor j. Then, a self-evident error would be received by process i but not by process j. Thus, if process i discards the erroneous value before voting, the two processes will vote using different sized multisets so that V would not be identical for all processes. Thus, a self-evident error must be tranformed into a symmetric error by assigning a default value before voting.

Intersection Convergence –   The conditions necessary to ensure that two processes i and j are Intersection Convergent can be derived as a variant on the completely connected system previously described. We begin with the following definitions:

a = The number of asymmetrically faulty processes in $\mathbf{P}_{i \cap j}$.

s = The number of symmetrically faulty processes in $\mathbf{P}_{i \cap j}$.

χ = $|\mathbf{P}_i| - |\mathbf{P}_{i \cap j}| = |\mathbf{P}_j| - |\mathbf{P}_{i \cap j}|$, the number of processes whose values are receivable by either i or j, but *not* by both.

A synchronous MSR algorithm can be Intersection Convergent only if [7]:

$$\tau \geq a + s + \chi, \tag{4.12}$$

$$\sigma \begin{cases} \geq & 1 \ : \ [a + \chi] = 0 \\ \geq & 2 \ : \ [a + \chi] > 0 \end{cases}, \tag{4.13}$$

$$V \geq 2\tau + max\,([a + \chi] + 1, \sigma). \tag{4.14}$$

Furthermore, the convergence rate of an algorithm is:

$$C \ = \ \frac{\gamma_{a+\chi}}{\sigma}. \tag{4.15}$$

Equations (4.6) – (4.9) become (3.1) – (3.4), respectively, using:

$$\alpha = a + \chi \quad \text{and} \quad \tau_0 = a + s + \chi. \tag{4.16}$$

Union Convergence – Union Convergence requires convergence within $\rho\,(\mathbf{U}_{i \cup j})$ rather than $\rho\,(\mathbf{U}_{i \cap j})$. Accordingly, Union Convergence is possible under less restrictive conditions than Intersection Convergence. We retain the previous definitions that a and s are the number of asymmetric and symmetric faults, respectively in $\mathbf{P}_{i \cap j}$. We then define:

f = The maximum number of faults in either $\{\mathbf{P}_i \backslash \mathbf{P}_{i \cap j}\}$ or $\{\mathbf{P}_j \backslash \mathbf{P}_{i \cap j}\}$, regardless of their modes.

It has been shown that a synchronous MSR algorithm can be Union Convergent only if [7]:

$$\tau \geq a + s + f, \tag{4.17}$$

$$\sigma \begin{cases} \geq & 1 \ : \ [a + \chi] = 0 \\ \geq & 2 \ : \ [a + \chi] > 0 \end{cases}, \tag{4.18}$$

$$V \geq 2\tau + max\,([a + \chi] + 1, \sigma). \tag{4.19}$$

Furthermore, the convergence rate of an algorithm is:

$$C \ = \ \frac{\gamma_{a+\chi}}{\sigma}. \tag{4.20}$$

Equations (4.11) – (4.14) become (3.1) – (3.4), respectively, using:

$$\alpha = a + \chi \quad \text{and} \quad \tau_0 = a + s + f. \tag{4.21}$$

4.2   Asynchronous Systems

What makes a system *asynchronous* is that no bounds on process operation exist and a process might have to wait "forever" to receive data from all non-faulty processes [3]. It is thus impossible to differentiate between a slow non-faulty process and a "dead" process. Accordingly, asynchronous systems possess two important characteristics: First, since a process can not distinguish between a slow process and a faulty process, a non-faulty process may never receive values from all N processes. Therefore, a non-faulty process accepts the first $N - \tau$ values received before voting on a new value. Depending on the distribution of communication delays, the voting multisets of the non-faulty processes may not be identical *even in a fault-free system*. Second, no fault may be considered *benign*, since any value transmitted may not be included by all processes in their voting multisets. Thus, an error is not immediately evident to *all* non-faulty processes.

4.2.1   Convergence with Complete Connectivity

It has been shown that in the worst case, an asynchronous system containing a asymmetric faults, has exactly the same convergence and fault-tolerance properties as a synchronous system containing $(a + \tau)$ asymmetric faults [1]. Consequently, an asynchronous MSR algorithm can be convergent only if:

$$\tau \geq a + s, \tag{4.22}$$

$$\sigma \begin{cases} \geq 1 & : (a + \tau) = 0 \\ \geq 2 & : (a + \tau) > 0 \end{cases}, \tag{4.23}$$

$$V \geq 2\tau + max\,(a + \tau + 1, \sigma). \tag{4.24}$$

Furthermore, the convergence rate of an algorithm is

$$C = \frac{\gamma_{a+\tau}}{\sigma}. \tag{4.25}$$

Equations (4.16) – (4.19) become (3.1) – (3.4), respectively, using:

$$\alpha = a + \tau \quad \text{and} \quad \tau_0 = a + s. \tag{4.26}$$

4.2.2   Local Convergence with Partial Connectivity

In an asynchronous partially connected system, non-faulty process i votes on a new value when it receives the first $|\mathbf{P}_{inj}| - \tau$ values. Since each non-faulty process is required to send one value for each round, each non-faulty process will eventually receive $V = |\mathbf{P}_{inj}| - \tau$ values and start its voting algorithm.

Intersection Convergence – An asynchronous MSR algorithm can be Intersection Convergent only if [2]:

$$\tau \geq a + s + \chi, \tag{4.27}$$

$$\sigma \begin{cases} \geq & 1 \quad : [a + \tau + \chi] = 0 \\ \geq & 2 \quad : [a + \tau + \chi] > 0 \end{cases}, \tag{4.28}$$

$$V \geq 2\tau + \max\,(a + \tau + \chi + 1, \sigma). \tag{4.29}$$

Furthermore, the convergence rate is:

$$C \;=\; \frac{\gamma_{a+\tau+\chi}}{\sigma}. \tag{4.30}$$

Equations (4.21) – (4.24) become (3.1) – (3.4), respectively, using:

$$\alpha = a + \tau + \chi \quad \text{and} \quad \tau_0 = a + s + \chi. \tag{4.31}$$

Union Convergence – An asynchronous MSR algorithm can be Union Convergent only if [2]:

$$\tau \geq a + s + f, \tag{4.32}$$

$$\sigma \begin{cases} \geq & 1 \quad : [a + \tau + \chi] = 0 \\ \geq & 2 \quad : [a + \tau + \chi] > 0 \end{cases}, \tag{4.33}$$

$$V \geq 2\tau + \max\,(a + \tau + \chi + 1, \sigma). \tag{4.34}$$

Furthermore, the convergence rate of the algorithm is:

$$C \;=\; \frac{\gamma_{a+\tau+\chi}}{\sigma}. \tag{4.35}$$

Equations (4.26) – (4.29) become (3.1) – (3.4), respectively, using the definitions:

$$\alpha = a + \tau + \chi \quad \text{and} \quad \tau_0 = a + s + f. \tag{4.36}$$

4.3 Summary

Equations (3.1) – (3.4) specified general fault-tolerance and convergence rate criteria for MSR voting algorithms in any system environment. Then, the environment parameters α and τ_0 were defined for each specific system environment. Table 1(a) summarizes the definitions of α for the different environments. It should be noted that the value of α in partially connected systems is independent of whether Intersection Convergence or Union Convergence is desired. It is also noteworthy that the set of effective asymmetric errors in the asynchronous partially connected case is the union of that in the synchronous partially connected case, and the asynchronous completely connected case. Table 1(b) summarizes the definitions of τ_0 for the different types of convergence. The most interesting point is that τ_0 is the same for synchronous and asynchronous systems.

CONNEC.	SYNC.	ASYNC.
COMPLETE	a	$a + \tau$
PARTIAL	$a + \chi$	$a + \tau + \chi$

(a) Asymmetric (α)

CONV. TYPE	DEF.
COMPLETE	$a + s$
INTERSEC.	$a + s + \chi$
UNION	$a + s + f$

(b) Malicious (τ_0)

Table 1: Effective Numbers of Asymmetric and Malicious Faults

5   Conclusions

This paper has presented a unifying model for fault-tolerant convergent voting algorithms in different system environments. This model spans synchronous and asynchronous systems, as well as completely connected and partially connected systems. Within partially connected systems, the model spans both intersection convergence and union convergence.

Equations (3.1) – (3.4) define the fault-tolerance and convergence criteria for any MSR voting algorithm. These equations apply in any of the six known system environments, given the proper definitions of environment parameters α and τ_0. The expressions for α and τ_0 listed in Table 1 show a simple and logical progression as one moves from one system to another. These results show that the different environments are merely special cases of a general convergent voting process. Thus, analysis of a particular voting algorithm in the simplest system (synchronous, completely connected) can easily be extrapolated to any of the other system environments.

Continuing research on convergent voting is focused in two main areas. The first area involves the process of global convergence in partially connected systems. In many system applications, the goal of convergent voting is to achieve Approximate Agreement on a global level, i.e. to reduce the range of values held by *all* non-faulty processes. It is known that local Union Convergence is a prerequisite to global convergence. The quantitative relationships between local convergence rates and global convergence rates is currently being studied in the context of synchronous systems [7].

The second area of continuing research involves the exploitation of *omission* faults. A major constraint in existing MSR algorithms (and other classes of convergent voting algorithms) is that a "missing" value must be replaced by a default value before voting. This constraint exists because current analyses assume that all processes start with a voting multiset of the same size ($V_i = V_j$). Therefore, a process can not simply discard a value which is *a priori* erroneous (e.g. out of bounds) unless it is assured that all other processes do likewise. Thus, an error can not be classified as benign unless a Byzantine Agreement algorithm is first executed to ensure that all processes agree that it is indeed an error. Under these restrictions, a fault can be considered benign only after global, Byzantine-safe diagnosis has been performed. In this context, diagnosis is the process of transforming a malicious fault into a benign fault. The requirement that all processes use the same value of V also makes the benign fault mode unusable in partially connected systems.

Analysis of the properties of MSR algorithms is being conducted under the assumption that $V_i \neq V_j$. Thus, a voting process can *unilaterally* discard a missing value or obvious error before voting. Results indicate that algorithms with this property can have significantly improved fault-tolerance if spontaneous benign faults are possible.

References

[1] Azadmanesh, M.H., and R.M. Kieckhafer, *Asynchronous Approximate Agreement With Mixed-Mode Faults*, University of Nebraska, Dept. of Computer Science and Engineering, Technical Report Series, No. 93–19, June 1993.

[2] Azadmanesh, M.H., and R.M. Kieckhafer, "Asynchronous Approximate Agreement in Partially Connected Systems", *Journal of Computing and Information*, (to appear).

[3] D. Dolev, N.A. Lynch, S.S. Pinter, E.W. Stark, and W.E. Weihl, "Reaching Approximate Agreement in the Presence of Faults," *Proc. 3rd Symp. on Reliability in Dist. Software and Database Systems*, Oct 1983.

[4] D. Dolev, N.A. Lynch, S.S. Pinter, E.W. Stark, and W.E. Weihl, "Reaching Approximate Agreement in the Presence of Faults," *JACM*, V. 33, No. 3, pp. 499-516, Jul 1986.

[5] R.M. Kieckhafer, C.J. Walter, A.M. Finn, and P.M. Thambidurai, "The MAFT Architecture for Dist. Fault-Tolerance", *IEEE Trans. Comput.*, V. C-37, No. 4, pp. 398-405, Apr, 1988.

[6] R.M. Kieckhafer, and M.H. Azadmanesh, "Reaching Approximate Agreement With Mixed Mode Faults", *IEEE Trans. Par. and Dist. Sys.*, V. 5, No. 1, Jan 1994.

[7] R.M. Kieckhafer, and M.H. Azadmanesh, "Low Cost Approximate Agreement in Partially Connected Networks", *Journal of Computing and Information*, V. 3, No. 2, 1993.

[8] L. Lamport, and P.M. Melliar-Smith, "Synchronizing Clocks in the Presence of Faults", *JACM*, Vol. 32, No. 1, pp. 52-78, Jan 1985.

[9] C.L. Liu, *Elements of Discrete Mathematics, 2nd Ed.*, New York, McGraw-Hill, 1985.

[10] J. Lundelius, and N. Lynch, "A New Fault-Tolerant Algorithm for Clock Synchronization", 3rd *Symp. on Principles of Dist. Computing*, pp. 75-87, Aug 1984.

[11] F.J. Meyer, and D.K. Pradhan, "Consensus with Dual Failure Modes", *Proc. 17th Fault-Tolerant Computing Symp.*, pp. 48-54, Jul 1987.

[12] F.B. Schneider, *Understanding Protocols for Byzantine Clock Synchronization*, Report No. 87–859, Dept of Computer Science, Cornell University, Aug 1987.

[13] P.M. Thambidurai, and Y.K. Park, "Interactive Consistency with Multiple Failure Modes", *Proc. Seventh Reliable Dist Systems Symp.*, Oct 1988.

Specification and Verification
of Distributed Protocols

Specification and Verification of Behavioral Patterns in Distributed Computations

Özalp Babaoğlu* and Michel Raynal †
*Department of Mathematics
University of Bologna
40127 Bologna, Italy
ozalp@dm.unibo.it
†IRISA
Campus de Beaulieu
35042 Rennes Cedex, France
raynal@irisa.fr

Abstract

The ability to specify and verify dynamic properties of computations is essential for ascertaining the dependability of distributed systems engaged in critical applications. In this paper, we consider properties that can be encoded as general Boolean predicates over global system states. We introduce two global predicate classes called *simple sequences* and *interval-constrained sequences* for specifying desirable states in some causality-preserving order along with intervening undesired states. Our formalism is simpler than more traditional proposals and permits concise and intuitive expression of many interesting system properties. Algorithms are given for verifying formulas belonging to these predicate classes in an on-line and observer-independent manner during distributed computations. We illustrate the utility of our results by applying them to examples drawn from program testing, debugging and dynamic reconfiguration in distributed systems.

Keywords: Distributed debugging, program verification, temporal logics, computation tree logic (CTL), non-stable global predicates, causal sequences.

1   Introduction

Detecting when the state of a distributed computation satisfies a certain predicate constitutes a fundamental abstraction in the design of distributed algorithms for critical applications. A large class of problems including error reporting, process control, performance monitoring, dynamic reconfiguration and load balancing can be solved by defining an appropriate set of notification or control actions guarded by the appropriate global predicates that encode critical system properties.

The problem of detecting predicates defined over a single global state has been extensively studied (see [1] for a survey). The case where the predicate is stable leads to particularly simple and efficient solutions based on distributed snapshots [5]. Informally, a distributed snapshot algorithm gathers and pieces together a collection of local states so as to guarantee that the resulting global state is consistent — one that could have been constructed by an idealized external observer. Being stable, the predicate evaluating to true in the snapshot state is sufficient for concluding that it has been "detected."

Detection of non-stable predicates cannot be based on snapshots, even when they are applied repeatedly [14]. No matter how frequently taken, a sequence of snapshots may have gaps that correspond to exactly those global states in which the (non-stable) predicate holds. Any reasonable definition of "detection" for the case of non-stable predicates must be based on *observations* [19,6,16,10,1]. A further complexity in detecting non-stable predicates is due to the so-called "relativistic effect" which results in multiple observations for the same computation. Modal operators have been proposed so as to make detection of non-stable predicates independent of any particular observation [6,16,10].

While predicates (stable or not) over a single global state are able to capture many interesting system properties, they inherently lack notions of logical time or relative order. To be able to characterize dynamic *behavioral patterns* of distributed computations, simple predicate specifications must be extended to include a temporal component [3,15,19,12]. In other words, specifying and verifying dynamic properties require reasoning about sequences, rather than single instances of global states.

In this paper we consider behavioral pattern specifications that admit arbitrary predicates over global states as building blocks. We introduce the syntactic classes *SS* (simple sequence) and *ICS* (interval-constrained sequence) of global predicates that define sets of global states related through the notion of causality-preserving sequencing [13]. Our formalism can be viewed as an application of branching time temporal logics [8,4] to distributed computations. This relationship is further explored in Section 5. The syntax and semantics of class *ICS* is tailored for specifying and verifying properties that are of interest to dependable computing. The main contribution of our work lies in the development of on-line algorithms for verifying the satisfaction of formulas by an underlying computation. The verification takes place at a monitor internal to the system and concurrently with the actual computation. As such, our detection algorithms can be viewed as on-line versions of model checkers in various logic formalisms.

2 System Model

2.1 Asynchronous Distributed Systems

We adopt the terminology and notation of [1]. A distributed system is a collection of sequential *processes* $p_1, p_2, \ldots, p_n$ that communicate by exchanging messages. We assume that communication is reliable and that it incurs finite but arbitrary delays. Processes do not share state and do not have access to a global clock. Furthermore, no bounds exist on the relative speeds of processes. The system thus described corresponds to the well-known *asynchronous* model.

2.2 Distributed Computations

Informally, a distributed computation describes a single execution of a distributed program by a collection of processes. The activity of each sequential process is modeled as a sequence of *events*. A large portion of the total activity of a process may be irrelevant with respect to a given application. Thus, we implicitly consider only those events that are "relevant" in the sense that they can affect the system property under consideration [6].

The *local history* of process p_i during the computation is a sequence of events $h_i = e_i^1 e_i^2 \ldots$. The labeling of the events of process p_i is such that e_i^1 is the first event executed, e_i^2 is the second event executed, etc. Let $h_i^k = e_i^1 e_i^2 \ldots e_i^k$ denote an initial prefix of local history h_i containing the first k events. We define h_i^0 to be the empty sequence. The *global history* of the computation is a set $H = h_1 \cup \cdots \cup h_n$ containing all events that are executed.[1]

Formally, a *distributed computation* is a partially ordered set γ defined by the pair $(H, \rightarrow)$ where $\rightarrow$ is the binary causal precedence relation defined on events [13]. It is common to depict distributed computations using an equivalent graphical representation called a *space-time diagram* as shown in Figure 1 (a).

2.3 Global States, Observations and Lattice Structures

Let σ_i^k denote the local state of process p_i immediately after having executed event e_i^k and let σ_i^0 be its initial state before any events are executed. The *global state* of a distributed computation

---

[1]Sometimes we are interested in local histories as *sets* rather than *sequences* of events. Since all events of a computation have unique names in the canonical labeling, h_i as a set contains exactly the same events as h_i as a sequence. We use the same symbol to denote both when the appropriate interpretation is clear from context.

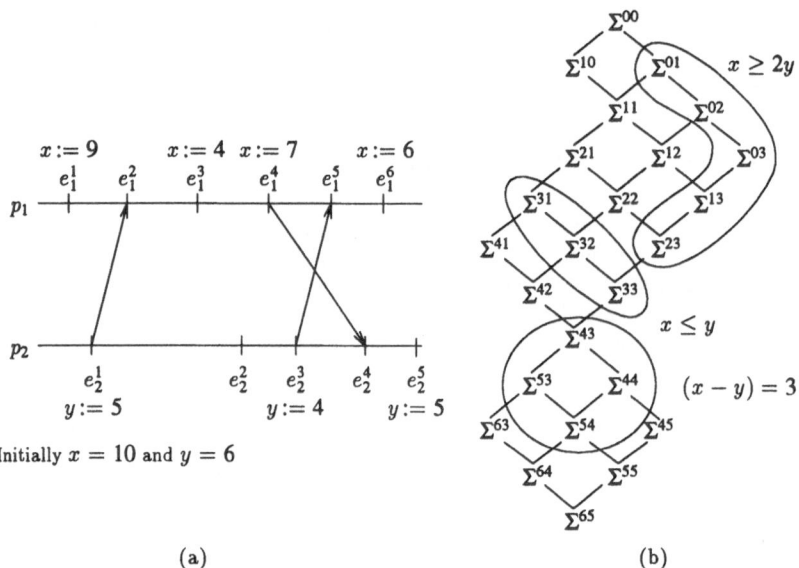

Figure 1: (a) Space-Time Diagram Representation of a Distributed Computation (b) Lattice of Consistent Global States

is an n-tuple of local states $\Sigma = (\sigma_1^{c_1}, \ldots, \sigma_n^{c_n})$, one for each process. Implicitly, global state Σ defines a *cut* of the global history as the subset $C = h_1^{c_1} \cup \cdots \cup h_n^{c_n}$ containing all events whose effects are reflected in Σ.

A global state is *consistent* if the cut associated with it is left closed under the causal precedence relation. In other words, global state Σ of computation γ is consistent if the cut C associated with Σ is such that for all events e and e' of the global history H we have $(e \in C) \wedge (e' {\rightarrow} e) \Rightarrow e' \in C$. Intuitively, consistent global states correspond exactly to those that could be constructed by an idealized observer external to the computation.

Excluding the possibility of simultaneous events, an actual execution of the distributed program results in a total ordering of γ called a *run*. Any total order of γ that is consistent with causal precedence and is constructed using mechanisms that are entirely internal to the system constitutes a *sequential observation* (observation for short) of the computation. Observations can be expressed as sequences of consistent global states rather than sequences of events. In particular, an observation $\Omega = e^1 e^2 \ldots$ of computation γ is equivalent to the sequence of global states $\Sigma^0\Sigma^1\Sigma^2\ldots$ where Σ^0 denotes the initial global state $(\sigma_1^0,\ldots,\sigma_n^0)$ and each global state Σ^i of the observation is obtained from the previous state Σ^{i-1} by some process executing the single event e^i in γ. For two such global states of observation Ω, we say that Σ^{i-1} *leads to* Σ^i *in* Ω. Let $\leadsto_\Omega$ denote the transitive closure of the leads-to relation in a given observation Ω. We say that Σ' is *reachable from* Σ if there exists some observation Ω such that $\Sigma \leadsto_\Omega \Sigma'$ and we drop the subscript from $\leadsto_\Omega$.

The set of all consistent global states of a computation along with the *leads-to* relation define a *lattice*. The lattice consists of n distinct axes, with one axis for each process. Let $\Sigma^{k_1\cdots k_n}$ be a shorthand for the global state $(\sigma_1^{k_1},\ldots,\sigma_n^{k_n})$ and let $k_1 + \cdots + k_n$ be its *level*. Figure 1 (b) illustrates the lattice of global states associated with the distributed computation of Figure 1 (a). Algorithms for on-line construction of the lattice can be found in [6,16,1,7].

3 Global Predicates

Global predicate classes can be established by providing syntax rules that define the set of well-formed formulas and describing the semantics associated with them. In other words, global predicate classes can be viewed as small languages for specifying system properties while detecting a global predicate can be viewed as checking a formula in a model, namely a distributed computation. As noted earlier, these formulas can be seen as an application of temporal logics to the analysis of distributed programs. They are tailored for the specification of properties that are of interest for dependable distributed computing. Section 4 will show how to detect them in an on-line manner.

3.1 Simple Predicates

Syntactically, the simplest global predicate class we consider is called *SP*, for *simple predicates*, and corresponds to general Boolean expressions defined over a single global state. Formulas of *SP* may reference any state variable of any process, the constants *true* and *false*, and follow the usual syntax rules for Boolean expressions. We let $\varphi(\Sigma)$ denote the value of predicate φ when evaluated in global state Σ. Suppose the annotated Figure 1 corresponds to a distributed computation where two variables x and y are maintained by two processes p_1 and p_2, respectively. Figure 1 illustrates three simple predicates defined over the global states of this computation, which can be expressed as pairs of (x, y) values, and identifies regions of the lattice satisfying each of them.

Note that the global predicates of [10,12] belong to a class $SP' \subset SP$ where Boolean expressions are limited to those that can be expressed as conjunctions or disjunctions of local predicates (those naming state variables of a single process only). There are many interesting system properties, including the following, that cannot be expressed in this restricted class SP' but have compact and natural expressions in SP:

1. Pressure level at all valves is balanced.

2. Message delay along the route from A to B is less than 5msec.

3. No more than 100 total users logged in to machines A, B and C.

With respect to observations, the semantics of the class SP is defined as follows:

Definition 1 *Let φ be a predicate in SP. Observation Ω satisfies φ, denoted $\Omega\models\varphi$, if and only if there exists global state $\Sigma \in \Omega$ such that $\varphi(\Sigma)$ is true.*

Stable predicates are members of the syntactic class SP with the following additional semantic requirement:

Definition 2 *Simple predicate φ is stable in computation γ if and only if*

$$\forall \Sigma, \Sigma' \in \gamma : \ \varphi(\Sigma) \wedge (\Sigma \leadsto \Sigma') \Rightarrow \varphi(\Sigma').$$

In other words, if a stable predicate is true in some global state of the computation, then it must be true in all global states reachable from it.

Detection of non-stable predicates poses problems even if we base them on observations rather than snapshots. It could be that the observation satisfies the predicate but the run does not, or vice versa. To make detection of non-stable predicates observer independent, Cooper and Marzullo have proposed that the class SP be augmented with two modal operators [6]. We adopt their proposal in order to define the semantics of the class SP with respect to computations.

Definition 3

1. *Distributed computation γ satisfies* **Pos** *φ, denoted $\gamma\models$**Pos** φ, if and only if there exists an observation Ω of γ such that $\Omega\models\varphi$.*

2. *Distributed computation γ satisfies* **Def** *φ, denoted $\gamma\models$**Def** φ, if and only if for all observations Ω of γ it is the case that $\Omega\models\varphi$.*

Note that the above definitions of **Pos** and **Def** have been called *weak* and *strong* formulas, respectively, by Garg and Waldecker in [10].

3.2 Simple Sequences

In order to include behavioral pattern specifications for distributed computations, global predicates must include a temporal component. One such proposal is due to Miller and Choi [15] who introduce the notion of *linked predicates* to add causal sequencing to a set of local predicates. We extend this idea by considering sequences of global predicates by composing instances of the class *SP*. The first extension we consider is called *SS*, for *simple sequences*, and is defined by the syntax rule

$$SS ::= SP \mid SS\,;SP$$

where *SP* is as defined in the previous Section. In other words, formulas in *SS* are semicolon-separated sequences of simple predicates, each defined over a single global state. Referring to Figure 1, the global predicate $(x \geq 2y); (x \leq y); (x - y = 3)$ is an instance of *SS*. Obviously, the class *SS* includes *SP* as sequences of length one.

The semantics of *SS* with respect to observations is as follows:

Definition 4 *Let* $\Phi = \varphi_1; \varphi_2; \ldots; \varphi_m$ *be a formula in SS. Observation* Ω *satisfies* Φ*, denoted* $\Omega \models \Phi$*, if and only if there exist m distinct global states* $\Sigma^{i_1} \ldots \Sigma^{i_m}$ *of* Ω *such that*

1. $\Sigma^{i_1} \leadsto_\Omega \Sigma^{i_2} \leadsto_\Omega \cdots \leadsto_\Omega \Sigma^{i_m}$ *and*

2. $\varphi_1(\Sigma^{i_1}) \wedge \varphi_2(\Sigma^{i_2}) \wedge \cdots \wedge \varphi_m(\Sigma^{i_m})$.

In other words, for the formula to be satisfied by observation Ω, each of the component predicates must hold in distinct global states of Ω, and each such global state must be observed in an order consistent with the syntactic position of its respective component predicate.

The semantics of *SS* with respect to computations remain exactly as given in Definition 3 with the introduction of modal operators **Pos** and **Def** and the above semantics with respect to observations. For example, the formulas

$$\gamma \models \textbf{Pos}\ (x \geq 2y); (x \leq y); (x - y = 3)$$
$$\gamma \models \textbf{Def}\ (x \leq y); (x - y = 3)$$
$$\neg(\gamma \models \textbf{Pos}\ (x \leq y); (x \geq 2y); (x - y = 3))$$

are all true for γ corresponding to the computation of Figure 1.

3.3   Negation of Predicates with Respect to Observations

Negation of a simple predicate φ with respect to a global state Σ has the obvious natural interpretation: $\neg\varphi$ holds in Σ if and only if φ does not hold in Σ. This definition, which we call *state negation*, can be extended to negation with respect to observations in one of several ways. We adopt the following interpretation since it captures our intuition for negating predicates when they are interpreted over sequences.

Definition 5 *Let Φ be a predicate. The* interval negation *of Φ with respect to observation Ω, denoted $[\Phi]$, is a predicate such that $\Omega\models[\Phi]$ if and only if $\neg(\Omega\models\Phi)$.*

Note that the negation is applied to the "satisfies" relation and not to the predicate itself. As a consequence, interval negation of a predicate is defined over sequences of global states even when the predicate is defined over a single state. For instance, consider $\Phi = \varphi$ for some φ belonging to the class *SP*. By Definition 1, we have $\Omega\models\varphi$ if and only if there exists some global state of Ω in which φ holds. Thus, for $\Omega\models[\varphi]$ to hold, our interpretation requires $\neg\varphi$ to hold in *all* global states of observation Ω.

Our notation $[\Phi]$ for denoting the interval negation of global predicate Φ suggests the fact that it is defined over an interval. Implicitly, the interval in question is the entire observation. Note that for Φ belonging to *SS*, the construct $\neg\Phi$ is meaningless while $[\Phi]$ is well defined. For Φ belonging to *SP*, we reserve the notation $\neg\Phi$ to denote state negation as opposed to interval negation.

3.4   Interval-Constrained Sequences

There are cases when the behavioral pattern specification must describe not only sequences of desired states, it must also describe undesirable states during intervals [12,21]. For example, to demonstrate the "low contention" property of a resource management strategy, we might be interested in verifying that during intervals while the resource is requested and acquired by a particular process, the total number of outstanding requests does not exceed some threshold. Note that the intervals of interest with respect to the undesired states are dynamic and cannot be established a priori. An interval is defined dynamically whenever two adjacent predicates are satisfied by two global states in the correct order.

The class *SP* extended with interval negation as discussed in the previous Section is sufficient for specifying undesired states over entire observations. What is lacking is the possibility to restrict negation dynamically to subintervals of an observation. We can achieve this by composing predicates from *SP* extended with interval negation to define *interval-constrained negation* as follows. We denote the beginning and end of an interval through simple predicates written to the left and right, respectively, of the negation. Thus, the interval-constrained negation $\varphi_1; [\theta]\varphi_2$ is satisfied by an observation if there exists a global state Σ^1 in which φ_1 holds, a later global state Σ^2 in which φ_2 holds, and θ does not hold in the global states in between. Note that Σ^1 and Σ^2 delimit the interval but do not belong to it. Thus, the interval can be empty if Σ^1 directly leads to Σ^2. Moreover, one or both of φ_1 and φ_2 may be missing, in which case the interval is implicitly extended to the initial state or to the final state of the observation, respectively.

We formalize these ideas by defining a new global predicate class called *ICS*, for *interval-constrained sequence*, with syntax rules

$$
\begin{aligned}
ICS &::= [SP] \mid COMB \mid COMB \,;\, [SP] \\
COMB &::= UNIT \mid COMB \,;\, UNIT \\
UNIT &::= [SP]\, SP \mid SP
\end{aligned}
$$

where *SP* is as before. The syntactic category *COMB* defines sequences with elements of *UNIT* as the building blocks. In their most general form, formulas in *ICS* are sequences of interval-constrained negations composed as alternations of two subsequences: those that describe desired conditions and their potential causal order (written as simple predicates) and those that describe undesired intervening conditions (written as interval-constrained negations). Note that the class *ICS* includes *SS* and (unconstrained) interval negation as special cases. Again referring to Figure 1, global predicates

$$(x \geq 2y); (x \leq y); (x - y = 3)$$

$$[x - y = 3]$$
$$[x \geq 2y](x \leq y); [true](x - y = 3)$$

are all instances of *ICS*.

In defining the semantics of *ICS* with respect to observations, we consider the most general formula belonging to this class:

Definition 6 *Let* $\Phi = [\theta_1]\varphi_1; \ldots; [\theta_m]\varphi_m; [\theta_{m+1}]$ *be a formula in ICS. If any of the* θ_i *in* Φ *are missing, we assume they are defined as the constant* false. *Observation* Ω *satisfies* Φ, *denoted* $\Omega \models \Phi$, *if and only if*

1. *If* $m = 0$, *then* $\forall \Sigma \in \Omega : \neg\theta_1(\Sigma)$,

2. *If* $m \neq 0$, *then there exist* m *distinct global states* $\Sigma^{i_1} \ldots \Sigma^{i_m}$ *of* Ω *such that*

   (a) $\Sigma^{i_1} \leadsto_\Omega \Sigma^{i_2} \leadsto_\Omega \cdots \leadsto_\Omega \Sigma^{i_m}$,

(b) $\varphi_1(\Sigma^{i_1}) \wedge \cdots \wedge \varphi_m(\Sigma^{i_m})$,

(c) $\forall i : i < i_1 : \neg\theta_1(\Sigma^i)$,

(d) $k = 2, \ldots, m : \forall i : i_{k-1} < i < i_k : \neg\theta_k(\Sigma^i)$ and

(e) $\forall i : i > i_m : \neg\theta_{m+1}(\Sigma^i)$.

With respect to the desired sequence, the semantics of *ICS* is the same as that of *SS*. For the class *ICS*, however, there is the additional requirement that between each pair of global states satisfying predicates φ_{i-1} and φ_i, there can be no intermediate states where θ_i holds. Furthermore, the prefix of the observation up to the state satisfying φ_1 must not include any state satisfying θ_1 and the suffix of the observation beyond the state satisfying φ_m must not include any state satisfying θ_{m+1}.

The class *ICS* happens to be quite expressive and can be used to specify many behavioral patterns that are of interest for the analysis of dependability properties in distributed systems. In particular, we can easily express notions such as atomicity, stability and invariance. For instance, in terms of observations, atomicity of distributed actions can be expressed as transforming one global state to another with no intermediate global states that are observable. In other words, the state in which the atomic action is applied and the resulting state must be adjacent to each other. In terms of *ICS* formulas, states that constrain the interval $[true]$ must be adjacent since only the empty sequence satisfies $[true]$. By inserting empty interval requirements at selected points in an *ICS* formula, we can express interesting specifications. Here are some *ICS* formulas and their meaning:

- $\varphi_1; [true]\varphi_2 \equiv$ Predicates φ_1 and φ_2 hold in adjacent global states, in that order.

- $[true]\varphi \equiv$ Predicate φ holds in the initial global state.

- $\varphi; [true] \equiv$ Predicate φ holds in the final global state.

- $[\varphi] \equiv$ Predicate $\neg\varphi$ is an invariant for the observation.

As before, we define the semantics of *ICS* with respect to computations by augmenting the class with the modal operators **Pos** and **Def**. The resulting semantics remain exactly as given in Definition 3 with the appropriate interpretation of satisfaction with respect to observations. For γ corresponding to the computation of Figure 1, the following formulas are all true:

$$\gamma \models \textbf{Pos}\,[x \geq 2y](x \leq y); [true](x - y = 3)$$
$$\gamma \models \textbf{Def}\,(x \leq y); [x \geq 2y](x - y = 3)$$
$$\neg(\gamma \models \textbf{Pos}\,(x \geq 2y); [x \leq y](x - y = 3))$$

The class *ICS* augmented with modal operators allows us to restate the stability property of simple predicates very concisely.

Definition 7 *Simple predicate φ is stable in computation γ if and only if*

$$\gamma \models \mathbf{Def}\,[\varphi](true); [\neg\varphi].$$

In other words, every observation of the computation consists of two segments — an initial prefix where the predicate is false, followed by a (potentially empty) suffix where the predicate is true.

3.5 Duality of Modal Operators

As noted in [1], modal operators **Pos** and **Def** are not duals of each other under state negation (i.e., with the syntax $\neg\varphi$). In other words, $\neg(\gamma \models \mathbf{Def}\,\varphi)$ is not equivalent to $\gamma \models \mathbf{Pos}\,\neg\varphi$ and $\neg(\gamma \models \mathbf{Pos}\,\varphi)$ is not equivalent to $\gamma \models \mathbf{Def}\,\neg\varphi$. As an example, consider γ corresponding to the distributed computation of Figure 1 where $\gamma \models \mathbf{Def}\,(x \le y)$ and $\gamma \models \mathbf{Pos}\,(x > y)$ are both true.

Under interval negation, which is on sequences of global states (i.e., with the syntax $[\Phi]$), the modal operators **Def** and **Pos** are indeed duals. Note that interval negation as given in Definition 5 remains valid also for our new class *ICS*. Thus, the following duality property is valid for all three global predicate classes *SP*, *SS* and *ICS*. For sake of brevity, we state results without proofs which can be found in [2].

Property 1 *If Φ is a formula in SP, SS or ICS, then for any computation γ we have*

$$\gamma \models \mathbf{Def}\,[\Phi] \equiv \neg(\gamma \models \mathbf{Pos}\,\Phi).$$

An immediate consequence of this result is that we need to develop detection algorithms only for formulas in *SS* and *ICS*, without negation. Formulas in these classes extended with interval negation can be detected indirectly through the duality property.

4 Detection Algorithms

In this section, we give an algorithm for verifying if a formula Φ belonging to the class *ICS* is satisfied by a distributed computation. Clearly, the same algorithm can also be used for detecting formulas belonging to the class *SS*, which is a special case of *ICS*. An equivalent algorithm optimized for the class *SS* is given in the Appendix. For a detailed description of the detection algorithms and their proofs, the reader is referred to [2].

Our algorithm operates on-line in the sense that it constructs the result by monitoring the computation as it evolves. It is no more intrusive than the algorithm used for constructing the lattice structure corresponding to the underlying computation. Any of the algorithms in [6,16,7] can be used for the on-line construction of the lattice used by our detection algorithm. Our algorithm needs to consider only two adjacent levels of the lattice at any given time. Thus, the overhead in space is only marginally greater than that of the algorithms used for the lattice construction.

```
       function Detect_ICS(Φ);
       var previous, current, verified: set of global states;
           path, p_path: array[0..m] of boolean;
           detected: boolean;
           u: integer;
       begin
1          previous := {Σ⁻¹};
2          path(Σ⁻¹)[0] := true;
3          for u := 1 to m do  path(Σ⁻¹)[u] := false od
4          verified := {};
5          current :={Σ⁰};
6          while (current ≠ {}) do
7              foreach Σ ∈ current do
8                  for u := 0 to m do  p_path[u] :=   ⋁      path(Σ′)[u] od
                                                    Σ′∈pred(Σ)
9                  path(Σ)[0] := p_path[0] ∧ ¬θ₁(Σ);
10                 for u := 1 to m do
11                     path(Σ)[u] := (p_path[u] ∧ ¬θ_{u+1}(Σ)) ∨ (p_path[u − 1] ∧ φ_u(Σ))
                   od
12                 if path(Σ)[m] then verified := verified ∪ {Σ} fi
               od
13             if verified ≠ {} then detected := true else detected := false fi
14             previous := current;
15             current :={global states directly reachable from those in previous};
16             verified := {};
           od
17         return (detected)
       end
```

Figure 2: Algorithm for Detecting Global Predicates in the Class *ICS*

The algorithm shown in Figure 2 detects **Pos** Φ for formulas belonging to the class *ICS*. With each global state Σ, algorithm *Detect_ICS* associates a Boolean array $path(\Sigma)$ of dimension $m+1$. For each $0 \leq u \leq m$, an element $path(\Sigma)[u]$ of this array is true if at least one observation starting with the initial state Σ^0 and ending at state Σ satisfies the prefix $[\theta_1]\varphi_1; \ldots; [\theta_u]\varphi_u; [\theta_{u+1}]$ of formula Φ. For notational convenience, we assume that a fictitious global state Σ^{-1} precedes the initial state Σ^0. The function $pred(\Sigma)$ returns a set of global states corresponding to the immediate predecessors of Σ. In other words, each member of $pred(\Sigma)$ leads to Σ. By definition, $pred(\Sigma^0) = \{\Sigma^{-1}\}$ and the predecessor of Σ^{-1} is undefined. The array associated with (fictitious) state Σ^{-1} is initialized to true for the prefix of length zero and false for all others.

The algorithm proceeds by trying to extend the prefix of the formula that is satisfied by increasing levels of the lattice. Formulas in *ICS* are not monotonic with respect to the length of observations (levels of the lattice). In other words, a prefix of the formula terminated with $[\theta_u]$ that is satisfied by an observation of length ℓ may cease to be satisfied when the observation is extended to length $\ell + 1$ since it may be impossible to avoid future undesired states (where θ_u holds). For this reason, we introduce the array *path* associated with each state so as to consider all possible values of the prefix up to u and a boolean variable *detected* that indicates if an observation terminating at the current level of the lattice satisfies the entire formula $\Phi = [\theta_1]\varphi_1; [\theta_2]\varphi_2; \ldots; [\theta_m]\varphi_m; [\theta_{m+1}]$.

The algorithm uses variables *previous*, *current* and *verified*, each a set of global states. We first compute *current* as the set of states directly reachable from those in *previous* (the set of states at the next level in the lattice). Note that *previous* is initialized to contain only the fictitious state Σ^{-1}. For each global state Σ under consideration, the prefix $[\theta_1]\varphi_1; \ldots [\theta_u]\varphi_u; [\theta_{u+1}]$ of the formula is satisfied, and consequently $path(\Sigma)[u]$ is true, if either some predecessor of Σ satisfies this prefix and Σ is not an undesired state (case $p\_path[u] \wedge \neg\theta_{u+1}(\Sigma)$ of line 11), or some predecessor of Σ satisfies the prefix $[\theta_1]\varphi_1; \ldots [\theta_{u-1}]\varphi_{u-1}; [\theta_u]$ and Σ is the next desired state (case $p\_path[u-1] \wedge \varphi_u(\Sigma)$ of line 11). If the entire formula is satisfied up to Σ, then the boolean *detected* becomes true. If there are no more levels in the lattice, the formula is satified if and only if *detected* has been set to true; if there are further levels in the lattice, they need to be examined since the formula is not monotonic with respect to the length of observations.

If the formula has the particular following form $\Phi = [\theta_1]\varphi_1; [\theta_2]\varphi_2; \ldots; [\theta_m]\varphi_m$ (i.e., the final $[\theta_{m+1}]$ is missing), then the detection algorithm can be simplified. Without the final invalidating predicate $[\theta_{m+1}]$, the entire formula Φ once again becomes monotonic with respect to the length of observations. Consequently the detection can be terminated as soon as at least one state Σ of the current level is such that $path(\Sigma)[m]$ is true. The algorithm can be modified by returning true immediately rather than setting the variable *detected* in line (13).

Property 2 *Given a formula* $\Phi = [\theta_1]\varphi_1; [\theta_2]\varphi_2; \ldots; [\theta_m]\varphi_m; [\theta_{m+1}]$ *belonging to the class ICS, function* Detect_ICS(Φ) *returns true for computation* γ *if and only if* $\gamma \models$**Pos** Φ.

5   Relation to Temporal Logics

The work reported in this paper has had distributed debugging as its practical motivation. Starting with the work of Cooper and Marzullo [6], we set out to explore ways in which their approach could be extended to capture larger classes of properties. The desire to include dynamic aspects of distributed computations has led us to consider sequences of states. Thus, it comes as no surprise that our proposal has strong parallels with other formalisms for reasoning about sequences. In particular, temporal logics which have emerged as particularly appropriate formal techniques for specifying and analyzing concurrent systems [4], parallel programs [17], protocols [9] and distributed systems [20,21] are closely related to our approach.

Temporal logics extend ordinary predicate logic with modal operators. Unlike predicate logic formulas, which have single states as models, temporal logic formulas have sets of possibly infinite sequences of states as their models. In terms of expressive power, our proposal adds nothing to a traditional temporal logic having the operators *next* and *until*. Every formula in *ICS* can be written as an equivalent, albeit very cumbersome, formula in this temporal logic. In this sense, *ICS* could be viewed as a stylized syntactic shorthand for temporal logic formulas appropriate for distributed computations. The two idioms of *ICS* — causality-preserving sequencing and interval negation — are designed so as to admit concise and intuitive formulas for specifying properties of distributed computations where these notions play a key role [21].

One distinction between our approach and that traditionally pursued by logicians who study temporal logic is how we determine the models in which a formula is to be evaluated. In our case, the model needs to be inferred from circumstantial evidence accumulated by monitoring the underlying computation. Neither the computation nor the program responsible for producing it is known. In contrast, traditional work in temporal logic assumes that a set of models is given or that a formal description (e.g., a state transition graph) is given to produce the set of models.

Interpreting our formulas over observations corresponds exactly to model checking in a *linear time* temporal logic. Interpreting our formulas over computations, on the other hand, corresponds [18] to model checking in *branching time* temporal logic [8]. There are strong parallels between our proposal and a particular branching time temporal logic known as *computation tree logic* (CTL) [4]. Our computations are the analogs of the state transition graphs of CTL and the lattice structure representing consistent global states is the analog of the tree structure in CTL representing possible program states. The CTL operators *inevitable* and *potential* [4,9] are the analogs of the modal operators **Def** and **Pos**, respectively, of Cooper and Marzullo [6].

Finally, our detection algorithms operate on-line with respect to the structure responsible for the model (the underlying computation). In that sense, they could be viewed as "on-the-fly" model checkers. Unlike traditional temporal logic applications where model checking is performed off-line, our approach constructs the relevant portions of the model (lattice of global states) as the underlying computation evolves. At no time is the model constructed or stored in its entirety. Only a current window of adjacents states is considered with dynamic programming principles being exploited to keep track of past states in a storage-efficient manner.

6   Conclusion

Two new classes of global predicates have been proposed to express dynamic behavior patterns in distributed computations. These classes admit Boolean expressions over global states as building blocks and include temporal specifications through causality-preserving sequencing and interval negation. The possibility to specify undesired states between desired ones adds significantly to the expressiveness of the proposal. Our formalism borrows ideas from temporal logics [8,4,21]. The originality of our work lies, on the one hand, in the syntactic simplicity of formulas expressing even complex properties, and on the other hand, in the design of the associated algorithms for on-line verification of such formulas. Our algorithms need access to only two adjacent levels at a time of the global state lattice structure to verify formulas. This space requirement is independent of the length of the formula being verified, so their cost is no more that the algorithm being used for the on-line construction of the lattice.

The cost of our detection algorithms may be reduced in several ways. One possibility is to restrict the class *SP*, from which classes *SS* and *ICS* are built, so as to admit efficient detection as is done in [23]. In this work, Tomlinson and Garg devise efficient mechanisms for detecting global predicates of the from $x + y < c$ where c is a constant and x and y are local variables at two processes. Unfortunately, the technique does not appear to be generalizable to more than two processes. Another possibility for reducing the cost of detecting general global predicates is to limit the size of the lattice being examined. It is well known that certain communication patterns (e.g., those resulting from remote procedure calls) in a computation will lead to very "lean" lattice structures avoiding exponential number of states. Or, we can avoid examining the entire lattice by employing certain heuristics in the detection algorithms for constructing the set *current* from the set *previous*. These heuristics could be based on the notion of "greedy" linear extensions associated with partial orders [22]. With incomplete explorations of the lattice structure, detection of $\gamma \models$**Def** Φ and $\neg(\gamma \models$**Pos** $\Phi)$ will take on probabilistic interpretations.

In the context of dependable computing, namely the construction of a debugging environment for distributed programs [11], an implementation of these algorithms is currently in progress.

Appendix

Detection Algorithm for the Class *SS*

The algorithm shown in Figure 3 detects if **Def** Φ is satisfied during a computation where

```
        function Detect_SS(Φ);
        var previous, current, verified: set of global states;
            prefix, u: integer;
        begin
1           previous := {Σ⁻¹};
2           prefix(Σ⁻¹) := 0;
        repeat
3               verified := {};
4               current := {global states directly reachable from those in previous};
5               foreach Σ ∈ current do
6                   u :=  min   (prefix(Σ'));
                        Σ'∈pred(Σ)
7                   if φ_{u+1}(Σ) then prefix(Σ) := u + 1
8                   else prefix(Σ) := u;
                    fi
9                   if (prefix(Σ) = m) then verified := verified ∪ {Σ} fi
                od
10              if ((verified ≠ {}) ∧ (verified = current)) then return (true) fi
11              previous := current − verified;
12          until (previous = {});
13          return (false)
        end
```

Figure 3: Algorithm for Detecting Global Predicates in the Class *SS*

$\Phi = \varphi_1; \varphi_2; \ldots; \varphi_m$ is a formula of length m belonging to the class *SS*. With each state Σ of the lattice, we associate an integer variable $prefix(\Sigma)$. Initially, only $prefix(\Sigma^{-1})$ is defined and is zero. In general, $prefix(\Sigma) = u$ indicates that all observations starting with the initial global state Σ^0 and ending at Σ satisfy a maximal prefix of length u of the formula Φ. Note that formulas in the class *SS* are monotonic with respect to the length of observations.

Given a formula $\Phi = \varphi_1; \varphi_2; \ldots; \varphi_m$ belonging to the class *SS*, verifying that a computation satisfies **Pos** Φ rather than **Def** Φ requires the following two simple syntactic modifications to algorithm *Detect_SS(Φ)*:

1. The min operator of line (6) is changed to max,

2. The test in line (10) is changed to "**if** $((verified \neq \{\}))$ **then return** (*true*) **fi** ".

Acknowledgements

We are grateful to Fred Schneider for his help in relating our work to temporal logics. The presentation has benefited from valuable comments by V. Garg, T. Jeron, K. Marzullo, M. Melliar-Smith and the anonymous referees. We thank M. Hurfin and M. Mizuno for pointing out an error in a previous version of the paper. This work has been supported in part by the Commission of European Communities under ESPRIT Programme Basic Research Project 6360 (BROADCAST). Babaoğlu was further supported by the Italian National Research Council and the Ministry of University, Research and Technology. Raynal was further supported by the CNRS under the grant Parallel Traces.

References

[1] Babaoğlu and K. Marzullo. Consistent global states of distributed systems: fundamental concepts and mechanisms. In S.J. Mullender, editor, *Distributed Systems*, chapter 4. ACM Press, 1993.

[2] Babaoğlu and M. Raynal. Specification and detection of behavioral patterns in distributed computations. Research Report UBLCS-93-11, Laboratory for Computer Science, University of Bologna, Italy, May 1993.

[3] P.C. Bates and J.C. Wileden. High-level debugging of distributed systems: the behavioral abstraction approach. *Journal of Systems and Software*, 4(3):255–264, December 1983.

[4] E.M. Clarke, E.A. Emerson, and A.P. Sistla. Automatic verification of finite state concurrent systems using temporal logic specifications. *ACM Transactions on Programming Languages and Systems*, 8(2):244–263, 1986.

[5] K.M. Chandy and L. Lamport. Distributed snapshots: Determining global states of distributed systems. *ACM Transactions on Computer Systems*, 3(1):63–75, February 1985.

[6] R. Cooper and K. Marzullo. Consistent detection of global predicates. In *ACM/ONR Workshop on Parallel and Distributed Debugging*, pages 163–173, Santa Cruz, California, May 1991.

[7] C. Diehl, C. Jard, and J.X. Rampon. Reachability analysis on distributed executions. In J.-P. Jouannaud M.-C. Gaudel, editor, *Proceedings of TAPSOFT*, number 668 in Lecture Notes on Computer Science, pages 629–643, Orsay, Paris, France, April 1993. Springer-Verlag.

[8] E.A. Emerson and J. Srinivasan. Branching time temporal logic. In G. Rozenberg J.W. de Bakker, W.-P. de Roever, editor, *Linear Time, Branching Time and Partial Order in Logics*

and Models of Concurrency, volume 354 of *Lecture Notes on Computer Science*, pages 123–172, Noordwijkerhout, The Netherlands, June 1988. Springer-Verlag.

[9] R. Gotzhein. Temporal logic and applications: a tutorial. *Computer networks and ISDN systems*, pages 203–218, 1992.

[10] V.K. Garg and B. Waldecker. Detection of unstable predicates in distributed programs. In *Proceedings of the 12th International Conference on Foundations of Software Technology and Theoretical Computer Science*, volume 652 of *Lecture Notes on Computer Science*, pages 253–264, New Delhi, India, December 1992. Springer-Verlag.

[11] M. Hurfin, N. Plouzeau, and M. Raynal. Debugging tool for distributed Estelle programs. *Journal of Computer Communications*, pages 328–333, May 1993.

[12] M. Hurfin, N. Plouzeau, and M. Raynal. Detecting atomic sequences of predicates in distributed computations. In *Proceedings of the ACM/ONR Workshop on Parallel and Distributed Debugging*, San Diego, California, May 1993.

[13] L. Lamport. Time, clocks, and the ordering of events in a distributed system. *Communications of the ACM*, 21(7):558–565, July 1978.

[14] F. Mattern. Efficient algorithms for distributed snapshots and global virtual time approximation. *Journal of Parallel and Distributed Computing*, 1993. To appear.

[15] B.P. Miller and J. Choi. Breakpoints and halting in distributed programs. In *Proceedings of the 8th IEEE International Conference on Distributed Computing Systems*, pages 316–323, San Jose, California, July 1988.

[16] K. Marzullo and G. Neiger. Detection of global state predicates. In *Proceedings of the Fifth International Workshop on Distributed Algorithms (WDAG-91)*, Lecture Notes on Computer Science. Springer-Verlag, Delphi, Greece, October 1991.

[17] A. Pnueli. Applications of temporal logic to the specification and verification of reactive sytems: a survey of current trends. volume 224 of *Lecture Notes on Computer Science*, pages 510–584. Springer-Verlag, 1986.

[18] F.B. Schneider. Personal Communication, May 1993.

[19] R. Schwarz and F. Mattern. Detecting causal relationships in distributed computations: In search of the Holy Grail. Technical Report SFB124-15/92, Department of Computer Science, University of Kaiserslautern, Kaiserslautern, Germany, December 1992.

[20] R.L. Schwartz and P.M. Melliar-Smith. Temporal logic specifications for distributed systems. In *Proceedings of the 2nd IEEE International Conference on Distributed Computing Systems*, pages 446–454, 1981.

[21] R.L. Schwartz, P.M. Melliar-Smith, and F.H. Vogt. An interval logic for high-level temporal reasonning. In *Proceedings of the ACM Symposium on Principles of Distributed Computing*, pages 173–186, Montreal, Canada, August 1983.

[22] M.M. Syslo. Minimizing the jump number for partially-ordered sets: a graph-theoretic approach, II. *Discrete Mathematics*, 63:279–295, 1987.

[23] A.I. Tomlinson and V.K. Garg. Detecting relational global predicates in distributed systems. In *Proceedings of the ACM/ONR Workshop on Parallel and Distributed Debugging*, San Diego, California, May 1993.

Specification and Verification of an Atomic Broadcast Protocol

Ping Zhou and Jozef Hooman
Department of Mathematics and Computing Science
Eindhoven University of Technology
5600 MB Eindhoven, The Netherlands.
wsinjh@win.tue.nl

Abstract

We apply a formal method based on assertions to specify and verify an atomic broadcast protocol. The protocol is implemented by replicating a server process on all processors in a network. First the requirements of the protocol are formally described. Next the underlying communication mechanism, the assumptions about local clocks, and the failure assumptions are axiomatized. Also the server process is represented by a formal specification. Then we verify that parallel execution of the server processes leads to the desired properties by proving that the conjunction of all server specifications and the axioms about the system implies the requirements of the protocol.

1 Introduction

Computing systems are composed of hardware and software components which can fail. Component failures can lead to unanticipated behaviour and service unavailability. To achieve high availability of a service despite failures, a key idea is to implement the service by a group of server processes running on distinct processors [6]. Replication of service state information among group members enables the group to provide the service even when some of its members fail, since the remaining members have enough information about the service state to continue to provide it. To maintain the consistency of these replicated global states, any state update must be broadcast to all correct servers such that all these servers observe the same sequence of state updates. Thus a communication service is needed for client processes to deliver updates to their peers. This communication service is called *atomic* broadcast. There are two sets of atomic broadcast protocols: *synchronous* ones, such as [3], and *asynchronous* ones, such as [5].

Synchronous atomic broadcast protocols assume that the underlying communication delays between correct processors are bounded. Given this assumption, local clocks of correct processors can be synchronized within certain bound [2]. Then the properties of synchronous atomic broadcast protocols are described in terms of local clocks as follows [3, 4]:

- Termination: every update whose broadcast is initiated by a correct processor at time T on its clock is delivered by all correct processors at time $T + \Delta$ on their own clocks, where Δ is a positive parameter called the *broadcast termination time*.

- Atomicity: if a correct processor delivers an update at time U on its clock, then that update was initiated by some processor and is delivered by each correct processor at time U on its own clock.

- Order: all correct processors deliver their updates in the same order.

In order to provide service despite the presence of faults, real-time systems often adopt fault-tolerance techniques. To achieve fault-tolerance, some kind of redundancy is introduced which will affect the timing behavior of a system. Hence it is a challenging problem to guarantee the correctness of real-time and fault-tolerant systems. We are interested in applications of formal verification methods to these systems. Since atomic broadcast service is one of the fundamental issues in fault-tolerance, we select an atomic broadcast protocol presented in [3, 4] which tolerates omission failures as our verification example. Henceforth, we use the term *atomic broadcast protocol* to refer to this protocol. An informal description of the protocol, an implementation, and an informal proof which shows that the implementation indeed satisfies the requirement of the protocol are presented in these papers. We follow the ideas of [4] as close as possible and compare our results in section 6.

The configuration of the service is illustrated in the figure 1. The atomic broadcast service is implemented by replicating a server process on all distributed processors in a network. Assume that there are n processors in the network. Pairs of processors are connected by links which are point-to-point, bi-directional, communication channels. The duration of message transmission along a link between correct processors is bounded. Each processor has access to a local clock. It is assumed that local clocks of correct processors are approximately synchronized. It is also assumed that only omission failures occur on processors and links. When a processor suffers an omission failure, it cannot send messages to other processors. When a link suffers an omission failure, the messages traveling along this link may be lost.

To send an update to its peers, a client process initiates the server process to broadcast that update. After such a request, each server process will perform a protocol and deliver that update to the client processes. To achieve the order property, there is a priority ordering among all

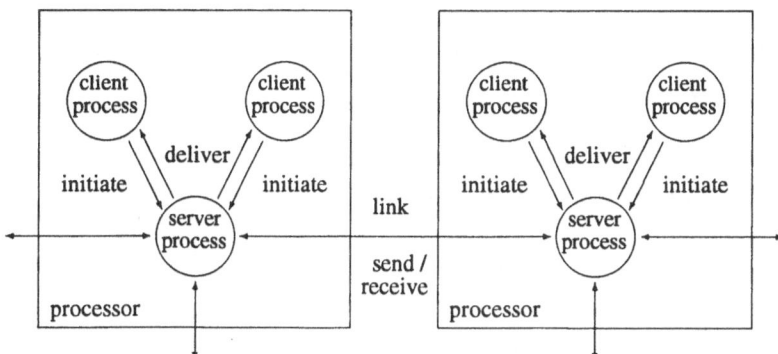

Figure 1: Atomic Broadcast Service Configuration.

processors. If two updates are initiated at different clock times, they will be delivered according to their initiation times. If they are initiated at the same clock time on different processors, they will be delivered according to the priority of their initiators.

In general, to formally verify a system, we need a proof theory which consists of axioms and rules about the system components. To be able to abstract from implementation details, it is often convenient to have a compositional verification method. Compositionality enables us to verify a system by using only specifications of its components without knowing any internal information of those components. Such compositional proof systems have been developed for non-real-time systems, e.g. [13], and real-time systems, such as [8] and [11].

In particular, if the system is composed of parallel components, the proof method should contain a *parallel composition rule*. Let $S(p)$ denote the atomic broadcast server process running on processor p, φ denote a specification written in a formal language based on first-order logic, and $S(p)$ **sat** φ denote that server process $S(p)$ satisfies specification φ. Under the condition of maximal parallelism (i.e., each process runs at its own processor), the parallel composition rule states that if server process $S(p_i)$ satisfies specification φ_i, and φ_i only refers to the interface of p_i, for $i = 1, \ldots, n$, then the parallel program $S(p_1)\| \cdots \|S(p_n)$ satisfies $\bigwedge_{i=1}^{n} \varphi_i$. This rule is formalized as follows.

Parallel Composition Rule

$$\frac{S(p_i) \text{ sat } \varphi_i, \varphi_i \text{ only refers to the interface of } p_i, \text{ for } i = 1, \ldots, n}{S(p_1)\| \cdots \|S(p_n) \text{ sat } \bigwedge_{i=1}^{n} \varphi_i}$$

We also need a *consequence rule* to weaken a specification and a *conjunction rule* to take the conjunction of specifications. Let S be any process.

Consequence Rule $\dfrac{S \text{ sat } \varphi, \varphi \rightarrow \psi}{S \text{ sat } \psi}$

Conjunction Rule $\dfrac{S \text{ sat } \varphi_1, S \text{ sat } \varphi_2}{S \text{ sat } \varphi_1 \wedge \varphi_2}$

We show that the verification of the protocol can be done compositionally by using specifications in which timing is expressed by local clock values. It proceeds as follows.

- In section 2, we specify the requirements of the protocol in a formal language based on first-order logic. We call this the *top-level specification* and denote it by ABS. Thus our aim is to prove
  $S(p_1) \| \cdots \| S(p_n)$ **sat** ABS.

- In section 3, we axiomatize the required assumptions about the system, including underlying communication mechanism, clock synchronization assumption, and failure assumptions. We denote the conjunction of all these axioms by AX.

- In section 4, we define the properties of the atomic broadcast server process running on processor p. We call this the *server process specification* and denote it by $Spec(p)$. It should only refer to the interface of p. We assume $S(p)$ **sat** $Spec(p)$.

- By the parallel composition rule,
  $S(p_1) \| \cdots \| S(p_n)$ **sat** $\bigwedge_{i=1}^n Spec(p_i)$.
  Using the conjunction rule, we obtain
  $S(p_1) \| \cdots \| S(p_n)$ **sat** $\bigwedge_{i=1}^n Spec(p_i) \wedge AX$.
  In section 5, we prove
  $\bigwedge_{i=1}^n Spec(p_i) \wedge AX \rightarrow ABS$.
  Hence the consequence rule leads to
  $S(p_1) \| \cdots \| S(p_n)$ **sat** ABS.

- We compare our results with [4] and conclude in section 6.

2   Top-Level Specification

In this section we formalize the top-level requirements of the atomic broadcast protocol. After the introduction of a number of notations, primitives, and abbreviations, we express termination by formula $TERM$ in section 2.1, atomicity by $ATOM$ in section 2.2, and order by formula $ORDER$ in section 2.3. Then the top-level specification of the atomic broadcast protocol, ABS, is the conjunction of these three properties, i.e., defined by

$ABS \equiv TERM \wedge ATOM \wedge ORDER.$

Let P (with elements $p, q, r, s, \ldots$) be a set of processor names and L (with elements $l, l_1, \ldots$) a set of link names. We assume that all processors and links have unique names. Let $G = P \cup L$. We assume that all real times range over a dense time domain called *RTIME*. Lower case letters, e.g. $t, u, v, \ldots$, are used to denote variables ranging over *RTIME*. Each processor has access to a local clock. Let C_p be a function which represents the value of the local clock of processor p, i.e., $C_p(t)$ is the value of the local clock of p at real time t. Suppose local clock values range over a domain called *CVAL*. We assume that, for any $T \in CVAL, T \geq 0$. We use capital letters, e.g. $T, U, V, \ldots$, to denote variables ranging over *CVAL*.

When a client process initiates a request to broadcasting update σ to a server process running on processor p, we call p the initiator of σ and say that p *initiates* σ. Similarly, when the server process of p delivers an update σ to client processes, we say that p *delivers σ to client processes*.

We define the following primitives:

- $correct(l)$: link l is correct (at any time).
- $correct(p)$ at t: processor p is correct at time t.
- $initiate(p, \sigma)$ at t: p finishes with receiving a request of broadcasting update σ from a client process at time t, i.e., p initiates σ at time t.
- $deliver(p, \sigma)$ at t: p starts to send update σ to client processes at time t.

Henceforth we use $\equiv$ to denote syntactic equality. For any primitive φ at t, we define the following abbreviations:

- $correct(p) \equiv \forall t : correct(p)$ at t, φ at$_p$ $T \equiv \exists t : \varphi$ at $t \wedge C_p(t) = T$
- φ by$_p$ $T \equiv \exists T_0 : \varphi$ at$_p$ $T_0 \wedge T_0 \leq T$
- φ in$_p$ $I \equiv \exists T \in I : \varphi$ at$_p$ T, where $I \subseteq CVAL$.

In [4], assumptions about the system are simplified. For instance, the message processing time on a correct processor is assumed to be zero. In this paper, however, we will take the time spent by a correct processor into account. Then the termination and atomicity properties can only be described by using an upper bound and an interval, respectively, instead of precise time points as in [4].

2.1 Termination

The termination property is stated as: every update whose broadcast is initiated by a correct processor s at clock value T will be delivered at all correct processors by clock value $T + D_1$ on their own clocks, where D_1 is a positive constant called the broadcast termination time.

This property is formally expressed as follows, with the usual convention that any free variable occurring in a formula is universally quantified.

$$TERM \equiv initiate(s, \sigma) \textbf{ at}_\textbf{s} \ T \land correct(s) \land correct(q) \rightarrow deliver(q, \sigma) \textbf{ by}_\textbf{q} \ T + D_1$$

2.2 Atomicity

Atomicity requires that, if a correct processor p delivers an update at clock value U, then that update was initiated by some processor s at some local time T and is delivered by all correct processors at some local clock value between $U - D_2$ and $U + D_2$, where D_2 is a positive constant and indicates the difference of delivery times of an update by two correct processors.

$$ATOM \equiv deliver(p, \sigma) \textbf{ at}_\textbf{p} \ U \land correct(p) \land correct(q) \rightarrow$$
$$\exists s, T : initiate(s, \sigma) \textbf{ at}_\textbf{s} \ T \land deliver(q, \sigma) \textbf{ in}_\textbf{q} \ [U - D_2, U + D_2]$$

Notice that the atomicity property does not follow from the termination property, because the initiator s need not be correct.

2.3 Order

The order property is expressed as follows: all correct processors deliver their updates in the same order. To formalize this we consider the sequence of updates of a processor. Let U be any clock value. If $\langle \sigma_1, \ldots, \sigma_k \rangle$ is a sequence of updates delivered by processor p before local time U, then there should exist a clock value V such that $\langle \sigma_1, \ldots, \sigma_k \rangle$ has also been delivered by any other processor q before local time V. Notice that U and V can be different. Furthermore, there is no reason to exclude the possibility that more than one update is delivered at the same time by a processor. This is modelled by using a *set* of sequences to represent the updates of a processor and to include all possible interleavings for simultaneous updates. We define the following abbreviation to express that no update has been delivered in a certain interval I.

- $\neg deliver(p) \textbf{ in}_\textbf{p} \ I \equiv \neg \exists \sigma : deliver(p, \sigma) \textbf{ in}_\textbf{p} \ I$.

Let $I\!N$ denote the set of all natural numbers (including 0) and $I\!N^+ = I\!N \setminus \{0\}$. Next define $List(p, U)$ to be the set of all possible sequences of updates delivered by p before local time U.

Definition 2.1 For any $p \in P$ and any $U \in CVAL$, define
$List(p, U) = \{ \langle \sigma_1, \ldots, \sigma_k \rangle \mid$ there exist $k \in I\!N^+$ and $U_1, \ldots, U_k \in CVAL$ such that
$\qquad\qquad U_1 \leq \ldots \leq U_k < U, deliver(p, \sigma_i) \textbf{ at}_\textbf{p} \ U_i$, for $i = 1, \ldots, k$,
$\qquad\qquad \neg deliver(p) \textbf{ in}_\textbf{p} \ (U_j, U_{j+1})$, for $j = 1, \ldots, k - 1$, and
$\qquad\qquad \neg deliver(p) \textbf{ in}_\textbf{p} \ [0, U_1).\}$

The order property is formalized as follows:

$ORDER \equiv correct(p) \land correct(q) \rightarrow \forall U \exists V : List(p, U) \subseteq List(q, V)$

By this property, we obtain that, for any pair of correct processors p and q,
$\forall U \exists V : List(p, U) \subseteq List(q, V)$ and, symmetrically, $\forall U' \exists V' : List(q, U') \subseteq List(p, V')$.
Hence p and q deliver their updates in the same order.

3 System Assumptions

Next we axiomatize the assumptions about the system, that is, properties of processors and links
in section 3.1, axioms about the underlying communication mechanism in section 3.2, properties
of local clocks in section 3.3, and the failure hypothesis in section 3.4. The conjunction of all
these axioms is denoted by AX.

3.1 Processors and Links

We first consider the topology of the network. Define the primitive.

- $link(l, p, q)$: l is a physical communication channel between p and q.

Define $Link(p) = \{l \mid \exists q : link(l, p, q)\}$, representing the set of links that connect p with other
processors. Two processors connected by a link are called neighbors. For any p, q, and l, if
$l \in Link(p)$, $l \in Link(q)$, and $p \not\equiv q$, then p and q are connected by l. This is expressed by the
following axiom.

Axiom 3.1 (Link) $l \in Link(p) \land l \in Link(q) \land p \not\equiv q \rightarrow link(l, p, q)$

We also assume that a link connects at most two processors.

Axiom 3.2 (Point-to-Point) $link(l, p, q) \land link(l, p, r) \rightarrow q \equiv r$

Let $FP = \{p \mid \neg correct(p)\}$ and $FL = \{l \mid \neg correct(l)\}$. Define $F = FP \cup FL$.
Thus F denotes the set of processors and links which are not always correct.

An important assumption about the network is that during any execution of the protocol all
correct processors remain connected via correct links. Recall that G is the set of all processors
and links, that is, $G = P \cup L$. Then we have $G \setminus F = \{p \mid correct(p)\} \cup \{l \mid correct(l)\}$,

denoting the set of correct processors and links. $G \setminus F$ can be considered as a graph in which processors are vertices and links are edges. We call $G \setminus F$ *connected* if and only if there exists a path between any two processors in $G \setminus F$. Now we can give the axiom for connectivity.

Axiom 3.3 (Connectivity) $G \setminus F$ is connected.

Given axiom 3.3, $d(p, q)$ is used to denote the *distance* between p and q and d to represent the diameter of $G \setminus F$.

3.2   Bounded Communication

To describe the communication mechanism, two primitives are introduced:

- $send(p, m, l)$ **at** t: p starts to send message m along l at time t.

- $receive(p, m, l)$ **at** t: p finishes with receiving m along l at time t.

The abbreviations defined in section 2 are also used for these two primitives. When $send(p, m, l)$ **at** t or $receive(p, m, l)$ **at** t holds, l must be a link of p. In terms of local clocks:

Axiom 3.4 (Connected Link)

$$send(p, m, l) \text{ } \mathbf{at_q} \text{ } T \text{ } \vee \text{ } receive(p, m, l) \text{ } \mathbf{at_q} \text{ } T \to l \in Link(p)$$

Two processors can send messages to each other if they are connected by a link. The duration of the transmission is bounded by two parameters γ and δ with $\gamma, \delta \in CVAL$, $\gamma > 0$, and $\gamma \leq \delta$. Let p and q be two correct processors connected by a correct link l. Let r be any correct processor to be used as reference. If p sends message m along link l at clock value U according to the clock of r, then q will receive m along l at some clock value in the interval $[U + \gamma, U + \delta]$ according to the clock of r.

Axiom 3.5 (Bounded Communication)

$$send(p, m, l) \text{ } \mathbf{at_r} \text{ } U \wedge correct(p) \wedge correct(q) \wedge link(l, p, q) \wedge$$
$$correct(l) \wedge correct(r) \to receive(q, m, l) \text{ } \mathbf{in_r} \text{ } [U + \gamma, U + \delta]$$

3.3 Clock Synchronization

Assume that clocks of correct processors are synchronized within a parameter ϵ.

Axiom 3.6 (Clock Synchronization)

$$correct(p) \text{ at } t \wedge correct(q) \text{ at } t \rightarrow |C_p(t) - C_q(t)| < \epsilon$$

Lemma 3.1 (Clock Synchronization) $correct(p) \wedge correct(q) \rightarrow |C_p(t) - C_q(t)| < \epsilon$

We also assume that local clocks are monotonic.

Axiom 3.7 (Monotonic Clock) $t_1 \leq t_2 \leftrightarrow C_p(t_1) \leq C_p(t_2)$

According to [7], an implicit assumption was used in [4], namely that any clock on a correct processor has a speed that differs from the speed of any other clock on a correct processor by a very small quantity ρ, $\rho \geq 0$. This drift ρ was neglected in [4] resulting in the following approximation: while a message travels between two processors the clocks of the two processors will keep their distance constant. We take this factor ρ into account and formalize this assumption as follows:

Axiom 3.8 (Relative Speed)

$$correct(p) \wedge correct(q) \wedge t_1 \leq t_2 \rightarrow$$
$$(1 - \rho)(C_p(t_2) - C_p(t_1)) \leq C_q(t_2) - C_q(t_1) \leq (1 + \rho)(C_p(t_2) - C_p(t_1))$$

3.4 Failure Assumptions

The protocol verified in this paper tolerates omission failures. When a processor suffers an omission failure, it cannot send out messages. More precisely, if a processor p is not correct at real time t, then p is not able to send any message m along any link l at time t. This is also called the *fail silence* property of processors.

Axiom 3.9 (Fail Silence) $\neg correct(p) \text{ at}_q T \rightarrow \neg send(p, m, l) \text{ at}_q T$

When a link suffers an omission failure, the messages entrusted on that link may be lost. But if a message has been received by a processor along a (possibly faulty) link, then that message should have been correctly transmitted by that link. For instance, the message is not corrupted and there are no timing errors. Therefore, if a processor q receives a message m along link l at clock value V, then there exists another processor p which has sent that message earlier along l at some time between $[V - \delta, V - \gamma]$ according to the clock of r.

Axiom 3.10 (Only Omission Failure)

$$receive(q, m, l) \ \textbf{at}_\textbf{r} \ V \wedge correct(r) \to \exists p \not\equiv q : send(p, m, l) \ \textbf{in}_\textbf{r} \ [V - \delta, V - \gamma]$$

4   Server Process Specification

In this section, we characterize $S(p)$, i.e., the atomic broadcast server process running on p. Notice that in the top-level specification only delivery of updates is important and thus a primitive $deliver(p, \sigma)$ at t is used. In the server process specification, however, information about the initiation time T and the initiator s of an update σ is needed to implement the top-level specification. Therefore we define a primitive

- $convey(p, <T, s, \sigma >)$ at t: p starts to send message $<T, s, \sigma >$ to client processes at real time t.

Then the relation between $deliver(p, \sigma)$ at t and $convey(p, <T, s, \sigma >)$ at t is clear:

$$deliver(p, \sigma) \ \text{at} \ t \leftrightarrow \exists s, T : convey(p, <T, s, \sigma >) \ \text{at} \ t$$

Assume that any correct processor can send a message to all its neighbors within $T_s \in CVAL$ time units and any correct processor can convey all the updates initiated at the same clock time to client processes within $T_c \in CVAL$ time units. Let $T_r \in CVAL$, $T_r \geq T_s$, be the time to ensure that all correct processors have received a message containing an update after it is initiated. These parameters will be used to determine the values of D_1 and D_2 occurring in the top-level specification.

Then the server specification is described by the following requirements.

- Initiation requirement.
  When p initiates an update σ at clock time T, it will send message $<T, p, \sigma >$ to all its neighbors within T_s. After T_r time units, assuming that then all correct processors have received that message, p will convey $<T, p, \sigma >$ to client processes. This is formalized by the following formula:
  $Start(p) \equiv initiate(p, \sigma) \ \textbf{at}_\textbf{p} \ T \to$
  $\qquad\qquad \forall l \in Link(p) : send(p, <T, p, \sigma >, l) \ \textbf{in}_\textbf{p} \ [T, T + T_s] \wedge$
  $\qquad\qquad convey(p, <T, p, \sigma >) \ \textbf{in}_\textbf{p} \ [T + T_r, T + T_r + T_c]$

- Relay requirement.
  When p receives a message $<T, s, \sigma >$, it will relay this message to all its links except the one along which it received this message. Later, as in the initiator's case, when its clock reaches $T + T_r$, p will convey $<T, s, \sigma >$ to client processes.

$Relay(p) \equiv receive(p, < T, s, \sigma >, l) \, \textbf{at}_\textbf{p} \, U \rightarrow$
$$\forall l_1 \in Link(p) \setminus \{l\} : send(p, < T, s, \sigma >, l_1) \, \textbf{in}_\textbf{p} \, [U, U + T_s] \, \wedge$$
$$convey(p, < T, s, \sigma >) \, \textbf{in}_\textbf{p} \, [T + T_r, T + T_r + T_c]$$

- Convey requirement.

  If processor p conveys a message $< T, s, \sigma >$ at clock time U, then we have that $U \in [T + T_r, T + T_r + T_c]$ and either p initiated σ itself at local clock time T or p is different from s and has received the message $< T, s, \sigma >$ at some clock value V.

  $Origin(p) \equiv convey(p, < T, s, \sigma >) \, \textbf{at}_\textbf{p} \, U \rightarrow$
  $$U \in [T + T_r, T + T_r + T_c] \wedge$$
  $$[\, (initiate(p, \sigma) \, \textbf{at}_\textbf{p} \, T \wedge p \equiv s) \vee$$
  $$(\exists l, V : receive(p, < T, s, \sigma >, l) \, \textbf{at}_\textbf{p} \, V \wedge p \not\equiv s)]$$

- Ordering requirement.

  If two messages are conveyed by processor p, then they will be conveyed in the order of the initiation times contained in these two messages. If these initiation times are the same, then they will be conveyed according to the priority of the initiators. Therefore it is assumed that there is a total order $\prec$ on the set of processor names P. This total order specifies a priority ordering among processors. A lexicographical ordering $\sqsubset$ on pairs $< T, s >$ is defined, for any two pairs (T_1, s_1) and (T_2, s_2), by
  $(T_1, s_1) \sqsubset (T_2, s_2)$ iff $(T_1 < T_2) \vee (T_1 = T_2 \wedge s_1 \prec s_2)$.
  Then the ordering requirement is formalized by the formula:
  $Sequen(p) \equiv convey(p, < T_1, s_1, \sigma_1 >) \, \textbf{at}_\textbf{p} \, V_1 \wedge convey(p, < T_2, s_2, \sigma_2 >) \, \textbf{at}_\textbf{p} \, V_2 \rightarrow$
  $$(V_1 < V_2 \leftrightarrow (T_1, s_1) \sqsubset (T_2, s_2))$$

The requirements mentioned above only hold for correct processors. Since omission failures are allowed, we still need to define the acceptable behaviour for faulty processors. This is captured by the following requirement, which holds for any arbitrary processor p.

- Failure requirement.

  When p sends a message $< T, s, \sigma >$ to a neighbor at local time U, there can be only two possibilities: either p initiated σ itself at local time T and $U \in [T, T + T_s]$ holds, or p received $< T, s, \sigma >$ at some local time V and $U \in [V, V + T_s]$ holds.

  $Source(p) \equiv send(p, < T, s, \sigma >, l) \, \textbf{at}_\textbf{p} \, U \rightarrow$
  $$(initiate(p, \sigma) \, \textbf{at}_\textbf{p} \, T \wedge U \in [T, T + T_s] \wedge p \equiv s) \vee$$
  $$(\exists l_1, V : receive(p, < T, s, \sigma >, l_1) \, \textbf{at}_\textbf{p} \, V \wedge$$
  $$U \in [V, V + T_s] \wedge p \not\equiv s)$$

Now we assume that server process $S(p)$ satisfies specification $Spec(p)$ with
$Spec(p) \equiv [correct(p) \rightarrow Start(p) \wedge Relay(p) \wedge Origin(p) \wedge Sequen(p)] \wedge Source(p)$

Axiom 4.1 (Server Process Specification) $S(p)$ **sat** $Spec(p)$

Thus the behavior of any processor p is specified by this axiom and the fail silence axiom 3.9.

5 Verification

As explained in section 1, our aim is to prove $\bigwedge_{i=1}^{n} Spec(p_i) \wedge AX \rightarrow ABS$. Thus we assume $\bigwedge_{i=1}^{n} Spec(p_i) \wedge AX$ and prove ABS, that is, $TERM$ in section 5.1, $ATOM$ in section 5.2, and $ORDER$ in section 5.3. Details of the proof can be found in [12], here we only present the main steps.

As a preparation, we first rewrite a part of the specification $Spec(p)$ to a more general form in which the clock values are measured on an arbitrary correct processor r.

Lemma 5.1 (Modified Server Specification)

$$correct(r) \rightarrow [correct(p) \rightarrow Forward(p,r)] \wedge NSource(p,r)$$

where $Forward(p,r)$ is a generalization of $Relay(p)$, defined by

$Forward(p,r) \equiv receive(p, < T, s, \sigma >, l)$ **at$_r$** $U \rightarrow$
$\qquad\qquad\qquad \forall l_1 \in Link(p) \setminus \{l\} : send(p, < T, s, \sigma >, l_1)$ **in$_r$** $[U, U + (1 + \rho)T_s]$

and $NSource(p,r)$ is a general form of $Source(p)$:

$NSource(p,r) \equiv send(p, < T, s, \sigma >, l)$ **at$_r$** $U \rightarrow$
$\qquad\qquad\qquad (initiate(p, \sigma)$ **at$_p$** $T \wedge U \in (T - \epsilon, T + T_s + \epsilon) \wedge p \equiv s) \vee$
$\qquad\qquad\qquad (\exists l_1, V : receive(p, < T, s, \sigma >, l_1)$ **at$_r$** $V \wedge$
$\qquad\qquad\qquad\qquad U \in [V, V + (1 + \rho)T_s] \wedge p \not\equiv s)$

5.1 Verification of Termination

In this section we prove the termination property of the protocol, starting with a few lemmas. The first lemma expresses that if a correct processor p receives a message $< T, s, \sigma >$ at time V measured on the clock of a correct processor r, then a correct neighbor q which is not s will receive $< T, s, \sigma >$ by $V + (1 + \rho)T_s + \delta$ measured on the clock of r.

Lemma 5.2 (Propagation)

$receive(p, < T, s, \sigma >, l_1)$ **at$_r$** $V \wedge correct(p) \wedge correct(q) \wedge link(l_2, p, q) \wedge$
$correct(l_2) \wedge q \not\equiv s \wedge correct(r) \rightarrow$
$\qquad\qquad \exists l : receive(q, < T, s, \sigma >, l)$ **by$_r$** $V + (1 + \rho)T_s + \delta$

If correct processor s initiates update σ at local time T, then any another correct processor q will receive $< T, s, \sigma >$ by time $T + d(s, q)((1 + \rho)T_s + \delta)$ measured on the clock of s.

Lemma 5.3 (Bounded Receiving)

$$initiate(s, \sigma) \text{ at}_s T \wedge correct(s) \wedge correct(q) \wedge q \not\equiv s \rightarrow$$
$$\exists l : receive(q, < T, s, \sigma >, l) \text{ by}_s T + d(s, q)((1 + \rho)T_s + \delta)$$

This lemma is proved by induction on $d(s, q)$, the distance between s and q, using lemma 5.2. By means of lemma 5.3 we can show that if a correct processor s initiates σ at local time T, then every correct processor q will convey $< T, s, \sigma >$ in the interval $[T + T_r, T + T_r + T_c]$ according to its own clock.

Lemma 5.4 (Convey)

$$initiate(s, \sigma) \text{ at}_s T \wedge correct(s) \wedge correct(q) \rightarrow$$
$$convey(q, < T, s, \sigma >) \text{ in}_q [T + T_r, T + T_r + T_c]$$

Finally we prove that the termination property $TERM$ follows from the axioms and lemmas given before.

Theorem 5.1 (Termination) If $D_1 \geq T_r + T_c$, then

$$initiate(s, \sigma) \text{ at}_s T \wedge correct(s) \wedge correct(q) \rightarrow deliver(q, \sigma) \text{ by}_q T + D_1$$

Proof: Assume that the premise of this theorem holds. By the convey lemma 5.4, we obtain $convey(q, < T, s, \sigma >) \text{ in}_q [T + T_r, T + T_r + T_c]$. As observed in section 4, this implies $deliver(q, \sigma) \text{ in}_q [T + T_r, T + T_r + T_c]$. Using $D_1 \geq T_r + T_c$, this leads to the required property $deliver(q, \sigma) \text{ by}_q T + D_1$. □

5.2 Verification of Atomicity

In this section, we prove the atomicity property of the atomic broadcast protocol, again using a number of lemmas. The first lemma states that if correct processor p receives message $< T, s, \sigma >$ at local time V, then that update σ was initiated by processor s at local time T.

Lemma 5.5 (Initiation)

$$receive(p, < T, s, \sigma >, l) \text{ at}_p V \wedge correct(p) \rightarrow initiate(s, \sigma) \text{ at}_s T$$

We define an abbreviation $Firstrec(p, < T, s, \sigma >, l)$ $\mathbf{at_r}$ V, which expresses that p receives $< T, s, \sigma >$ at time V measured on the clock of a correct processor r and p is one of the first correct processors which have received $< T, s, \sigma >$ according to the clock of r, that is, any other correct processor has not received $< T, s, \sigma >$ earlier.

$$Firstrec(p, < T, s, \sigma >, l) \ \mathbf{at_r} \ V \equiv$$
$$receive(p, < T, s, \sigma >, l) \ \mathbf{at_r} \ V \wedge correct(r) \wedge correct(p) \wedge$$
$$\forall p', l', V' : (correct(p') \wedge p' \not\equiv p \wedge receive(p', < T, s, \sigma >, l') \ \mathbf{at_r} \ V' \rightarrow V' \geq V)$$

The next lemma expresses that if p receives $< T, s, \sigma >$ at time V measured on the clock of a correct processor r, p is one of the first correct processors which have received $< T, s, \sigma >$, and s is faulty, then any processor q which is not p and has sent $< T, s, \sigma >$ earlier than V is a faulty processor.

Lemma 5.6 (Faulty Sender)

$$Firstrec(p, < T, s, \sigma >, l_1) \ \mathbf{at_r} \ V \wedge send(q, < T, s, \sigma >, l_2) \ \mathbf{at_r} \ U \wedge$$
$$p \not\equiv q \ \wedge \neg correct(s) \wedge U < V \rightarrow \neg correct(q)$$

The following lemma shows that if p receives $< T, s, \sigma >$ at time V measured on the clock of a correct processor r, p is one of the first correct processors which have received $< T, s, \sigma >$, and s is faulty, then $V < T + m((1 + \rho)T_s + \delta) + \epsilon$, where m is the maximum number of faulty processors in the network.

Lemma 5.7 (First Correct Receiving)

$$Firstrec(p, < T, s, \sigma >, l) \ \mathbf{at_r} \ V \wedge \neg correct(s) \rightarrow V < T + m((1 + \rho)T_s + \delta) + \epsilon$$

If p receives $< T, s, \sigma >$ at time V measured on the clock of a correct processor r and s is faulty, then any other correct processor q will receive $< T, s, \sigma >$ by time $V + d(p, q)((1 + \rho)T_s + \delta)$ measured on the clock of r.

Lemma 5.8 (Correct Receiving)

$$receive(p, < T, s, \sigma >, l') \ \mathbf{at_r} \ V \wedge \neg correct(s) \wedge correct(q) \wedge p \not\equiv q \rightarrow$$
$$\exists l : receive(q, < T, s, \sigma >, l) \ \mathbf{by_r} \ V + d(p, q)((1 + \rho)T_s + \delta)$$

Lemma 5.8 is proved using induction on $d(p, q)$. The next lemma expresses that if correct processor p learns of $< T, s, \sigma >$, then any correct processor q also learns of $< T, s, \sigma >$.

Lemma 5.9 (All Learn)

$$Learn(p, < T, s, \sigma >) \wedge correct(p) \wedge correct(q) \rightarrow Learn(q, < T, s, \sigma >)$$

If correct processor p conveys $< T, s, \sigma >$ at local time U, then any correct processor q conveys $< T, s, \sigma >$ in the interval $[T + T_r, T + T_r + T_c]$ on its own clock.

Lemma 5.10 (All Convey)

$$convey(p, < T, s, \sigma >) \textbf{ at}_\textbf{p} U \wedge correct(p) \wedge correct(q) \rightarrow$$
$$convey(q, < T, s, \sigma >) \textbf{ in}_\textbf{q} [T + T_r, T + T_r + T_c]$$

Finally we prove a theorem which shows that the atomicity property $ATOM$ follows from the axioms and lemmas given before.

Theorem 5.2 (Atomicity) If $D_2 \geq T_c$, then

$$deliver(p, \sigma) \textbf{ at}_\textbf{p} U \wedge correct(p) \wedge correct(q) \rightarrow$$
$$\exists s, T : initiate(s, \sigma) \textbf{ at}_\textbf{s} T \wedge deliver(q, \sigma) \textbf{ in}_\textbf{q} [U - D_2, U + D_2]$$

Proof: Assume that the premise of the theorem holds. From $deliver(p, \sigma)$ $\textbf{at}_\textbf{p}$ U, by definition, there exist s and T such that $convey(p, < T, s, \sigma >)$ $\textbf{at}_\textbf{p}$ U holds. By the server process specification axiom 4.1 and $correct(p)$, we have $Origin(p)$. By $Origin(p)$, we obtain $Learn(p, < T, s, \sigma >) \wedge U \in [T + T_r, T + T_r + T_c]$, i.e.,

$$initiate(p, \sigma) \textbf{ at}_\textbf{p} T \wedge p \equiv s \tag{1}$$

or

$$\exists l, V : receive(p, < T, s, \sigma >, l) \textbf{ at}_\textbf{p} V \wedge p \not\equiv s \wedge \tag{2}$$
$$U \in [T + T_r, T + T_r + T_c]. \tag{3}$$

If (1) holds, then $initiate(s, \sigma)$ $\textbf{at}_\textbf{s}$ T.
If (2) holds, the initiation lemma 5.5 leads to $initiate(s, \sigma)$ $\textbf{at}_\textbf{s}$ T.
Hence both cases lead to

$$\exists s, T : initiate(s, \sigma) \textbf{ at}_\textbf{s} T. \tag{4}$$

From $convey(p, < T, s, \sigma >)$ $\textbf{at}_\textbf{p}$ U, by the all convey lemma 5.10, we obtain $convey(q, < T, s, \sigma >)$ $\textbf{in}_\textbf{q}$ $[T + T_r, T + T_r + T_c]$.
By (3), we have $T \in [U - T_r - T_c, U - T_r]$.
Hence $convey(q, < T, s, \sigma >)$ $\textbf{in}_\textbf{q}$ $[U - T_c, U + T_c]$.
Thus, by definition, $deliver(q, \sigma)$ $\textbf{in}_\textbf{q}$ $[U - T_c, U + T_c]$. Since $D_2 \geq T_c$ this leads to $deliver(q, \sigma)$ $\textbf{in}_\textbf{q}$ $[U - D_2, U + D_2]$. Together with (4) this proves the theorem. □

5.3 Verification of Order

The order property of the atomic broadcast protocol will be proved in this section. We first prove a lemma which expresses that, for any correct processors p and q, if p conveys $< T, s, \sigma >$ at local time U, q conveys $< T, s, \sigma >$ at local time V, and no update is delivered by p in the interval $[0, U)$, then there is also no update delivered by q in the interval $[0, V)$.

Lemma 5.11 (First Delivery)

$$convey(p, < T, s, \sigma >) \textbf{ at}_\textbf{p} U \wedge convey(q, < T, s, \sigma >) \textbf{ at}_\textbf{q} V \wedge correct(p) \wedge$$
$$correct(q) \wedge \neg deliver(p) \textbf{ in}_\textbf{p} [0, U) \rightarrow \neg deliver(q) \textbf{ in}_\textbf{q} [0, V)$$

The next lemma expresses that, for any pair of correct processors p and q, if p conveys $< T_1, s_1, \sigma_1 >$ at local time U_1 and $< T_2, s_2, \sigma_2 >$ at local time U_2, q conveys $< T_1, s_1, \sigma_1 >$ at local time V_1 and $< T_2, s_2, \sigma_2 >$ at local time V_2, and there is no update delivered by p in the interval (U_1, U_2), then there is also no update delivered by q in the interval (V_1, V_2).

Lemma 5.12 (No Delivery)

$$convey(p, < T_1, s_1, \sigma_1 >) \textbf{ at}_\textbf{p} U_1 \wedge convey(p, < T_2, s_2, \sigma_2 >) \textbf{ at}_\textbf{p} U_2 \wedge$$
$$convey(q, < T_1, s_1, \sigma_1 >) \textbf{ at}_\textbf{p} V_1 \wedge convey(q, < T_2, s_2, \sigma_2 >) \textbf{ at}_\textbf{p} V_2 \wedge$$
$$correct(p) \wedge correct(q) \wedge \neg deliver(p) \textbf{ in}_\textbf{p} (U_1, U_2) \rightarrow$$
$$\neg deliver(q) \textbf{ in}_\textbf{q} (V_1, V_2)$$

This leads to the following theorem.

Theorem 5.3 (Order)

$$correct(p) \wedge correct(q) \rightarrow \forall U \exists V : List(p, U) \subseteq List(q, V)$$

6   Comparison and Conclusion

Comparing our paper with [4], the basic ideas of proving properties of the protocol are similar. For instance, in the algorithm of [4] a processor only relays a message to its neighbors if the message is received by the processor for the first time and it is not a "late message". Actually these two factors do not affect the correctness of the protocol. Adding them to the algorithm only improves the efficiency of the implementation. Thus the informal proof in [4] verifies the protocol without taking these factors into account. Our formal proof proceeds similarly, as can be seen from the $Relay(p)$ property.

We have observed that if an update σ is initiated by a processor s at local clock time T, then any correct processor p will receive the message $< T, s, \sigma >$ in less than $(d + m)((1 + \rho)T_s + \delta) + \epsilon$ time units measured on its own clock, where d is the maximal distance between two correct processors and m is the maximal number of faulty processors. Thus $T_r \geq (d + m)((1 + \rho)T_s + \delta) + \epsilon$. The corresponding time in [4] is $(d + m)\delta + \epsilon$, which equals our bound if we assume $T_s = 0$ and $\rho = 0$ as in [4]. Notice that the condition on T_r is only needed for the implementation of the server specification $Spec(p)$, not for the correctness proof of the protocol.

To prove the atomicity property, we have to show that if a correct processor p delivers σ at some time U, then σ was initiated by some processor s at some clock time T. This has not been proved in [4]. We proved it (in lemma 5.5) by means of available timing information, using a lower bound for message transmission delay between two correct processors.

There is an implicit assumption in [4] about the drift of local clocks. We have formalized this assumption in axiom 3.8. This axiom is used to formulate (in lemma 5.1) part of the server specification in terms of the local clock of any correct processor. Together with the other axioms about local clocks, this makes it possible to perform the verification in terms of local clock values. In contrast with most formal methods, see e.g. [1], there is no need to refer to global times during the protocol verification. This leads to a convenient and natural calculus.

There is quickly growing literature on the formal verification of real-time and fault-tolerant distributed systems. Closely related to our approach is the work on the proof checker EHDM and its successor PVS. Rushby and von Henke [9] use EHDM to check the proofs of Lamport and Melliar-Smith's interactive convergence clock synchronization algorithm. Mechanical verification of a generalized protocol for Byzantine fault-tolerant clock synchronization by using EHDM is described in [10]. In future applications of our approach we will certainly investigate the use of such an interactive proof checker.

Acknowledgement

Many thanks go to Flaviu Cristian for stimulating discussions and and valuable comments on preliminary versions of this paper.

References

[1] J.W. de Bakker, C. Huizing, W.-P. de Roever, and G. Rozenberg(Eds.). *Real-Time: Theory in Practice*. LNCS 600, 1991.

[2] F. Cristian, H. Aghili, and R. Strong. Clock synchronization in the presence of omission and performance failures, and processor joins. In *Global States and Time in Distributed Systems*. Z. Yang and T.A. Marsland (Eds.), 1993.

[3] F. Cristian, H. Aghili, R. Strong, and D. Dolev. Atomic broadcast: From simple message diffusion to Byzantine agreement. In *The 15th Annual International Symposium on Fault-Tolerant Computing*, pages 200 – 206. Ann Arbor, USA, 1985.

[4] F. Cristian, H. Aghili, R. Strong, and D. Dolev. Atomic broadcast: From simple message diffusion to Byzantine agreement. Technical Report RJ 5244, IBM Almaden Research Center, 1989.

[5] J.M. Chang and N. Maxemchuck. Reliable broadcast protocols. *ACM Trans. on Computer Systems 2(3)*, pages 251–273, 1984.

[6] F. Cristian. Synchronous atomic broadcast for redundant broadcast channels. *Real-Time Systems 2*, pages 195–212, 1990.

[7] F. Cristian. Comments. *Private Correspondence*, 1993.

[8] J. Hooman. *Specification and Compositional Verification of Real-Time Systems*. LNCS 558, Springer-Verlag, 1991.

[9] J. Rushby and F. von Henke. Formal verification of algorithms for critical systems. *IEEE Trans. on Software Engineering*, 19(1):13–23, 1993.

[10] N. Shankar. Mechanical verification of a generalized protocol for Byzantine fault tolerant clock synchronization. In *LNCS 600*, pages 217–236, 1992.

[11] P. Zhou and J. Hooman. A proof theory for asynchronously communicating real-time systems. In *Proc. of the 13th IEEE Real-Time Systems*, pages 177–186, 1992.

[12] P. Zhou and J. Hooman. Formal specification and compositional verification of an atomic broadcast protocol. Technical report CSN 94/05, Eindhoven University of Technology, To appear in *Real-Time Systems*, 1994.

[13] J. Zwiers. *Compositionality, Concurrency and Partial Correctness*. LNCS 321, Springer-Verlag, 1989.

Trace-Based Compositional Refinement of Fault Tolerant Distributed Systems*

Henk Schepers* and Jos Coenen†
*Philips Research Laboratories
Information & Software Technology
Specification, Design & Realisation Department
5656 AA Eindhoven
schepers@prl.philips.nl
†Department of Mathematics and Computing Science
Eindhoven University of Technology
5600 MB Eindhoven, The Netherlands
wsinjosc@win.tue.nl

Abstract

We present a trace-based compositional framework for the refinement of fault tolerant distributed systems. Important in such systems is the failure hypothesis that stipulates the class of failures that must be tolerated. In the formalism presented in this report, the failure hypothesis of a system is formalized as a relation between the system's normal behaviour (i.e., the behaviour that conforms to the specification) and its acceptable behaviour, that is, the normal behaviour together with the exceptional behaviour (i.e., the behaviour whose abnormality should be tolerated). We highlight two aspects of refinement of fault tolerant distributed systems. First we show how to classify the system that under a particular failure hypothesis should satisfy a given specification. In the second place we determine the least stringent failure hypothesis such that a given system still satisfies a particular specification.

---

*This research was supported through NWO project "Fault Tolerance: Paradigms, Models, Logics, Construction." under STW grant number NWI88.1517 and SION grant number 612-316-103.

1   Introduction

In this report we model the operations of a distributed system using a network of processes that communicate synchronously via directed channels. We focus on the formalization of fault tolerance in relation to concurrency. Therefore, we abstract from the internal states of the processes and concentrate on the input and output behaviour that is observable at their interface. In particular, we only describe the sequence of communications that are performed by the processes. So, in our theory we do not deal with the sequential aspects of processes and instead use a simple formalism to reason about the properties of networks of processes. We do not consider the timing of communications and the enabledness of a process to communicate. Furthermore, we restrict ourselves to the specification and verification of *safety* properties of fault tolerant systems. Safety properties are important for reliability because, in the characterization by Lamport [4], they express that "nothing bad will happen". The method is compositional to allow reasoning with the specifications of processes while ignoring their implementation details. Thus, our approach supports top-down design.

We express properties by means of a first order trace logic. To express that a process P satisfies a safety property ϕ we use a correctness formula of the form P **sat** ϕ. We use a special variable h to denote the trace, also called the history, of P. Such a history describes the observable behaviour of a process by recording the communications along its visible channels.

A fault may cause a process to behave abnormally, and a failure hypothesis divides such abnormal behaviour into exceptional and catastrophic behaviours. Relative to the failure hypothesis an exceptional behaviour exhibits an abnormality which should be tolerated. A catastrophic behaviour has an abnormality that was not anticipated. The exceptional behaviour together with the normal behaviour constitutes the *acceptable* behaviour.

In [6] a failure hypothesis is formalized as a relation between the normal and the acceptable behaviour of a system. The construct $P \wr \chi$ (read "P under χ") is introduced to indicate execution of process P under failure hypothesis χ. This construct enables us to specify *failure prone processes*. In [8] a trace-based compositional proof theory to verify safety properties of fault tolerant distributed systems is presented. There, a failure hypothesis χ of a process P is formalized as a predicate, whose only free variables are h and h_{old}. The interpretation is such that h_{old} represents a normal history of process P whereas h is an *acceptable* history of P with respect to χ. Such relations enable us to abstract from the precise nature of a fault and to focus on the abnormal behaviour it causes. To characterize the acceptable behaviour of a failure prone process FP with respect to a failure hypothesis χ, the following inference rule is given:

$$\frac{FP \ \textbf{sat} \ \phi}{FP \wr \chi \ \textbf{sat} \ \phi \wr \chi}$$

In practice, a designer is faced with the problem of constructing a system which, given a failure hypothesis that characterizes the circumstances assumed for the system, satisfies a given specification. Although the above failure hypothesis introduction rule can be used to obtain a specification of the acceptable behaviour from the specification of the normal behaviour, it cannot be used to identify the normal behaviour specification that results in the desired acceptable behaviour specification. Another problem one may encounter in practice concerns reusability: does a given system continue to satisfy its acceptable behaviour specification when the circumstances get worse? Then, the above inference rule is again not of much use.

Essentially, a failure hypothesis relates the abstract level at which a process behaves normally to a concrete level at which that process behaves acceptably. More precisely, the specification of the normal process behaviour can be seen as a *refinement* of the acceptable process behaviour because it is more restrictive. In this report we study the relationship between the compositional proof theory of [8] and the compositional refinement theory of [10]. One particular aim is to classify the processes that, given a particular failure hypothesis, satisfy a given specification. Also, we try to determine the least restrictive failure hypothesis under which a process still satisfies a given specification.

This report is organized as follows. Section 2 introduces the model of computation. In Section 3 we present assertions, failure hypotheses, and correctness formulae. Section 4 contains the compositional refinement theory. In Section 5 we illustrate our method by investigating a transmission medium that might corrupt messages and one that might be transiently stuck at zero. A conclusion and suggestions for future research can be found in Section 6.

2   Model of computation

The set of natural numbers (including 0) is denoted by $\mathbb{N}$. Let *VAR* be a nonempty set of program variables, *CHAN* a nonempty set of channel names, and let *VAL* be a denumerable domain of values ($VAL \supseteq \mathbb{N}$). We consider networks of processes that communicate synchronously via directed channels. Channels always connect exactly two processes. A channel via which a process communicates with its environment is called an *external* channel of that process. When processes are composed in parallel their joint channels are said to be the *internal* channels of that composite process.

We assume a programming language, e.g. CSP [2], which can be used to define such networks of processes. Besides sequential constructs, this language includes the construct $P_1 \parallel P_2$ to indicate parallel execution of processes P_1 and P_2, as well as the construct $P \setminus cset$ to hide communications along the channels from a set *cset* of internal channels.

Define $var(P)$ as the set of variables occurring in process P. Parallel processes do not share program variables, i.e., for $P_1 \parallel P_2$ we require $var(P_1) \cap var(P_2) = \emptyset$.

The set of visible, or observable, input channels of process P, notation $in(P)$, can be defined for sequential constructs. Then, $in(P_1 \parallel P_2) = in(P_1) \cup in(P_2)$ and $in(P \setminus cset) = in(P) - cset$.

The set $out(P)$ of observable output channels of process P is defined likewise.

Definition 1 (Channels of a process) The set of channels of a process P, notation $chan(P)$, is defined by $chan(P) = in(P) \cup out(P)$.   ◇

For $P_1 \| P_2$ we require $in(P_1) \cap in(P_2) = \emptyset$ and $out(P_1) \cap out(P_2) = \emptyset$ to guarantee that channels are unidirectional and connect exactly two processes. To guarantee that only communications via internal channels are hidden we require for $P \setminus cset$ that $cset \subseteq in(P) \cap out(P)$.

We denote a synchronous communication of value $\mu \in VAL$ along channel $c \in CHAN$ by a pair (c, μ), and define $ch((c, \mu)) = c$ and $val((c, \mu)) = \mu$. A history (also called a trace), typically denoted by θ, is a finite sequence of the form $\langle (c_1, \mu_1), \ldots, (c_n, \mu_n) \rangle$, where $n \in \mathbb{N}$, $c_i \in CHAN$, and $\mu_i \in VAL$, for $1 \leq i \leq n$. If $\theta = \langle (c_1, \mu_1), \ldots, (c_n, \mu_n) \rangle$, then the length of θ, notation $len(\theta)$, is defined by $len(\theta) = n$. Let $\langle \rangle$ denote the empty history, that is, the history of length 0. A history represents the communications of a process along its observable channels up to some point in an execution.

Example 2 (Transmission medium) Consider the first-in first-out transmission medium M that accepts messages via channel *in* and delivers them via channel *out*. Some possible histories of M are $\langle \rangle$, $\langle (in, 1) \rangle$, and $\langle (in, 1), (out, 1) \rangle$.   △

The concatenation of two histories $\theta_1 = \langle (c_1, \mu_1), \ldots, (c_k, \mu_k) \rangle$ and $\theta_2 = \langle (d_1, \nu_1), \ldots, (d_l, \nu_l) \rangle$, denoted by $\theta_1 {}^{\wedge} \theta_2$, is defined as $\langle (c_1, \mu_1), \ldots, (c_k, \mu_k), (d_1, \nu_1), \ldots, (d_l, \nu_l) \rangle$. We use $\theta^{\wedge}(c, \mu)$ as an abbreviation of $\theta^{\wedge} \langle (c, \mu) \rangle$. Let $TRACE$ be the set of traces.

Definition 3 (Projection) For a trace $\theta \in TRACE$ and a set of channels $cset \subseteq CHAN$, we define the *projection* of θ onto $cset$, denoted $\theta \uparrow cset$, as the sequence obtained from θ by deleting all records with channels not in $cset$. Formally,

$$\theta \uparrow cset = \begin{cases} \langle \rangle & \text{if } \theta = \langle \rangle, \\ \theta_0 \uparrow cset & \text{if } \theta = \theta_0{}^{\wedge}(c, \mu) \text{ and } c \notin cset, \\ (\theta_0 \uparrow cset)^{\wedge}(c, \mu) & \text{if } \theta = \theta_0{}^{\wedge}(c, \mu) \text{ and } c \in cset. \end{cases}$$

◇

Definition 4 (Hiding) Hiding is the complement of projection. Formally, the *hiding* of a set $cset$ of channels from a trace $\theta \in TRACE$, notation $\theta \setminus cset$, is defined as $\theta \setminus cset = \theta \uparrow (CHAN - cset)$.   ◇

Definition 5 (Prefix) The trace θ_1 is a *prefix* of a trace θ_2, notation $\theta_1 \preceq \theta_2$ if, and only if, there exists a trace θ_3 such that $\theta_1{}^{\wedge}\theta_3 = \theta_2$.   ◇

A set of traces is *prefix closed* if every prefix of a trace in the set is also an element of that set. For a process P defined using a programming language in which processes communicate synchronously via directed channels it is quite standard, using the above defined operations, to give the prefix closed set $\mathcal{H}[\![P]\!]$ of possible finite traces that can be observed up to any point in an execution of P (see for instance [9, 8]). Instead of the infinite executions of P this set contains all finite approximations, which is justified since we only deal with safety properties (see for instance [9]), and since (the semantics of) our programming language is such that an infinite trace represents a behaviour of the process if, and only if, all its prefixes do. The set $\mathcal{H}[\![P]\!]$ represents the normal behaviour of process P. In Section 3 we determine the set $\mathcal{H}[\![P\wr\chi]\!]$ representing the acceptable behaviour of P under the assumption χ, that is, the normal behaviour of failure prone process $P\wr\chi$.

3 Assertions, failure hypotheses, and correctness formulae

We use a correctness formula P **sat** ϕ to express that process P satisfies safety property ϕ. Informally, since we focus on the pattern of communications, such a correctness formula expresses that any sequence of communications performed by P satisfies ϕ.

Similar to the semantic denotation of traces in the previous section, we use in assertions communication record expressions such as (c, μ), with $c \in CHAN$ and $\mu \in VAL$. We have channel expressions, e.g. using the operator ch which yields the channel of a communication record, and value expressions, including the operator val which yields the value of a communication record. To reason about natural numbers, integer expressions include the length operator len. We use the empty trace, $\langle\rangle$, traces of one record, e.g. $\langle(c, \mu)\rangle$, as well as the concatenation operator $^\wedge$ and the projection operator $\uparrow$ to create trace expressions. Further, for a trace expression $texp$ and an integer expression $iexp$ we use $texp(iexp)$ to refer to record number $iexp$ of $texp$, provided $iexp$ is a positive natural number less than or equal to $len(texp)$. To refer to the communication history of a process we use a special variable h. For instance, a process which outputs value 2 along channel c satisfies the assertion $h\uparrow\{c\} = \langle\rangle \vee h\uparrow\{c\} = \langle(c, 2)\rangle$.

In assertions we furthermore use logical variables which serve as placeholders for arbitrary values. Let *IVAR*, with typical representative i, be the set of logical value variables ranging over $\mathbb{N}$, let *VVAR*, with typical representative v, denote the set of logical value variables ranging over *VAL*, and let *TVAR*, with characteristic element t, be the set of logical trace variables ranging over *TRACE*.

Example 6 (Transmission medium) The transmission medium M introduced in Example 2 satisfies the following specification:

$$M \quad \textbf{sat} \quad \forall i : 1 \leq i \leq len(h\uparrow out) \rightarrow val(h\uparrow out(i)) = val(h\uparrow in(i)).$$

△

For an assertion ϕ we define the set $chan(\phi)$ of channels such that $c \in chan(\phi)$ if a communication along c might affect the validity of ϕ. For instance, the validity of assertion $h = \langle\rangle$ is affected by any communication. On the other hand, the validity of assertion $(h\uparrow\{c\})^{\wedge}(d,7) = \langle(d,7)\rangle$ can only be changed by a communication along channel c, although d also occurs in the assertion. Rather than the channels occurring syntactically in ϕ, $chan(\phi)$ consists of the channels to which references to h in ϕ are restricted (cf. [9, 3]). Note that the value of a logical variable is not affected by any communication. In a compositional approach the specification ϕ of a process P should not impose restrictions on communications along channels other than those of P, that is, $chan(\phi) \subseteq chan(P)$.

For an assertion ϕ we also write $\phi(h)$ to indicate that ϕ has one free variable h. We use $\phi(texp)$ to denote the expression which is obtained from ϕ by replacing h by trace expression $texp$.

We use an environment γ to interpret logical variables. This environment maps a logical value variable $i \in IVAR$ to a value $\gamma(i) \in \mathbb{N}$, a logical value variable $v \in VVAR$ to a value $\gamma(v) \in VAL$, and a logical trace variable $t \in TVAR$ to a trace $\gamma(t) \in TRACE$. An assertion is interpreted with respect to a pair (θ, γ), where the value of h is obtained from θ (a formal definition is given in [8]). We write $(\theta, \gamma) \models \phi$ to denote that trace θ and environment γ satisfy assertion ϕ.

Example 7 (Satisfaction) Consider the assertion $t \preceq h \rightarrow t\uparrow\{c\} = \langle\rangle$. Since h obtains its value from θ, we have $(\theta, \gamma) \models t \preceq h \rightarrow t\uparrow\{c\} = \langle\rangle$ for any environment γ and trace θ such that if $\gamma(t) \preceq \theta$ then $\gamma(t)\uparrow\{c\} = \langle\rangle$.   △

Definition 8 (Validity of an assertion) An assertion ϕ is *valid*, notation $\models \phi$, if, and only if, for all θ and γ, $(\theta, \gamma) \models \phi$.   ◇

As mentioned in the introduction, a failure hypothesis χ of a process P is formalized as a predicate which represents a relation between the normal and acceptable histories of P. Such a predicate is expressed in a slightly extended version of the assertion language. This version contains, besides h, the special variable h_{old}. With respect to some failure hypothesis, h represents an acceptable history of process P, whereas h_{old} represents a normal history of P.

Example 9 (Corruption) Consider the transmission medium M presented in Example 2. To formalize corruption we require that

- with respect to the number of recorded *in* and *out* communications h_{old} and h are equally long,

- the order of *in* and *out* communications as recorded by h_{old} is preserved by h, and

- the ith input value as recorded by h equals the ith input value as recorded by h_{old}.

By not specifying the value part of an *out* record in h, we allow it to be any element of *VAL*. Formally,

$$\begin{aligned}
Cor \equiv \quad & len(h{\uparrow}\{in, out\}) = len(h_{old}{\uparrow}\{in, out\}) \\
& \wedge\ \forall i:\ 1 \le i \le len(h{\uparrow}\{in, out\}) \\
& \quad \to ch(h{\uparrow}\{in, out\}(i)) = ch(h_{old}{\uparrow}\{in, out\}(i)) \\
& \wedge\ \forall i:\ 1 \le i \le len(h{\uparrow}in) \to val(h{\uparrow}in(i)) = val(h_{old}{\uparrow}in(i)).
\end{aligned}$$

Observe that the predicate *Cor* holds for $h = h_{old} = \langle(in, 1), (out, 1)\rangle$. It also holds for $h = \langle(in, 1), (out, 9)\rangle$ and $h_{old} = \langle(in, 1), (out, 1)\rangle$. $\triangle$

Sentences of the extended language are called *transformation expressions*, with typical representative ψ. Let $chan(\psi)$ be the set of channels such that the validity of transformation expression ψ might be affected by a communication on c if $c \in chan(\psi)$. Since h and h_{old} play similar roles, this set consists of the channels to which references to h or h_{old} in ψ are restricted.

For a transformation expression ψ we also write $\psi(h_{old}, h)$ to indicate that ψ has two free variables h_{old} and h. We use $\psi(texp_1, texp_2)$ to denote the expression which is obtained from ψ by replacing h_{old} by $texp_1$, and h by $texp_2$. A transformation expression is interpreted with respect to a triple $(\theta_0, \theta, \gamma)$. Trace θ_0 gives h_{old} its value, and, as before, trace θ gives h its value, and environment γ interprets the logical variables of $IVAR \cup VVAR \cup TVAR$ (see [8] for details). We write $(\theta_0, \theta, \gamma) \models \psi$ to denote that traces θ_0 and θ, and environment γ satisfy transformation expression ψ.

Definition 10 (Validity of a transformation expression) A transformation expression ψ is *valid*, notation $\models \psi$, if, and only if, for all θ_0, θ, and γ, $(\theta_0, \theta, \gamma) \models \psi$. $\Diamond$

Definition 11 (Failure hypothesis) A *failure hypothesis* χ is a transformation expression which represents a reflexive relation on the normal behaviour, in order to guarantee that the normal behaviour is part of the acceptable behaviour:

- $\models \chi(h_{old}, h_{old})$.

As mentioned before, the semantics of a process contains the finite traces that can be observed up to any point in a normal execution. To maintain this property for acceptable behaviour, we require a failure hypothesis χ to preserve the prefix closedness:

- $\models (\chi(h_{old}, h) \wedge t \preceq h) \rightarrow \exists t_{old} \preceq h_{old} : \chi(t_{old}, t)$.

Furthermore, a failure hypothesis for a process FP does not impose restrictions on communications along those channels that are not in $chan(FP)$:

- $chan(\chi) \subseteq chan(FP)$.   ◇

As in [6], the construct $P \wr \chi$ indicates execution of process P under the assumption χ. This construct enables us to specify *failure prone processes*.

Example 12 (Corruption) As we have seen above, *Cor* holds for $h = h_{old} = \langle (in, 1), (out, 1) \rangle$. Hence, M's normal behaviour $\langle (in, 1), (out, 1) \rangle$ is contained in $\mathcal{H}[\![M \wr Cor]\!]$, the set representing the acceptable behaviour of M under *Cor*. Also, because *Cor* holds for $h = \langle (in, 1), (out, 9) \rangle$ and $h_{old} = \langle (in, 1), (out, 1) \rangle$, we obtain that the abnormal behaviour $\langle (in, 1), (out, 9) \rangle$ is an element of $\mathcal{H}[\![M \wr Cor]\!]$.   △

Using P to denote a process expressed in the programming language mentioned in Section 2, the syntax of our extended programming language is given by:

$$\textit{Failure Prone Process} \quad FP \ ::= \ P \ | \ FP_1 \| FP_2 \ | \ FP \backslash cset \ | \ FP \wr \chi$$

Since $chan(\chi) \subseteq chan(FP)$, we obtain $chan(FP \wr \chi) = chan(FP)$. As before, we define $chan(FP_1 \| FP_2) = chan(FP_1) \cup chan(FP_2)$, and $chan(FP \backslash cset) = chan(FP) - cset$. We require the specification ϕ of a failure prone process FP to satisfy $chan(\phi) \subseteq chan(FP)$.

Definition 13 (The traces of a failure prone process)   The trace semantics of a failure prone process is inductively defined as follows:

- $\mathcal{H}[\![FP_1 \| FP_2]\!] = \{ \theta \mid \text{ for } i = 1, 2, \theta \!\uparrow\! chan(FP_i) \in \mathcal{H}[\![FP_i]\!],$
  $$\text{and } \theta \!\uparrow\! chan(FP_1 \| FP_2) = \theta \ \},$$

- $\mathcal{H}[\![FP \backslash cset]\!] = \{ \theta \backslash cset \mid \theta \in \mathcal{H}[\![FP]\!] \}$, and

- $\mathcal{H}[\![FP \wr \chi]\!] = \{ \theta \mid \text{ there exists a } \theta_0 \in \mathcal{H}[\![FP]\!] \text{ such that,}$
  $$\text{for all } \gamma, (\theta_0, \theta, \gamma) \models \chi, \text{ and } \theta \!\uparrow\! chan(FP) = \theta \ \}.$$

This definition ensures that if $\theta \in \mathcal{H}[\![FP]\!]$ then $\theta \!\uparrow\! chan(FP) = \theta$.   ◇

The syntactic construct $FP \wr \chi$ mixes, in effect, process terms with failure hypotheses. In such a *mixed terms* formalism it is convenient to interpret an assertion, just as a process term, as a set of computations, rather than by means of truth values [9, 5]. In our case we interpret an assertion as a set of traces:

- $\llbracket \phi \rrbracket_{cset} = \{ \theta \mid \theta \setminus cset = \langle \rangle \wedge \exists \gamma : (\theta, \gamma) \models \phi \}$.

A transformation expression is interpreted as a set of pairs of traces:

- $\llbracket \psi \rrbracket_{cset} = \{ (\theta_0, \theta) \mid \theta_0 \setminus cset = \langle \rangle \wedge \theta \setminus cset = \langle \rangle \wedge \exists \gamma : (\theta_0, \theta, \gamma) \models \psi \}$.

Definition 14 (Valid correctness formula)

$$FP \;\; \mathbf{sat} \;\; \phi \text{ if, and only if, } \mathcal{H}\llbracket FP \rrbracket \subseteq \llbracket \phi \rrbracket_{chan(FP)}.$$

$\diamond$

4 Compositional refinement

Compositional refinement can be defined in terms of the relational composition $X \,\mathbf{;}\, R$, the weakest precondition $[R]X$, and the leads-to relation $X \rightsquigarrow Y$ [10]. But this definition is complex due to its generality and its formulation in terms of parameterized, i.e. higher order, processes. In this report, we restrict the definition of these operators on the mixed terms formalism of CSP-like processes and first order assertional trace specifications. We only consider those cases which are needed to develop our refinement theory.

Definition 15 (Composition operator) For the set $X \subseteq TRACE$ and the set $R \subseteq TRACE^2$ the composition $X \,\mathbf{;}\, R$ is defined as follows:

$$X \,\mathbf{;}\, R = \{ \theta \mid \exists \theta_0 : \theta_0 \in X \wedge (\theta_0, \theta) \in R \}.$$

$\diamond$

Definition 16 (Weakest precondition operator) For the set $X \subseteq TRACE$ and the set $R \subseteq TRACE^2$ the weakest precondition for X with respect to R, notation $[R]X$, is defined as follows:

$$[R]X = \{ \theta \mid \forall \theta_0 : (\theta, \theta_0) \in R \rightarrow \theta_0 \in X \}.$$

Definition 17 (Leads-to operator) For the sets $X, Y \subseteq TRACE$ the leads-to relation $X \rightsquigarrow Y$ is defined as follows:

$$X \rightsquigarrow Y = \{ (\theta_0, \theta) \mid \theta_0 \in X \rightarrow \theta \in Y \}.$$

$\diamond$

The following two lemmas relate the above defined operators.

Lemma 18 Let $X, Y \subseteq TRACE$ and $R \subseteq TRACE^2$, then

$$X \, \mathbf{;} \, R \subseteq Y \text{ if, and only if, } X \subseteq [R]Y.$$

Proof.

$$X \, \mathbf{;} \, R \subseteq Y$$
$$\Leftrightarrow \quad \forall \theta : ((\exists \theta_0 : (\theta_0 \in X \wedge (\theta_0, \theta) \in R)) \rightarrow \theta \in Y)$$
$$\Leftrightarrow \quad \forall \theta : (\forall \theta_0 : ((\theta_0 \in X \wedge (\theta_0, \theta) \in R) \rightarrow \theta \in Y))$$
$$\Leftrightarrow \quad \forall \theta : (\forall \theta_0 : (\theta_0 \in X \rightarrow ((\theta_0, \theta) \in R \rightarrow \theta \in Y)))$$
$$\Leftrightarrow \quad \forall \theta_0 : (\theta_0 \in X \rightarrow \forall \theta : ((\theta_0, \theta) \in R \rightarrow \theta \in Y))$$
$$\Leftrightarrow \quad X \subseteq [R]Y$$

$\square$

Lemma 19 Let $X, Y \subseteq TRACE$ and $R \subseteq TRACE^2$, then

$$X \, \mathbf{;} \, R \subseteq Y \text{ if, and only if, } R \subseteq X \rightsquigarrow Y.$$

Proof.

$$X \, \mathbf{;} \, R \subseteq Y$$
$$\Leftrightarrow \quad \forall \theta : (\exists \theta_0 : (\theta_0 \in X \wedge (\theta_0, \theta) \in R) \rightarrow \theta \in Y)$$
$$\Leftrightarrow \quad \forall \theta : (\forall \theta_0 : ((\theta_0 \in X \wedge (\theta_0, \theta) \in R) \rightarrow \theta \in Y))$$
$$\Leftrightarrow \quad \forall \theta : (\forall \theta_0 : ((\theta_0, \theta) \in R \rightarrow (\theta_0 \in X \rightarrow \theta \in Y)))$$
$$\Leftrightarrow \quad R \subseteq X \rightsquigarrow Y$$

$\square$

For assertions ϕ and ξ, and transformation expression ψ, the above operators can be expressed in the first order assertion language of the previous section as follows:

- $[\![\phi]\!] \, \mathbf{;} \, [\![\psi]\!]$ is expressed by $\exists t : (\phi(t) \wedge \psi(t, h))$ (abbreviated as $\phi \wr \psi$);

- $[[\psi]][\phi]$ is expressed by $\forall t : (\psi(h,t) \to \phi(t))$;

- $[\phi] \leadsto [\xi]$ is expressed by $\phi(h_{old}) \to \xi(h)$.

The inference rule for introducing failure hypotheses given in [8] is reformulated below as Theorem 20. Remember that a failure hypothesis only refers to a subset of the channels of the process.

Theorem 20 (Failure hypothesis introduction)

$$\text{If } FP \ \text{ sat } \ \phi \text{ then } FP \wr \chi \ \text{ sat } \ \phi \wr \chi.$$

Proof.

$$
\begin{aligned}
& FP \ \text{ sat } \ \phi \\
\Leftrightarrow \quad & \mathcal{H}[\![FP]\!] \subseteq [\![\phi]\!]_{chan(FP)} \\
\Rightarrow \quad & \mathcal{H}[\![FP]\!] \, \mathring{,} \, [\![\chi]\!]_{chan(FP)} \subseteq [\![\phi]\!]_{chan(FP)} \, \mathring{,} \, [\![\chi]\!]_{chan(FP)} \\
\Leftrightarrow \quad & \mathcal{H}[\![FP \wr \chi]\!] \subseteq [\![\exists t : (\phi(t) \wedge \chi(t,h))]\!]_{chan(FP)} \\
\Leftrightarrow \quad & FP \wr \chi \ \text{ sat } \ \exists t : (\phi(t) \wedge \chi(t,h))
\end{aligned}
$$

$\square$

Next we investigate how given a failure hypothesis χ and an assertion ϕ we can find a specification for FP such that $FP \wr \chi \ \text{ sat } \ \phi$. In this context, a trace h of FP is characterized by the fact that any trace t with $\chi(h,t)$ conforms to ϕ.

Theorem 21 (Failure hypothesis elimination)

$$FP \wr \chi \ \text{ sat } \ \phi \text{ if, and only if, } FP \ \text{ sat } \ \forall t : (\chi(h,t) \to \phi(t)).$$

Proof.

$$
\begin{aligned}
& FP \wr \chi \ \text{ sat } \ \phi \\
\Leftrightarrow \quad & \mathcal{H}[\![FP \wr \chi]\!] \subseteq [\![\phi]\!]_{chan(FP)} \\
\Leftrightarrow \quad & \mathcal{H}[\![FP]\!] \, \mathring{,} \, [\![\chi]\!]_{chan(FP)} \subseteq [\![\phi]\!]_{chan(FP)} \\
\Leftrightarrow \quad & \mathcal{H}[\![FP]\!] \subseteq [[\chi]\!]_{chan(FP)}][\![\phi]\!]_{chan(FP)} \\
\Leftrightarrow \quad & \mathcal{H}[\![FP]\!] \subseteq [\![\forall t : (\chi(h,t) \to \phi(t))]\!]_{chan(FP)} \\
\Leftrightarrow \quad & FP \ \text{ sat } \ \forall t : (\chi(h,t) \to \phi(t))
\end{aligned}
$$

$\square$

Suppose ξ_{FP} is a *precise* specification for process FP, i.e. $\theta \in \mathcal{H}[\![FP]\!]$ if, and only if, $\theta \in [\![\xi_{FP}]\!]_{chan(FP)}$. The following theorem identifies the weakest, that is the least restrictive, class χ of failure hypotheses is, such that $FP \wr \chi$ **sat** ϕ for a suitable specification ϕ. Remember that, like failure hypothesis χ, the specification ϕ should only refer to a subset of $chan(FP)$.

Theorem 22 (Failure hypothesis isolation)

$$FP \wr \chi \quad \textbf{sat} \quad \phi \text{ if, and only if, } \chi \to (\xi_{FP}(h_{old}) \to \phi(h)).$$

Proof.

$$
\begin{array}{ll}
& FP \wr \chi \quad \textbf{sat} \quad \phi \\
\Leftrightarrow & \mathcal{H}[\![FP \wr \chi]\!] \subseteq [\![\phi]\!]_{chan(FP)} \\
\Leftrightarrow & \mathcal{H}[\![FP]\!] \, \mathring{,} \, [\![\chi]\!]_{chan(FP)} \subseteq [\![\phi]\!]_{chan(FP)} \\
\Leftrightarrow & [\![\xi_{FP}]\!]_{chan(FP)} \, \mathring{,} \, [\![\chi]\!]_{chan(FP)} \subseteq [\![\phi]\!]_{chan(FP)} \\
\Leftrightarrow & [\![\chi]\!]_{chan(FP)} \subseteq [\![\xi_{FP}]\!]_{chan(FP)} \rightsquigarrow [\![\phi]\!]_{chan(FP)} \\
\Leftrightarrow & \chi \to (\xi_{FP}(h_{old}) \to \phi(h))
\end{array}
$$

$\square$

5   Examples

In this section we illustrate the use of the failure hypothesis elimination and isolation theorems by investigating a transmission medium that might corrupt messages and one that might be transiently stuck at zero.

5.1   A transmission medium that might corrupt messages

Consider the specification of the transmission medium M given in Example 6 and the failure hypothesis Cor discussed in Example 9. Using failure hypothesis introduction theorem 20 we obtain:

$$
\begin{aligned}
M \wr Cor \quad \textbf{sat} \quad \exists t : \quad & \forall i : 1 \leq i \leq len(t \upharpoonright out) \to val(t \upharpoonright out(i)) = val(t \upharpoonright in(i)) \\
& \wedge \, len(h \upharpoonright \{in, out\}) = len(t \upharpoonright \{in, out\}) \\
& \wedge \, \forall i : 1 \leq i \leq len(h \upharpoonright \{in, out\}) \\
& \qquad \to ch(h \upharpoonright \{in, out\}(i)) = ch(t \upharpoonright \{in, out\}(i)) \\
& \wedge \, \forall i : 1 \leq i \leq len(h \upharpoonright in) \to val(h \upharpoonright in(i)) = val(t \upharpoonright in(i)).
\end{aligned}
$$

Because there is no relationship any more between the values input and those output, the strongest property of $M \wr Cor$ is:

$$M \wr Cor \quad \textbf{sat} \quad len(h{\uparrow}out) \leq len(h{\uparrow}in) \leq len(h{\uparrow}out) + 1,$$

which no longer specifies a transmission medium. By failure hypothesis elimination theorem 21 we know that

$$CM \wr Cor \quad \textbf{sat} \quad \forall i : 1 \leq i \leq len(h{\uparrow}out) \rightarrow val(h{\uparrow}out(i)) = val(h{\uparrow}in(i))$$

if, and only if,

$$
\begin{aligned}
CM \quad \textbf{sat} \quad \forall t : (\ & len(t{\uparrow}\{in, out\}) = len(h{\uparrow}\{in, out\}) \\
& \wedge \forall i : 1 \leq i \leq len(t{\uparrow}\{in, out\}) \\
& \quad \rightarrow ch(t{\uparrow}\{in, out\}(i)) = ch(h{\uparrow}\{in, out\}(i)) \\
& \wedge \forall i : 1 \leq i \leq len(t{\uparrow}in) \rightarrow val(t{\uparrow}in(i)) = val(h{\uparrow}in(i))) \\
& \rightarrow \forall i : 1 \leq i \leq len(t{\uparrow}out) \rightarrow val(t{\uparrow}out(i)) = val(t{\uparrow}in(i)).
\end{aligned}
$$

However, this implication is equivalent to **false** because the premise may hold even if not $\forall i : 1 \leq i \leq len(t{\uparrow}out) \rightarrow val(t{\uparrow}out(i)) = val(t{\uparrow}in(i))$, and therefore such a CM cannot be implemented. A possible way to deal with corruption is to use coding. An encoding function transforms a dataword into a codeword which contains some redundant bits. Thus the set of datawords is mapped into only a small fraction of a much larger set of codewords. The codewords some dataword is mapped into are called valid, and the encoding ensures that it is very unlikely that due to corruption one valid codeword is changed into another.

Using the function *Valid* with the obvious interpretation we formalize the detectable corruption hypothesis as follows:

$$
\begin{aligned}
DetCor \equiv \ & len(h{\uparrow}\{in, out\}) = len(h_{old}{\uparrow}\{in, out\}) \\
& \wedge \ \forall i : 1 \leq i \leq len(h{\uparrow}\{in, out\}) \\
& \quad \rightarrow ch(h{\uparrow}\{in, out\}(i)) = ch(h_{old}{\uparrow}\{in, out\}(i)) \\
& \wedge \ \forall i : 1 \leq i \leq len(h{\uparrow}in) \rightarrow val(h{\uparrow}in(i)) = val(h_{old}{\uparrow}in(i)) \\
& \wedge \ \forall i : 1 \leq i \leq len(h{\uparrow}out) \\
& \quad \rightarrow \quad val(h{\uparrow}out(i)) = val(h_{old}{\uparrow}out(i)) \\
& \qquad\quad \vee \neg Valid(val(h{\uparrow}out(i))).
\end{aligned}
$$

Now, we seek CM such that

$$CM \wr DetCor \quad \textbf{sat} \quad \forall i : \; 1 \le i \le len(h{\uparrow}out)$$
$$\rightarrow (\; Valid(val(h{\uparrow}out(i)))$$
$$\rightarrow val(h{\uparrow}out(i)) = val(h{\uparrow}in(i)) \;).$$

Using failure hypothesis elimination theorem 21 once more we obtain:

$$CM \quad \textbf{sat} \quad \forall t : (\quad len(t{\uparrow}\{in, out\}) = len(h{\uparrow}\{in, out\})$$
$$\wedge \, \forall i : 1 \le i \le len(t{\uparrow}\{in, out\})$$
$$\rightarrow ch(t{\uparrow}\{in, out\}(i)) = ch(h{\uparrow}\{in, out\}(i))$$
$$\wedge \, \forall i : 1 \le i \le len(t{\uparrow}in) \rightarrow val(t{\uparrow}in(i)) = val(h{\uparrow}in(i))$$
$$\wedge \, \forall i : 1 \le i \le len(t{\uparrow}out)$$
$$\rightarrow \quad val(t{\uparrow}out(i)) = val(h{\uparrow}out(i))$$
$$\vee \, \neg Valid(val(t{\uparrow}out(i))) \;)$$
$$\rightarrow (\, \forall i : 1 \le i \le len(t{\uparrow}out)$$
$$\rightarrow (\; Valid(val(t{\uparrow}out(i)))$$
$$\rightarrow val(t{\uparrow}out(i)) = val(t{\uparrow}in(i)) \;) \;).$$

5.2   A transmission medium that might be transiently stuck at zero

Suppose the medium M presented in Example 2 might be transiently stuck at zero. What is a suitable failure hypothesis *StuckAtZero* such that

$$M \wr StuckAtZero \quad \textbf{sat} \quad \forall i : 1 \le i \le len(h{\uparrow}out)$$
$$\rightarrow \quad val(h{\uparrow}out(i)) = val(h{\uparrow}in(i))$$
$$\vee \, val(h{\uparrow}out(i)) = 0 \; ?$$

Using failure hypothesis isolation theorem 22 we can classify *StuckAtZero* as follows:

$$StuckAtZero \rightarrow (\quad \forall i : 1 \le i \le len(h_{old}{\uparrow}out)$$
$$\rightarrow val(h_{old}{\uparrow}out(i)) = val(h_{old}{\uparrow}in(i))$$
$$\wedge \, len(h_{old}{\uparrow}out) \le len(h_{old}{\uparrow}in) \le len(h_{old}{\uparrow}out) + 1$$
$$\rightarrow \forall i : 1 \le i \le len(h{\uparrow}out)$$
$$\rightarrow \quad val(h{\uparrow}out(i)) = val(h{\uparrow}in(i))$$
$$\vee \, val(h{\uparrow}out(i)) = 0 \;).$$

A natural candidate is the following:

$$StuckAtZero \equiv \quad len(h{\uparrow}\{in, out\}) = len(h_{old}{\uparrow}\{in, out\})$$
$$\wedge \, \forall i : 1 \le i \le len(h{\uparrow}\{in, out\})$$
$$\rightarrow ch(h{\uparrow}\{in, out\}(i)) = ch(h_{old}{\uparrow}\{in, out\}(i))$$
$$\wedge \, \forall i : 1 \le i \le len(h{\uparrow}in) \rightarrow val(h{\uparrow}in(i)) = val(h_{old}{\uparrow}in(i))$$
$$\wedge \, \forall i : 1 \le i \le len(h{\uparrow}out)$$
$$\rightarrow \quad val(h{\uparrow}out(i)) = val(h_{old}{\uparrow}out(i))$$
$$\vee \, val(h{\uparrow}out(i)) = 0.$$

6   Conclusions and future research

We have presented a trace-based compositional framework for the refinement of fault tolerant distributed systems. In this formalism the failure hypothesis of a process is formalized as a relation between the normal and acceptable observable input and output behaviour of that process. Two interesting applications, namely the classification of the processes that, given a particular failure hypothesis, satisfy a given specification, and the determination of the least restrictive failure hypothesis such that a given process still satisfies a given specification, have been illustrated.

Finding a logic to express failure hypotheses more elegantly, e.g. using the classification of failures that appears in [1], is a subject of future investigation. Also, further research should indicate how our framework can be adapted to allow a failure hypothesis to be formalized as a transformation on sequences of states rather than communication sequences.

Another interesting extension of the research described in this report is the introduction of time to the refinement theory. This enables one to reason about the properties of fault tolerant real-time systems. Clues for such an extension can be found in [7] where time is introduced to the proof theory of [8]. Then, the characterization that safety properties express that "nothing bad will happen" and liveness properties express that "eventually something good will happen" is no longer appropriate, as indeed mentioned in [4]. Consider, for instance, a transmission medium that accepts messages via a channel *in* and relays them to a channel *out*. The real-time property "after a message is input to the medium via *in* it is output via *out* within 5 seconds" is a safety property, because it can be falsified 5 seconds after an *in* communication. Note, however, that it expresses that something must happen. Hence, by adding time, the class of safety properties also includes real-time properties.

Acknowledgements

We would like to thank Job Zwiers for his helpful comments during the initial stage of this research. The referees provided useful remarks.

References

[1] F. Cristian. Understanding Fault Tolerant Distributed Systems. *Communications of the ACM* **34**(2) (1991) 56 – 78.

[2] C.A.R. Hoare. Communicating Sequential Processes. *Communications of the ACM* **21**(8) (1978) 666–677.

[3] J. Hooman. Specification and Compositional Verification of Real-Time Systems. *Lecture Notes in Computer Science* **558** (Springer-Verlag, 1992).

[4] L. Lamport. What Good is Temporal Logic. In R.E. Manson (ed.). *Information Processing* (North-Holland, 1983) 657–668.

[5] E.R. Olderog. Nets, Terms, and Formulas. *Cambridge Tracts in Computer Science* **23** (Cambridge University Press, 1991).

[6] H. Schepers. Tracing Fault Tolerance. In *Proc. 3rd IFIP Int. Working Conference on Dependable Computing for Critical Applications*, Dependable Computing and Fault Tolerant Systems **8** (Springer-Verlag, 1993) 91–110.

[7] H. Schepers and R. Gerth. A Compositional Proof Theory for Fault Tolerant Real-Time Distributed Systems. In *Proc. 12th Symp. on Reliable Distributed Systems* (IEEE Computer Society Press, 1993) 34–43.

[8] H. Schepers and J. Hooman. A Trace-Based Compositional Proof Theory for Fault Tolerant Distributed Systems. *Eindhoven University of Technology*, 1993. To appear in *Theoretical Computer Science*; an extended abstract appeared in *Proc. Parallel Architectures and Languages Europe (PARLE) '93*, Lecture Notes in Computer Science **694** (Springer-Verlag, 1993) 197–208.

[9] J. Zwiers. Compositionality, Concurrency and Partial Correctness. *Lecture Notes in Computer Science* **321** (Springer-Verlag, 1989).

[10] J. Zwiers, J. Coenen, and W.-P. de Roever. A Note on Compositional Refinement. In *Proc. 5th BCS–FACS Refinement Workshop*, 5th Refinement Workshop (Springer-Verlag, 1992) 342–366.

Design Techniques
for Robustness

A Modular Robust Binary Tree

Nasser A. Kanawati, Ghani A. Kanawati, Jacob. A. Abraham
Computer Engineering Research Center
The University of Texas at Austin
Austin, Texas 78758
U.S.A.
kanawati@cerc.utexas.edu

Abstract

This paper presents a new robust binary tree designed using the theory of robust data structures. A basic module composed of three elements is replicated as necessary to form the robust tree, allowing a tree to be built up in a modular fashion while preserving its robust characteristics. The proposed structure is shown to be able to detect 2 errors (*2-detectable*) or correct 1 error (*1-correctable*) in any module of the tree under a generic fault model. The advantages of the tree when compared with other proposed trees presented in the literature are presented. Results of fault/error injection experiments on an implementation of the proposed structure in C++, including error coverage and performance overhead, are also provided.

Index terms: Robust data structures, modular design, binary trees, error detection and correction, locality

This research was sponsored in part by the office of Naval research under contract N00014-89-K-0098

1 Introduction

Complex computational and data base systems are being increasingly used in applications where there is a high cost of failure, and such systems have to meet requirements of high reliability in addition to high performance. These systems have to be designed with the capability to detect errors (due to faults in the system) concurrently with normal operation in order to preserve the integrity of the results, particularly since it has been observed that the majority of the faults occurring in a system are transient in nature. In order to reduce the overhead of achieving reliable operation, recent techniques have used information about the computations being performed [1], [2], and have been implemented at the system architecture level.

Since computer systems utilize data structures to represent and manage their data, a mechanism to concurrently detect and correct errors in these systems is to use robust data structures which are designed with the inherent redundancy for this purpose. In their fundamental paper, Taylor, Morgan, and Black [1] have presented the foundation of robust data structures which employ redundancy in order to detect and correct errors in their structural components. Since then, several techniques and improvements have been presented in the literature for achieving robust data structures. Black and Taylor [3] introduced the concept of *locality* for detection and correction of a number of errors, provided these errors are sufficiently separated from each other. Li, Chen and Fuchs [4] confined this locality to a checking window. Similarly, Kant and Ravichandran [5] presented the *neighborhood* of an element node as those elements that need to be modified to accommodate a new element or delete an existing element. Procedures for the construction of of error correction routines tends to be a very difficult problem, and Taylor [1] has presented guidelines for these procedures.

Concurrent detection and correction algorithms, however, tend to incur a performance penalty. Proper selection of redundant structural information (referred to in the paper as attributes) would yield effective robustness at low performance overhead. For example, proper selection of redundant pointers will yield a locally detectable and correctable data structure [5].

In this paper we present a new robust binary tree architecture, called the Modular Robust Binary (**MRB**) tree. The proposed tree is 2-detectable or 1-correctable under the **SCL** error model as well as under the generic fault model (**GF**)[1] applied to each module of the tree, which is the locality in this case. The robust tree is able to detect faults that create errors in the attributes of the tree structure. The faults leading to these errors are those occurring in the logic blocks performing the address calculation of pointer and integer values of the robust tree. The capability of the proposed tree to tolerate errors in its structural components are evaluated analytically.

Our main contributions are the relatively low performance overhead of the MRB tree, 162% for traversing the tree [2], and our utilization of fault and error injection to study the effectiveness of its error detection capabilities. The capability of the proposed tree (and other trees in the literature) to tolerate errors in its structural components is evaluated analytically. Our experiments have also shown that a variety of faults in the processor and memory led to detectable errors in the tree data structure. Section 2 presents the design of the proposed robust tree. The implementation of the robust tree is illustrated in Section 3. In Section 4 we present the results of our experiments.

---

[1] The SCL error model and the GF fault model are summarized in Section 1.1.

[2] To the knowledge of the authors, no performance overhead was reported for other robust binary trees in the literature.

1.1   Robust Binary Trees

The binary tree is a widely used data structure in software systems and applications. Robust binary trees presented in the literature are based on the threaded binary tree. The literature includes the chained and threaded binary tree (CT-tree) [1], the CTB and the mod(2) CTB trees, [1], [6], the *robust binary tree* [7], and the *three pointer tree* [8].

The above trees utilize global detection and correction techniques for achieving overall system reliability. This process, however, degrades overall system performance when used concurrently with normal tree operations since the whole tree has to be traversed for the purpose of error detection. The other trees presented in the literature, [4], [5], [10] as well as the proposed robust binary tree, consider the benefits of local detection and correction of errors. In these trees, during normal tree operations (search, insert, delete, traverse), only a small number of nodes that lie within a defined locality are involved in the detection and correction process. The AVL tree [10] is an example where the complexity of insert and delete operations remains logarithmic. However, erroneous correction results when two or more errors occur in the same node. Li *et al* [4] have presented a locally detectable and correctable tree structure where detectability and correctability within a defined checking window are a function of the window size. Performing error detection and correction does not degrade the performance of the tree in [4] since it does not involve nodes that lie off the traversed path.

Most of the robust behavior of trees can be characterized by utilizing the so called Single Component Global (**SCG**) or Single Component Local (**SCL**) error models presented by Taylor [12]. The single component error model assumes that each error affects one component in a node element. This error model may be applied to the whole structure or to a small locality of the structure. Another error model which has been proposed is the generic fault model **GF** [5], which describes, under faults, the behavior of structures that are tolerant to errors in their components in every defined locality. In this model, a fault may tend to damage a particular component type and may produce damage to the same component type in multiple nodes.

2   The Modular Robust Binary (MRB) Tree

A simple binary tree has three basic attributes. These are: 1) a pointer to a left child node, 2) a pointer to a right child node, and 3) a key which is used for ordered insert. Our objective was to construct a robust binary tree that is 2-detectable or 1-correctable against any of the above attributes becoming erroneous. In order to accomplish this objective, the design of the MRB tree was based on a modular approach. Basic units, or modules, of three nodes (elements) are interconnected in a systematic way to form the desired binary tree. Relationships between the

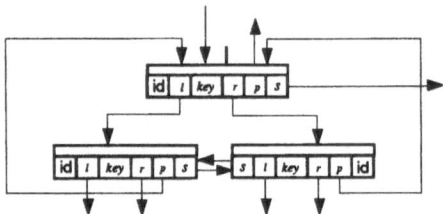

Figure 1: Basic module of the MRB tree.

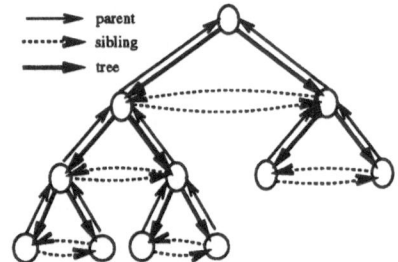

Figure 2: **MRB**, The modular robust binary tree.

three nodes are developed to establish the dependability of the tree at every module. [3] The MRB modular approach provides the capability of concurrently manipulating different units of the tree. Figure 1 shows the data structure for the basic unit of our design. In this figure, l is left child pointer, r is right child pointer, p is parent pointer, s is sibling pointer, and id is the identifier for every tree node. The proposed MRB tree is shown in Figure 2.

The basic module is composed of three nodes, a parent node, a left child node, and a right child node with two disjoint paths between each pair of the nodes. Taylor [13] and Kant [5] have shown that a circular doubly linked list with a forward pointer, a backward pointer, an identifier, and a counter for the number of elements in the list is at least 2-detectable or 1-correctable. In the MRB tree the forward and backward pointers are replaced by the tree (child), parent and sibling pointers. The count is implicit in such an arrangement and is always equal to three. Thus, the structural components considered in the design of the MRB tree are: the tree pointers (left child and right child), a parent pointer, a sibling pointer, an identifier, a key, and an implicit count equal to three. The identifier is used to indicate the type of a node. The MRB tree has three types of nodes. These are: a branch node, a leaf node, and a ghost node. (The definition of a ghost node will be given below.) Nodes for every data structure are distinguished from nodes

---

[3]It was shown in [1] that a storage structure which is 2-detectable and 1-correctable is sufficient for most typical applications.

of other data structures by tagging the identifier of every node for that particular data structure. Hence, if an application utilizes two binary trees, each tree will have a selected tag value in its identifier field. The selection of the tag value was experimentally found to have an impact on the measured fault coverage as will be presented later. The *key* is part of the semantics in every node and is employed for ordered insert. The MRB tree is designed to accept only those elements whose key values are greater than zero. The neighborhood as well as the window [4] for the MRB tree is the basic module which has a fixed size of three nodes.

In Figure 2, a node has either no children or two children. During insert procedures, a *ghost* node is constructed for every newly added element to maintain a consistent doubly linked three-node structure. A node, as mentioned before, is flagged by an identifier field to be a ghost, a branch or a leaf node. Leaf nodes as well as ghost nodes have their tree pointers set to NULL in order to differentiate them from branch nodes. A ghost node is further differentiated from a leaf node by having its *key* set to zero. The ghost node is converted later to a leaf node if its position was selected by the *search* routine in the process of insertion. Delete routines are modified accordingly to accommodate the ghost nodes. Hence, if a node is to be deleted and has a ghost sibling, both nodes are deleted.

2.1 Attributes of the MRB Tree

We adopted methods developed by Kant [5] in formulating the logical rules that describe the relationships among the original and redundant attributes of the binary tree data structure. The following rules are designed to logically relate three attributes among the set of selected attributes utilized in the construction of the robust tree. These attributes are: a left-child pointer, a right-child pointer, a parent pointer, a sibling pointer, an identifier, and a *key*. When formulating any of these rules, one rule or more should evaluate to *FALSE* if any of the involved attributes are faulty. In special cases when two or more of these attributes are faulty, the rules may still be satisfied. This situation is observed for *compensating faults* which occur when two or more attributes are faulty simultaneously, yet the rule that relates these attributes is satisfied. The design of the MRB tree has included provisions for detecting two compensating faults.

2.1.1 Logical Rules of the MRB Tree Attributes

The logical rules that relate the different attributes of the MRB tree are as follows:

a) $right\_child(node) = sibling(left\_child(node))$.

b) $left\_child(node) = sibling(right\_child(node))$.

c) $parent(sibling(left\_child(node))) = node$.

d) $parent(sibling(right\_child(node))) = node.$

e) $left\_child(node) = sibling(sibling(left\_child(node))).$

f) $right\_child(node) = sibling(sibling(right\_child(node))).$

g) $ident(left\_child(node)) \neq GHOST \implies key(left\_child(node)) \leq key(node).$

h) $ident(right\_child(node)) \neq GHOST \implies key(right\_child(node)) > key(node).$

i) $parent(left\_child(node)) = node \implies ident(node) = BRANCH.$

j) $parent(right\_child(node)) = node \implies key(node) \neq NULL.$

In the set of rules above, *node* is the parent node in every module. Rule (h), states that for any node in the binary tree, if the identifier field of its right child indicates that the right child is not a ghost node, then the *key* value of the right child is greater than the *key* value of the node itself. The *key* attribute is protected by adding two extra copies for each *key* attribute. Hence, in the set of rules above, when the attribute *key* appears in a rule, a macro that employs a software triple modular redundancy scheme is invoked. If one of the copies disagrees with the other two copies, its value is masked. On the other hand, if all copies disagree, then the macro returns with a FALSE value. Rule (h) evaluates to *TRUE* only when a node *key* value is less than the *key* value of its right child and that right child is not a ghost node. This result is consistent with the ordered insert procedure and the property of non-ghost nodes. Another example is rule (i), which relates three attributes in a module. These are: left-child pointer of the parent node, parent pointer of the left child node, and the identifier of the parent node. This rule evaluates to *TRUE* since it states that the identifier type of a parent node may only be that of a branch type.[4]

An error which swaps the left child pointer and the right child pointer is detected by rules (g) and (h). If neither pointer points to a ghost node, then rules (g) and (h) are both violated and all other rules evaluate to *TRUE*. Note that swapping child pointers leaves the right child *key* value less or equal to the *key* value of its parent, whereas the left child *key* value becomes greater than its parent *key* value. On the other hand, if the left child node was a ghost node and an error that swaps the child pointers occurred, then the right child node becomes the ghost node. In this case rule (h) evaluates to *TRUE* since it concerns itself with what should hold if the right child node is not a ghost node. Rule (g), however, evaluates to *FALSE* since the parent node is not pointing to a ghost node any more (indicated by the tag in the tree pointer of the parent node) and the *key* value of the left child node is greater than the *key* value of the parent node. Similar arguments hold when the right child node was a ghost node and swapping occurred. In this situation, rule (g) evaluates to *TRUE* and rule (h) evaluates to *FALSE*.

---

[4]Note that a parent always branches to two child nodes.

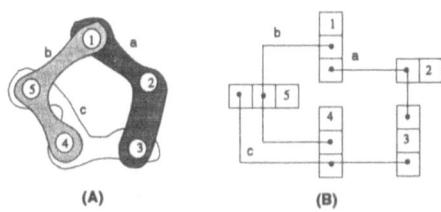

Figure 3: An example of a rules hypergraph

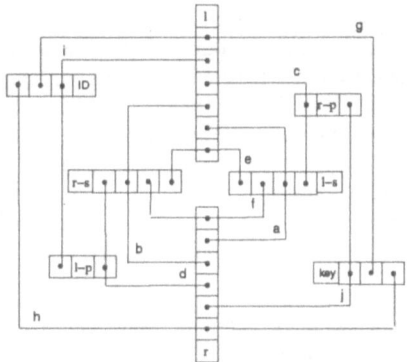

Figure 4: Rules hypergraph model for the proposed MRB tree

2.1.2 Rules Hypergraph of the MRB Tree Attributes

A *rules hypergraph* is consequently obtained which relates the set of attributes in a module in order to show the correctability of the MRB tree. A rules hypergraph is a hypergraph $G(V,E)$ where V is a set of attributes represented by the vertices in the hypergraph, and E is the set of rules represented by the hyperedges connecting these attributes. In Figure 3, an example of a hypergraph is presented along with its translation to a modified hypergraph notation used later for presenting the rules hypergraph of the MRB tree. In Figure 3A, hyperedge (a) connects items (1,2,3), (b) connects (1,4,5) and (c) connects (3,4,5). In the rules hypergraph, items (1) through (5) represent the attributes of the structure whereas hyperedges (a) through (c) represent the logical rules among these attributes. As the number of attributes, logical rules, and number of attributes per rule increase, the complexity of the hypergraph increases. We have adopted a new notation for representing the hypergraph in the Figure 3(A), which is shown in Figure 3(B). In Figure 3(B), a continuous path is drawn to connect items (1), (2), and (3) representing hyperedge (a) in Figure 3(A).

Figure 4 presents the rules hypergraph of the MRB tree using this new hypergraph notation. In this figure, eight attributes, the six pointers shown in Figure 1 (l = left-child pointer, r = right-child pointer, l-p = left-child parent pointer, r-p = right-child parent pointer, l-s = left-child sibling pointer, r-s = right-child sibling pointer), an identifier (ID) and a key item are represented as eight nodes. Hyperedges (*a - j*) are the logical rules developed for the MRB tree. Every three attributes (vertices) in the hypergraph are represented by a hyperedge and are shown to form a continuous path in the figure. As an example, a path is drawn among three attributes, left-child pointer (l), left-child sibling pointer(l-s), and right-child sibling pointer(r-s) representing rule (e).

In order to represent faulty attributes in the rules hypergraph, the vertices representing these attributes and all the hyperedges that are incident on these vertices are removed. An exception is when faulty attributes compensate each other. In this case, the vertices representing the compensating faulty attributes and the hyperedges connecting these vertices are maintained, whereas all other hyperedges that are incident on only one of these vertices are removed. The rules hypergraph for the proposed robust tree shows that any single faulty attribute leaves the hypergraph distinct and non-empty. In addition, the resulting hypergraph, due to a single faulty attribute, is distinct from any hypergraph with two faulty nodes, including compensating faulty nodes. This result is a necessary condition for correcting a single faulty attribute. In addition, any two faulty attributes,[5] leave the hypergraph non-empty, which is a necessary condition for detection [5].

In Figure 5, the sibling pointer of the left child is faulty. Thus vertex (l-s) and all hyperedges passing through (l-s) are removed. Rules (*b,d,g,h,i,j*) are not violated and hence will be used to correct the faulty sibling pointer of the left child.

Figures 6 and 7 show two cases for two faulty pointers. In Figure 6 the right child pointer and the sibling pointer of the left child are both faulty. In this case, the vertices (attributes) and hyperedges (rules) (*a,b,c,d,e,f,h,j*) connecting these nodes are disconnected. Figure 7, on the other hand, depicts the case where these pointers are faulty and are compensating. This situation is also illustrated in Figure 8. As shown in Figure 7, hyperedges *a*, and *f* connecting these faulty attributes remain intact, whereas other edges connecting each of the two pointers to other attributes are removed.

2.1.3   Analysis of Dependability Properties of MRB

In the following, we will evaluate the detectability and correctability properties of the proposed tree.

Definition(1): A structure is K-connected if there are K disjoint sets of paths between any pair of its node elements [13]. Two disjoint paths exist for the MRB tree. These can be traversed

---

[5]including masking faulty attributes

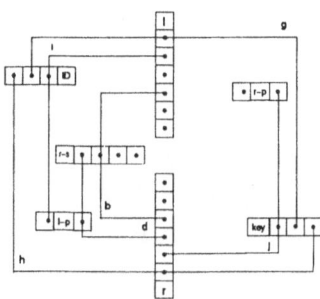

Figure 5: Rules hypergraph when left child sibling pointer is faulty.

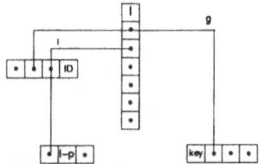

Figure 6: Rules hypergraph when left child sibling pointer and right child pointer are faulty.

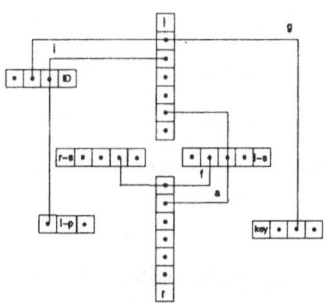

Figure 7: Rules hypergraph when left child sibling pointer and the right child pointer are faulty and are compensating each other.

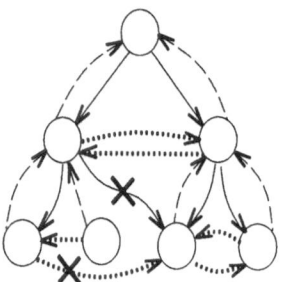

Figure 8: Right child pointer and the sibling of the left child pointer are faulty and are compensating.

either clockwise or counterclockwise starting from the parent node.

We adopt the definition of Li, Chen and Fuchs [4] to determine the detectability of each module in the tree. This is similar to the concept of changes defined by Taylor, Morgan and Black [13] in determining the distance between two data structures in a global context. If N changes are necessary to go from one correct structure to another, then the detectability is N - 1.

Definition (2): For an operation in the binary tree with the module defined as its size of locality, the local distance in this module (*dist*) is the minimum number of erroneous attributes in the module that would result in the *Detect* (defined in the next section) routine returning "no error detected" when applying the rules relating these attributes. Note that three or more changes are required to move from one correct module into another.

Definition (3): The error detectability of a module in an MRB tree for a given operation is $D = \max(dist) - 1$; hence for the MRB tree, $dist = 2$, (Definition (2)).

Definitions (2) and (3) are similar to the ones presented by Li, Chen and Fuchs [4].

Theorem (1): The MRB tree is 2-detectable or 1-correctable.

Proof: The general correction theorem presented by Taylor et al [13] states that if a storage structure has $r + 1$ disjoint paths, an identifier field, and is *2r-detectable* then that structure is r correctable.

These results lead to the following properties of the MRB tree:

1. The tree is composed of systematically connected modules.

2. The tree is locally 2-detectable since three or more changes are required to move from one correct module into another (*Definition (3)*). The MRB tree, however, does not attempt

to correct the attribute *key* using these established logical relations since its value cannot be inferred from these relations, rather it votes among the three redundant copies of each key attribute.

3. Since there are 2-disjoint paths in the module (Definition (1)), an identifier, and the tree was pre-established to be 2-detectable, the tree is then 1-correctable in that module, hence it is locally 1-correctable in that module or window (Theorem (1)). In addition, the constructed rules hypergraph shows that all single errors violate at least one rule which leaves the hypergraph model distinct.

4. Single errors can be corrected in constant time in two directions. During a forward move, tree pointers are used to traverse down the tree, whereas, during a backward move, a parent pointer is used to traverse towards the root of the tree.

5. It has the feature of backward traversal without using a stack.

In Table 1, we tabulate characteristics of existing robust binary trees. These include the CTB [14], the AVL tree [10], and the proposed MRB tree. The MRB tree allows insertion of leaf nodes only. In the MRB tree, *insert* and *delete* operations are similar to a normal binary tree implementation with special provisions to construct and delete sibling *ghost* nodes if necessary. A disadvantage of this scheme is that if the *insert* routine is presented with a file of sorted keys, a ghost node would be constructed for each inserted node in the tree.

Table 1: Characteristics of selected robust trees

Tree Features	Trees		
	CTB [14]	AVL [10]	MRB
detectability	2	1	2
correctability	1	1	1
locality	no	yes	yes
error model	SCG	SCL	SCL+GF
No. of fields per node	2m+3	2+3	2+5
types of pointers	3	2	3
additional storage	one header	one header	two headers
types of nodes	3	1	3
backward traversal	no	no	yes

3 Implementation

In order to experimentally evaluate the properties and overhead of the proposed tree, an implementation of the robust data structure was coded in C++ [15]. All nodes of the tree are objects of the same class *NODE*. A *NODE* class object contains an identifier, two tree pointers, a parent pointer, a sibling pointer, three copies for the key, and a pointer to a structure encapsulating the semantic information for every object.

The MRB tree has a header which is the root of the tree. If this header is faulty, the whole structure will be lost. A software triple modular redundancy fault tolerance scheme was employed to mask an error in the header or any of its two backups.

A detection routine (*Detect*) which applies the ten logical rules was implemented. *Detect* is invoked whenever a node in the tree is accessed by any routine (method) of the binary tree data structure. The MRB data structure routines considered were:

- *Search* and *Traverse*. *Detect* was placed inside the WHILE loops traversing the tree. Search is called to find an object location prior to insertion or deletion. These assertions are executed at every level in the structure. The overhead in performance for search and traverse routines varies from $O(log\ n)$ for a balanced tree to $O(n)$ for the sparsest tree (note that for the sparsest tree, $n/2$ of the total nodes are ghost nodes).

- *Insert*. *Detect* was applied before inserting a new node or before changing a ghost node to a leaf node.

- *Delete*. Two cases are possible. If a node is a leaf node (has no children), then *Detect* is applied before and after deletion. On the other hand, if a node is a branch node, then *Detect* is applied before and after a node replacement. [6]

If any of the logical rules evaluates to *FALSE*, a correction routine is invoked to locate the error as well as to apply correction procedures. Table 2 shows the syndrome table constructed in order to isolate the distinct errors. A syndrome table is like a hardware maintenance dictionary. Each row corresponds to an attribute or attributes in error. Each column represents the rule that connects the different attributes at different rows. If an attribute x is faulty, all logical rules that relate x to other attributes evaluates to *FALSE*, indicated by a value of 1 in the table. The result of applying the logical rules are compressed into a code word *signal*, shown in Table 2. These signals are passed to the correction routines to correct the faulty attribute. Since our design considered only 2-detectability or 1-correctability and only eight attributes were involved, construction of the syndrome table was not costly.

In Table 2, signal (0x157) indicates that the left-child pointer is in error. Upon receiving this signal, the correction routine sets the left-child pointer value equal to the sibling pointer of the

---

[6]In this case the deleted node is replaced by a leaf node.

right-child node. The correction routine handles signals (0x035) through (0x108) similarly. Signal (0x1c0) (indicating that attribute *ident* is faulty), is treated differently. Since the constructed rules do not distinguish among the identifiers of the parent and its two children [7], there is not a separate *ident* for the parent. The correction routine applies the following algorithm to correct the faulty identifier:

Algorithm *Correction of identifier fields*
begin

> // Applied rules have already indicated that the node has two children
> *node → identifier* = BRANCH
> // Correction of left-child-node identifier
> **if** (*left-child-node → key* = 0) **and** (*left-child-pointer* = *ghost-pointer*)
> //Note that least significant bits of every pointer are used to indicate the type of the node
> > (normal or ghost) the pointer is pointing at
>
> > > *left-child-node → identifier* = GHOST
>
> **else if** (*left-child-node → key* ≠ 0)
> //then test whether the left child node is a branch node or a leaf node
>
> > > **if** (*left-child-node → children-pointers* = 0)
> > > > **then** *left-child-node → identifier* = LEAF
> > > > **else** *left-child-node → identifier* = BRANCH
>
> // end Correction of left-child-node identifier
> // Correction of right-child-node identifier
> //same as Procedure applied to the left-child-node identifier
> **Call** *Detect*
>
> > **if** *Passed*
> >
> > > **print** "an error was corrected"
> >
> > **else**
> >
> > > **print** "more than one error was found"
> > > **exit**

end

Finally signals (0x2c0) through (0x300) are not passed to the correction routine since the MRB tree is designed to correct only one error. As mentioned earlier, when the attribute *key* is faulty, indicated by signal (0x2c0), no correction is applied.

---

[7]In the constructed rules, *ident* in rule *g* refers to the identifier of the left-child node, in rule *h* it refers to the identifier of the right-child node, and in rule *i* it refers to the identifier of parent node.

Attributes in error	Rules										signal
	j	i	h	g	f	e	d	c	b	a	
l	0	1	0	1	0	1	0	1	1	1	x157
l-s	0	0	0	0	1	1	0	1	0	1	x035
r-p	1	0	0	0	0	0	0	1	0	0	x204
r	1	0	1	0	1	0	1	0	1	1	x2ab
r-s	0	0	0	0	1	1	1	0	1	0	x03a
l-p	0	1	0	0	0	0	1	0	0	0	x108
ident	0	1	1	1	0	0	0	0	0	0	x1c0
key	1	0	1	1	0	0	0	0	0	0	x2c0
...	.	.	.	.	.	.	.	.	.	.	...
l&l-s	0	1	0	1	1	1	0	1	1	1	x177
l&l-s/C	0	1	0	1	1	0	0	0	1	0	x162
l&r-p	1	1	0	1	0	1	0	1	1	1	x357
l&r-p/C	1	1	0	1	0	1	0	0	1	1	x353
...	.	.	.	.	.	.	.	.	.	.	...
l-s&r-p	1	0	0	0	1	1	0	1	0	1	x235
l-s&r-p/C	1	0	0	0	1	1	0	0	0	1	x231
...	.	.	.	.	.	.	.	.	.	.	...
r-p&r	1	0	1	0	1	0	1	1	1	1	x2af
r-p&r/C	0	0	1	0	1	0	1	1	1	1	x0af
...	.	.	.	.	.	.	.	.	.	.	...
r&r-s	1	0	1	0	1	1	1	0	1	1	x2bb
r&r-s/C	1	0	1	0	0	1	0	0	0	1	x291
...	.	.	.	.	.	.	.	.	.	.	...
r-s&l-p	0	1	0	0	1	1	1	0	1	0	x13a
r-s&l-p/C	0	1	0	0	1	1	0	0	1	0	x132
...	.	.	.	.	.	.	.	.	.	.	...
l-p&ident	0	1	1	1	0	0	1	0	0	0	x1c8
l-p&ident/C	0	0	1	1	0	0	1	0	0	0	x0c8
...	.	.	.	.	.	.	.	.	.	.	...
ident&key	1	1	1	1	0	0	0	0	0	0	x3c0
ident&key/C	1	1	0	0	0	0	0	0	0	0	x300

Table 2: Syndrome table showing rules and attributes. /C = compensating faults

4 Empirical Results

Although the robust tree has been shown analytically to be 2-detectable or 1-correctable for any error in its set of pointers and its identifier, its behavior under various types of faults and for different fault durations needs to be experimentally evaluated. These experiments would measure the effectiveness of the robust tree in contributing to the overall error detection capabilities of a system. Selection of fault models for these experiments should reflect possible physical failures in the system hardware [16]. Examples of physical faults are those occurring in external address and data buses, internal data buses, memory, memory select circuitry, Program Counter (PC) and addressing circuitry, and internal registers of a processor.

Towards that end, we employed fault/error injection experiments to evaluate the detection and correction schemes, and to obtain realistic performance overheads. Fault/error injection has been recognized as the best approach to evaluate the behavior and performance of complex systems under faults and to obtain statistics on parameters such as coverages and latencies. There are several advantages in adopting the fault/error injection approach for evaluating these systems. These advantages include: 1) the effects and latencies of errors can be determined when executing realistic programs, 2) the overhead of recovery algorithms under permanent faults as well as transient errors can be evaluated, 3) the effects of errors occurring during recovery process can be studied, and 4) analytical models can be refined by utilizing data such as fault coverage and recovery coverage.

Experiments were conducted on a SUN4 workstation. FERRARI (Fault and ERRor Automatic Real-time Injector), a software implemented fault injector, was utilized [17]. In FERRARI, the system is exercised under injected faults/errors to measure the effectiveness of its detection and correction capabilities and error detection latency. FERRARI emulates hardware faults and errors by modifying the executable program modules. Table 3 lists several error models FERRARI is capable of injecting. In this table, inserting a data line error, for example, models several actual faults in the processor hardware. These include faults in the external and internal data lines, faults in the PC and its internal registers, faults in address calculating circuitry, and faults in memory.

Results for every run are logged. Each run records the location of the error (virtual address), the affected bit, and the affected register, if any. The monitor appends status flags for every run indicating whether the error was latent (did not lead to a failure during the lifetime of program execution or was overwritten), detected or has led to a failure. In addition the monitor logs the identity of the error detection mechanism if the error was detected as well as the latency period for detecting the error. The collection and analysis module collects these results along with the associated status flags and determines the count for each incident. At the end of the experiment, the module determines the count and percentages (with respect to coverage, latency, error detection mechanism) for each run.

Over 60,000 error injection experiments were conducted to measure the effectiveness of the

Error Model	Description
1	address line error resulting in executing a different instruction
2	address line error resulting in executing two instructions
3	address line error when a data operand is fetched
4	address line error when an operand is stored
5	data line error when an opcode is fetched
6	data line error when an operand is loaded
7	data line error when an operand is stored
8	condition code flags

Table 3: Error models used in the experiment

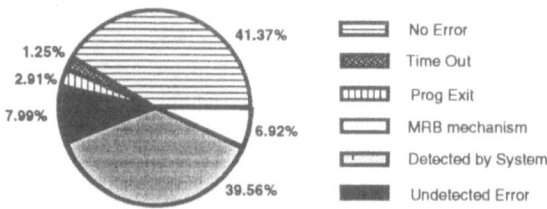

Figure 9: Distribution of detected errors, undetected errors, latent errors, and timeouts for the MRB tree.

error detection capabilities of the MRB tree. In our fault and error injection experiments, we were interested in intermittent faults of a short duration since they are hard to detect. Permanent faults, on the other hand, eventually cause detectable errors, as has been observed in previous studies [18]. The selected error models are listed in Table 3.

In Figure 9, we present the distribution of detected errors, undetected errors, latent errors, and timeouts for the MRB tree against all the faults in Table 3. "Time out" in Figure 9 is the percentage of runs when task execution entered an infinite loop due to the injected fault. The "NO ERROR" part in Figure 9 is the case were errors are overwritten or where the errors did not lead to a failure. "MRB Mechanism" in the same figure is the coverage due to the MRB tree. "Prog Exit" is a combination of several programming robustness features, such as checking status of I/O operations when opening and closing files. In Figure 9, over 39% of injected errors were detected by the SUN4 built-in error detection mechanisms such as, illegal instruction, segmentation fault, bus error, bad system calls, etc. Of the remaining, the MRB tree detected 6.9% and "Prog Exit" detected 2.9%. Note that the error detection mechanisms of the SUN4 system were triggered before any other error detection techniques. Finally, 7.99% of the injected errors were not detected by any of the error detection techniques. The origin of these "Undetected" errors will be explained later.

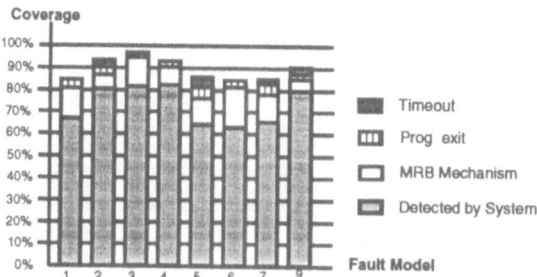

Figure 10: Error coverage of the MRB tree as a function of the injected fault model.

Figure 10 shows the error coverage results. In this figure, latent errors were excluded from coverage calculations. Over 60% of the detected errors were trapped by the system error detection mechanisms. Although the MRB tree error detection mechanism is a data value checking technique, it was still effective for other errors caused by the injected faults. The effectiveness of the MRB tree error detection, however, was the highest when errors were injected into the data bus when operands were either fetched or stored (error models 6 and 7). The percentage of errors that were not detected varied between 15.0% for error model 1 and 4.0% for error model 3.

For the purpose of measuring performance penalties of the MRB tree, I/O operations as well as any computation not related to the robust structure operation were excluded from the timing measurements. This situation would measure the worst case performance penalty incurred by the addition of redundancy as well as invoking the detection routine, *Detect* (which applies the assertion rules relating the attributes of the tree), for every traversal of the tree. System timing routines were utilized to measure execution times before adding redundancy to a binary tree, and after adding redundancy and invoking the *Detect* routine. Performance evaluation involved the following measurements.

- Traversal of the tree without key comparisons. The performance overhead was measured to be 162% over execution time of an ordinary binary tree with no added redundancy.

- Construction of the tree which combines two routines, insert and search for a position. Performance overhead was measured to be 96%.

- Deletion of elements of the tree which combines two routines, finding the selected element and deleting the element. Here overhead was measured to be 91%.

Detect is invoked inside the loops (WHILE loops) of search and traverse routines, after inserting an element, before and after deleting an element, and at entry and exit points of all routines. Note that for insert and delete routines, *Detect* is invoked only when inserting or deleting an element respectively. Maintaining high error coverage incurred the above overhead percentages for every data structure routine. Note that when I/O operations were included in the timing measurements, the overhead for reading, inserting, traversing, and writing 100 elements (integers) dropped to less than 10%.

Using FERRARI to locate errors, it was found that a significant percentage of the errors that were detected neither by the system nor by the concurrent detection mechanism were those that originated inside I/O handling routine libraries. Examples of faults leading to these errors are data line faults occurring while reading or writing to an I/O buffer. In order to increase error coverage, though at a higher overhead cost, we experimented by introducing two-phase input reading (Mode1) whereby data to be inserted in the tree was read twice and was compared for any discrepancies. We applied the same procedure for the output stage in Mode2 whereby the output of the MRB is written to a file and is later read to compare the data set values to those stored in memory. Finally, we combined Mode1 and Mode2 into Mode3, thus performing two phase input read and two phase output write. Table 4 shows the coverages for these modes as a function of the injected error model.

In Table 4, Mode0 refers to the ordinary MRB tree. Error coverages for modes Mode1, Mode2, and Mode3 were higher than for an ordinary MRB tree. In one case, for Mode1 and with respect to models 3, and 4, error coverage was lower than the coverage for Mode0. When these undetected errors were examined, they were found to be those that corrupted the arguments passed between the robust tree routines and the operating system library functions. In these situations, a checksum mechanism can be added to detect these errors. Hence, the checksum of the values of the arguments to be passed among routine calls are calculated and are appended to the argument list. The called function would thus calculate the checksum for the original arguments and compare it to the passed checksum argument. Inclusion of this method in the MRB tree running Mode3 will increase the coverage of errors produced by the errors we have injected in our experiments.

Application	Fault Model							
	1	**2**	**3**	**4**	**5**	**6**	**7**	**8**
Mode0	84.8	93.5	98.3	94.0	84.6	83.6	82.0	91.4
Mode1	91.2	97.8	94.6	93.4	85.0	88.6	83.9	93.1
Mode2	97.7	98.0	98.1	83.6	92.9	95.1	93.0	98.4
Mode3	98.7	98.0	99.2	99.2	93.7	97.4	93.9	99.0

Table 4: Fault coverage for the MRB tree for four modes of I/O mechanisms.

In Figure 11, we present the performance overhead as a function of error coverage for the previous modes, Mode0 through Mode3. In this experiment, insertion and traversal including I/O operations of 100 elements of integers was timed. As shown in the figure, obtaining higher coverage was achieved at the expense of higher performance overhead. The overhead using multiple I/O operations, however, is still considered much lower than replicated systems (over

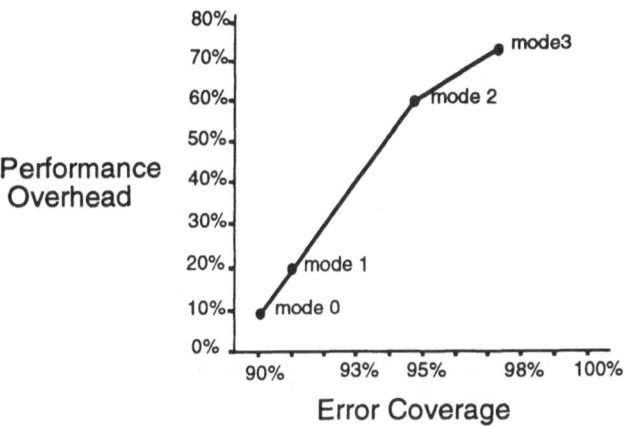

Figure 11: Fault coverage versus performance overhead for four I/O modes for the MRB tree.

100%).

5 Conclusion

In this paper we have presented a new robust binary tree structure. The MRB tree has a fixed window size which is its basic unit module. These unit modules are replicated as required to form the binary tree structure. The added redundancy allows the tree to detect 2 errors or correct a single structural error in every module. The *Traverse* routine showed the highest performance overhead penalty since it invokes the *Detect* routine for every recursive walk through the tree. However, even this routine incurred only a 162% performance overhead. Fault and error injection experiments utilizing FERRARI have shown the effectiveness of the robust tree in contributing to the overall error detection capabilities of a system against a variety of error models.

References

[1] D.J. Taylor, D.E. Morgan and J.P. Black, "Redundancy in data structures: Improving software fault tolerance," *IEEE Trans. on Software Eng.*, vol. SE-6, no. 6, pp. 585-594, Nov. 1980.

[2] K. Huang and J.A. Abraham,"Algorithm-based fault tolerance for matrix operations," *IEEE Trans. on Computers*, vol. C-33, no. 6, pp. 518-528, June 1984.

[3] J.P. Black, and D.J. Taylor, "Local correctability in robust storage structures," Tech. Rep. CS-84-44, Dep. Comput. Sci., Univ of Waterloo, Dec. 1985.

[4] C.-C. J. Li, P. P. Chen, and W. K. Fuchs, "Local concurrent error detection and correction in data structures using virtual backpointers", *IEEE Trans. on Computers*, vol. 38, no. 11, pp. 1481-1492, Nov. 1989.

[5] K. Kant, and A. Ravichandran,"Synthesizing robust data structures – An introduction," *IEEE Trans. on Computers*, vol. 39, no. 2, pp. 161-173, Feb. 1990.

[6] S.C. Seth, and R. Muralidhar, "Analysis and design of robust data structures," in *Dig FTCS-15*, pp. 14-19, June 1985.

[7] M. Sampaio, and J.P. Sauve, "Robust trees," in *Dig FTCS-15*, pp. 23-28, June 1985.

[8] K. Yoshihara, Y. Koga, and T. Ishihara, "A robust data structure scheme with checking loops," in *Dig FTCS-13*, pp. 241-248, June 1983.

[9] E. Reingold and W. Hansen, "Data Structures", *Little Brown Computer Series, library of congress*, 1983

[10] I.Davis, "A locally correctable AVL tree," in *Dig FTCS-17*, pp. 85-88, July 1987.

[11] D.J. Taylor, D.E. Morgan and J.P. Black, "A locally correctable B-Tree implementation," *The Computer J.*, vol. 29, no. 3, pp. 269-276, Nov. 1986.

[12] D.J. Taylor , "Error models in robust data structures," in *Dig FTCS-20*, pp. 416-422, June 1990.

[13] D.J. Taylor, D.E. Morgan and J.P. Black, "Redundancy in data structures: Some theoretical results," *IEEE Trans. on Software Eng.*, vol. SE-6, no. 6, pp. 595-602, Nov. 1980.

[14] J.P. Black, D.J. Taylor, and D.E. Morgan , "A robust B-tree implementation," in *Proc. 5th Int. Conf. Software Eng.*, pp. 63-70, March 1981.

[15] S.B. Lippman, "C++ Primer," Addison Wesley. 1989.

[16] K. A. Hua and J. Abraham, "Design of systems with concurrent error detection using software redundancy", *Proc. ACM/IEEE Fall Joint Computer Conference*, Dallas, Texas, Nov. 1986.

[17] G. A. Kanawati N. A. Kanawati and J.A. Abraham, "FERRARI: A Tool for The Validation of System Dependability Properties", *Dig FTCS-22*, Boston, July 1992, pp. 336-344.

[18] P. Banerjee, J.T. Rahmeh, C. Stunkel, V.S. Nair, K. Roy, V. Balasubramanian and J.A. Abraham,"Algorithmic-based fault tolerance on hypercube multiprocessor," *IEEE Trans. on Computers*, vol. 39, no. 9, pp. 1132-1146, Sept. 1990.

Secondary Storage Error Correction Utilizing the Inherent Redundancy of the Stored Data

Randy Rowell and V.S.S. Nair*
BNR Inc.
Research Triangle Park, North Carolina 27709
U.S.A. rowell@bnr.ca
*Computer Science and Engineering Department
Southern Methodist University
Dallas, Texas 75206
U.S.A.
nair@seas.smu.edu

Abstract

Data records in secondary storage have some inherent redundancy of information. This redundancy cannot precisely be predicted as in the case of typical error correction scheme's artificial redundancy. However, the redundancy can be exploited to provide error correction with some degree of confidence. To that end, we develop a statistical error location and correction (SELAC) approach. We use simple and weighted checksum schemes for error detection and present algorithms for single and multiple error correction using SELAC. An implementation of SELAC will be described with an elaborate study of its error correction capabilities. A conspicuous aspect of SELAC is that it will not cost any processor time and storage overhead until after an error is encountered unlike the classical schemes using SEC-DED and DEC-TED codes.

1   Introduction

As technology advances, computer memories are becoming larger and larger. Any physical storage device is subject to hardware faults that lead to loss of data. It is possible, with some storage of redundant information to allow a certain amount of error correction in order to minimize the impact of hardware failures. However, the expense of this approach is often costly, especially as the number of bits possible to correct gets larger and larger.

The data that is typically stored on secondary storage devices is either the result of some program output, or the data to be used as program input. In almost all cases, any useful file will have some structure that is used to carry meaning to some user or program. However, contemporary storage and retrieval treat the data independent from its semantic content.

The field of Information Theory gives the following definition of quantity of information:

$$I(E) = \log \frac{1}{P(E)}$$

where E is an element of communication and $P(E)$ is the probability of the communication element E occurring. If the logarithm is base two, then $I(E)$ is the quantity of information in bits. An observation of the types of data stored in various primary and secondary memory storage systems in today's computers will show that the amount of information present is significantly less than the amount of information that could be present.

We can reason that many data records have some inherent redundancy of information. This redundancy cannot precisely be predicted as in the case of a typical error correction scheme's artificial redundancy. However, this redundancy can be exploited to provide error correction with some degree of confidence. This paper explores the the degree of confidence that can be attained.

As an example, a large, text oriented file stored on a typical software development workstation takes 281096 bytes. Running a simple, off the shelf, compression algorithm reduces the same file to 86167 bytes. This simple one minute experiment demonstrates that such a file has a theoretical redundancy of at least three times as much storage as needed. It is unlikely that memory cost will ever become so dominant that all items put onto secondary storage will be tightly compressed. This experiment demonstrates that there is great potential to exploit the redundant nature of stored data for its own error correction.

A great advantage of error correction based on the data's inherent redundancy is that the amount of effort to be expended on error correction can be determined after the error has occurred when the cost of the error is known. Also, such error correction does not require a memory system specially designed for correction. This approach to error correction may be more or less effective depending on the error detection scheme employed by the memory system.

This paper is organized into the following sections:

- **Previous Work** will discuss the theoretical basis for the error correction approach.

- **Statistical Error Location and Correction (SELAC)** will describe the statistical error correction approach based on the theory discussed in the preceding section.

- **Experimental Results** will give results of the experiments discussed is this paper. The results will show how the nature of the data affects error correction success by showing different results for text oriented files, loadable binary files, Pascal source files, Pascal listing files, and other types of files. It will also show differing results due to the differing nature of errors and different types of checksums.

- **Comparison Verses Existing Error Correction Schemes** will put the results into some perspective by giving quantitative comparisons with accepted error correction schemes.

- **Further Direction** gives direction for further study. There is vast potential for further study, some of which may lead to viable error correction improvements on certain domains.

2   Previous Work

It was pointed out earlier that almost all data stored on a secondary storage device is very highly structured and therefore somewhat predictable. There is an inherent redundancy in almost all such data. In order to exploit this information, it is necessary to analyze the data in the stored file to figure out the nature of the data. A large part of this project focuses on the analysis of computer data for the purpose of later error correction on that data.

To see how we are going to do this correction, consider the following stream of text that could be stored on a secondary storage device:

Don't fire until you sge the whites of their eyes!

If told there is an error in the above sentence, almost any English speaker can spot it. For those of you who are stumped, the word *sge* should be *see*. An English speaker determines this based on the context of the sentence and the fact that sge is not a word.

For computer error correction however, we have to exercise a little caution and make a few simplifications. It is a very difficult problem to make a computer correct the above sentence based on the context of the sentence. Furthermore, most things stored on a computer storage device are a little more cryptic than English. It is always possible that someone intended to write the sentence as it is above (for example, I as I am writing this paper). An error correcting program has to have some means of ensuring that it doesn't make corrections when no correction is necessary.

A simple statistical method to determine which positions are the most likely to be erroneous can be developed from the information theory discussed in [1]. In this discussion on the structure of language, the author gives several statistical approximations to English based on random sequences of the characters in the alphabet. Several levels of approximation are given, the more accurate approximations are based on the probability distributions of characters that are calculated conditionally upon the preceding characters. In the sequence of characters given below, all the characters are equally likely to appear at any point.

That writer calls this the *Zeroth approximation to English*.

zewrtzynsadxesyjrqy wgecijj obvkrbqpozbymbuawvlbtqcnikfmp kmvuugbsaxhlh-
sie m

A better approximation can be given by taking into account the relative frequencies of each character as is given below in the *First Approximation to English*:

ai ngae itf nnr asaev oie baintha hyroo poer setrygaietrwco ehduaru eu c ft nsrem
diy eese f o sris r unnashor

This sequence only takes into account their relative frequencies. It is possible to take into account what is called in this paper, the conditional frequencies which is based on the set of conditional probabilities. For example, what is the distribution of characters that follow an A, that follow a B, that follow a C, and so on. This information can be used to construct a *Second approximation to English*. If one takes into account the conditional probabilities of characters following any pair of characters, it is possible to construct a *Third Approximation to English*. The third approximation from [1] is given below:

ianks can ou ang rler thatted of to shor of to havemem a i mand and but whissitably
thervereer eights takillis ta

As you can see, this sequence has a remarkable resemblance to English.

3   Statistical Error Location and Correction (SELAC)

In this section we discuss a statistical error location and correction method based on the previous section and propose algorithms to test the viability of the theory. A program will be implemented to run experiments in statistical error correction based on the information theory above. The program will be referred to in this paper as SELAC, which stands for Statistical Error Locator and Corrector. SELAC will have two primary functional areas. The first is the statistical analysis of the nature of data in a test file and the second is error location and correction based on this analysis. Error detection will be bases on conventional methods as discussed in [3]. Error location and correction will not be attempted unless an error has been detected.

In the current implementation, the analysis consists of collecting the data necessary to generate an *N*th approximation to the characters in the input file. The value of *N* can be configured to any value desired, but for practical needs of good performance and good results, *N* will typically be three in all experiments documented in this paper. Higher values would likely get slightly better results, but experiments would take too long to run. Lower values would likely get significantly worse results.

In order to generate the *N*th approximation, it is necessary to collect a count of character occurrences conditional upon the previous *N* - 1 characters occurring. For example, SELAC collects a history of all characters following the sequence **AB** and separately collects a history of all characters following the sequence **AC**. For better statistical utilization, SELAC will keep these statistics to enable it to examine the conditional transitional probabilities in both the forward and the backward direction in the file. For example, in the sequence **ABC**, SELAC considers the probability of **AB** being followed by a **C** as well as the probability of **BC** being preceded by an **A**.

The process of error location cannot be separated from the process of error correction. In fact, the error location process actually only produces a list of candidate error locations within a data block and the error correction process actually only produces a list of candidate error corrections for any given error location. SELAC will use these candidate lists to find a location and a correction that together it believes to be the valid correction. The strength in this approach comes from using the statistical analysis to generate good guesses at the error location and good guesses at the error correction. Both of these processes will be discussed more later in this paper.

3.1 Simplifying Assumptions

There are several complications that could come up that are not dealt within this project. For example, if a file with a bad record is small, it may not be possible to gather enough statistics on the file to provide much help in error correction. It is possible to use other files that have similar contents for statistical analysis, however this project does not go that far. Even for a very large file, it may be a little more complicated if errors show up very early in the file because there is not much statistical information available in the earlier records of the file. Ideally, one could read the blocks that appear later in the file, then go back and correct the error. This project, however, will only allow errors after a predetermined number of blocks have been processed in the beginning of the file. The predetermined number of blocks is chosen almost arbitrarily. The significance of this choice needs more investigation. In order to evaluate the error correction process, after a certain number of blocks are read, errors are inserted into every single block of the file. This simplifying assumption is primarily an implementation short cut, although it is important to realize that SELAC will be ineffective in correcting errors on some files if the statistics on similar data cannot be collected.

This project assumes that all data blocks are stored along with a checksum value that will be discussed a little later. In any storage medium, it is quite possible that the checksum value itself could become corrupted. In that case, the error correction approach taken by this project will not be able to correct the error. However, since the chance of a corrupted checksum is small compared to the chance of corrupted data due to the large disparity in number of bits in each, this project will ignore the chance of a checksum error. It is possible to add functionality that would attempt to correct errors in the checksum also, but this project will not implement that functionality. Because of this simplification, the correctability results gathered for this paper are slightly better than they would have been had this assumption not been made.

3.2   Error Detection

There are several ways available for detecting errors on secondary storage devices. This project will consider two of the checksum schemes discussed in [3], simple checksums and weighted checksums. The first scheme, adds up all the characters giving them an equal weight of 1. It stores the sum in an M-bit checksum field where M can be configured to any value and will differ throughout these experiments. Any overflow from the checksum addition is simply masked off and discarded. There is a danger in this equal weight checksum approach. A single bit error may for example change a bit from a 1 to a 0. Since many bits in a data block share the same place value within their respective octets, an attempted error correction that changes another bit from a 0 to a 1 has a very good chance of producing what will be called a *false correction*.

To help circumvent the above weakness, the weighted checksum scheme gives each position in the block a prime number weight to be multiplied by the ASCII (or non-ASCII) character value. Position number 0 has a weight of 3, position number 1 has a weight of 5, position number 2 has a weight of 7, and so on. Prime weights were chosen because intuitively it will minimize the chance of a two bit error being accepted as a correct block and thus reduce the chance of a *false correction*. This checksum scheme masks off any overflow beyond the M bit checksum.

In the actual experiments, the choice of checksum schemes from above made very little difference in the success of the error correcting algorithm. The number of bits in the checksum is varied in different runs of these experiments.

Because of the limitations of the checksums, both schemes are susceptible to what is called in this paper a *false correction*. A false correction is when the error corrector thinks it found a correction, but it actually has not. Instead, it has changed the data block to some other incorrect block. It is also possible for an error to occur and the error detection scheme fail to detect the error. This scenario will be called a *failed detection*.

3.3 Error Location

When a block is known to contain at least one error, the first thing necessary is to find the symbol which contains the error. It is not possible to know for sure what symbol contains the error, so we developed a set of heuristics based partially on the discussion above. These heuristics are used to generate a matrix with a column for each symbol in the data block and a row for each of the 14 heuristics that are defined. Each entry in the matrix is a numerical value between zero and one, higher values suggesting a higher likelihood of an error in or near the symbol in the data block. Some entries can only have zero or one, in the case of true/false heuristics. Other entries can take on any value between zero and one.

Consider the following character window:

Ther$fore

The character sequence **he** may frequently be followed by an **r**. However, the character sequence **er** is unlikely to be commonly followed by a **$**. Based on this reasoning, we could guess with some confidence that **$** is an incorrect character. But we might also reason that the **e** or the **r** could be incorrect characters instead of the **$**. So the following heuristics are used to capture many of the highly suspect cases that can occur.

- ZFIRST: The character in question has never occurred before. This is a very strong heuristic.

- ZSECOND: The character in question is never followed by the next character.

- ZTHIRD: The character in question is followed by the next character, but the two are never followed by the third character.

- ZFOURTH: The same idea, but for the fourth following character.

- ZBSECOND: The character in question has never followed the preceding character before.

- ZBTHIRD: The character in question has never followed the two preceding characters before.

- ZBFOURTH: The character in question has never followed the three preceding characters before.

- ZRATIO: A number based on the ratio of this character verses all characters in the file so far. Precisely it is the reciprocal of the probability of seeing the character in this position divided by the total number of possible symbols. Thus, the more likely the character, the lower the number for this heuristic.

- ZB2RATIO: A number based on the ratio of this character verses all other characters that follow the preceding character. Precisely it is calculated just like ZRATIO except that it is based on a conditional probability rather than just a probability of the single character.

- ZB3RATIO: A number based on the ratio of this character verses all other characters that follow the two preceding characters. Precisely it is calculated just like ZRATIO except that it is based on a conditional probability rather than just a probability of the single character.

- ZB4RATIO: A number based on the ratio of this character verses all other characters that follow the three preceding characters. Precisely it is calculated just like ZRATIO except that it is based on a conditional probability rather than just a probability of the single character.

- RB2RATIO: Same as ZB2RATIO except that it is calculated by using reverse order conditional probabilities. Precisely it is calculated just like ZRATIO except that it is based on a conditional probability rather than just a probability of the single character.

- RB3RATIO: Same as ZB3RATIO except that it is calculated by using reverse order conditional probabilities. Precisely it is calculated just like ZRATIO except that it is based on a conditional probability rather than just a probability of the single character.

- RB4RATIO: Same as ZB4RATIO except that it is calculated by using reverse order conditional probabilities. Precisely it is calculated just like ZRATIO except that it is based on a conditional probability rather than just a probability of the single character.

ZFIRST – ZBFOURTH are set to 0 for false and 1 for true. The rest are given numerical values greater than zero. The most important characteristic about these heuristics based on ratios is that they be monotone increasing as the likelihood of the symbol appearing within its respective scope decreases. The weights of these heuristics are automatically tuned by SELAC as it gathers statistics on an input file. The formula used for each of these heuristics based on ratio is:

$$\frac{Input\_Characters}{Matching\_Characters * Number\_of\_Symbols}$$

Table 1. Heuristic Indicators for Example Sequence

	Wght	T	h	e	r	$	f	o
Zfirst	10000	0	0	0	0	0	0	0
Zsecond	0.4	0	0	0	1	1	0	0
Zthird	0.1	0	0	1	0	1	0	0
Zfourth	0.1	0	0	0	0	0	0	0
Zbsecond	800	0	0	0	0	1	1	0
Zbthird	20	0	0	0	0	1	1	1
Zbfourth	1	0	0	0	0	0	0	0
Zratio	0.1	0.355	0.195	0.087	0.150	3.91	0.206	0.122
Zb2ratio	0.1	0.186	0.064	0.024	0.025	1	1	0.040
Zb3ratio	0.1	0.035	0.014	0.012	0.021	1	1	1
Zb4ratio	0.1	0	0	0	0	0	0	0
Rb2ratio	0.1	0.098	0.035	0.022	1	1	0.033	0.029
Rb3ratio	0.1	0.009	0.021	1	1	1	0.025	0.022
Rb4ratio	0.1	0	0	0	0	0	0	0
Total	–	0.068	0.032	0.113	0.062	821.3	820.2	20.12

Table 1 shows the heuristic indicators for the given example sequence. The $ is intended to be an error, and as you can see, it has been selected by the heuristics as the top candidate for the error location. The row labeled "Total" is calculated by summing the product of the heuristic indicators and the corresponding heuristic weight Both the heuristic indicators and the heuristic weights in this table are examples intended to illustrate the concept and do not correspond to an actual run of SELAC.

The table shows the initial value of the heuristic weights. The weights should be tuned so as to maximize the chance of finding the erred location early in the search. It is a potentially very difficult problem to tune these weights. Instead of tuning them before program execution, SELAC tunes them as it reads through data blocks with no errors. For each block, it simulates an error and tunes each weight by plus or minus ten percent in order to maximize the difference in the erred location and other locations. Errors are simulated by flipping a single bit in a random location in the input file.

When a block is found to contain a real error, these heuristics are calculated for each character in the block and a numerical score is assigned to each one. The top 20 of these characters are considered as candidates for correction. The number 20 is chosen somewhat arbitrarily and can easily be changed for other experiments.

3.4   Error Correction

When a symbol is determined to be a candidate for correction, it must be determined what symbol will be tried as a replacement. It is possible to try every single possible symbol, however, that is expensive and may not be beneficial. A predetermined number of symbols will be tried for each error correction candidate. This number depends on which candidate number is being analyzed and how many errors are currently being tried for correction. For example, if a heuristic search determines a given character is the most likely of all characters to be erred, it is worth trying more alternative symbols as corrections than if the heuristic search determines that the given character is only the 20th most likely character of being in error. The error corrector could be configured to try 25 times to correct an error when the character is most likely in the block to be incorrect, however only try five times to correct an error when the character is the 20th most likely in the block to be incorrect.

SELAC can be configured to try and correct any number of errors. However, when two errors are close to each other in the data block, it is possible that the statistical corrector will be confused. To circumvent this problem, error correction of multiple errors is done recursively. For each attempt at correcting a single error, the SELAC recursively recalculates all the heuristics with the assumption that the attempted correction is a legitimate correction. Multiple error correction is discussed more completely in "Handling of Multiple Errors". Because of exponential growth of processing time as a result of the recursive search, the search effort at different levels of the recursive search is configurable independent of the other levels. In other words, it is possible to configure SELAC to search less extensively for the first of two suspected error corrections than when it searches for a single error correction.

In selecting the symbol to be tried as a replacement, it is desirable to use the most specific statistics available. For example, for the following fragment of data:

> *Hello therw, professor.*

in trying to find a character to replace the **w**, the fact that it follows the characters **er** should be used before the fact that it follows just an **r**. It may also be more useful to consider the fact that it is followed by a comma and a space before considering the fact that it is following an **r**. All of this of course depends what the actual statistics that the file in question determine. In general, the experiments try characters as replacements starting at the most common character of the most specific level to the least common character at the least specific level.

At first glance, this implementation might resemble a glorified spell checker. However, a spell checker cannot generally correct a misspelled word once found. Also, a spell checker is given ahead of time a strict vocabulary which does not change. SELAC effectively configures its vocabulary as it examines a file, although in reality it does not have a vocabulary, but a statistical summary of character sequences. SELAC also does not need to delineate words in order to find corrections. Thus, SELAC can work just as easily on any language or any file with highly structured data.

The following algorithm shows the basic steps in correcting a single error:

Algorithm 1 *Single Error Correction Algorithm*

1 Generate List of Candidate Error Locations
    1.1 Generate Matrix of Heuristic Indicators
    1.2 Multiply by Weights to Generate Candidate Strengths
    1.3 Sort to Find Top Candidate Locations
2 For Each Top Candidate Location (Configurable Number)
    2.1 While Symbols Remaining And Limit Not Exceeded
        (Configurable Limit)
        2.1.1 Find Next Candidate Symbol for Replacement
        2.1.2 Generate Checksum with Candidate Replacement
        2.1.3 If Checksum Matches Expected Checksum
            – Return "found correction"
    2.2 End While
3 End For
4 Return "give up"

3.4.1 Handling of Multiple Errors

SELAC will correct P errors, where P is configurable. In order to handle a single error, it will keep trying replacements as dictated by the algorithm described above. For each symbol it attempts to use as a replacement, it recalculates the checksum of the new block until it either finds a corrected block with the same checksum as given, or it runs out of replacements to try. If it does find a block with a matching checksum, there is a chance that it is a *false correction*. SELAC will determine this and keep performance statistics accordingly.

To handle multiple errors, it replaces the first symbol, then recursively does the process all over again for a second or third symbol. As a part of this recursion, it reevaluates all of the heuristics and recalculates the rankings of the most likely candidates. This recalculation is necessary since two errors close together could be confusing and make it difficult to correct both errors.

Since it is presumed that single errors will be much more common than double errors, and double errors more common than triple errors, the algorithm will completely try all single error corrections before doing any double error corrections. Likewise, it will completely try all double error corrections before trying any triple error corrections. It may seem that this choice is a mere program efficiency issue, however it is not. It is possible, for example, for a false double error correction to be incorrectly performed when there is only a single error if double error corrections are attempted before all single error correction attempts.

Algorithm 2 *Multiple Error Correction Algorithm*

1 Generate List of Candidate Error Locations
  1.1 Generate Matrix of Heuristic Indicators
  1.2 Multiply by Weights to Generate Candidate Strengths
  1.3 Sort to Find Top Candidate Locations
2 For Each Top Candidate Location (Configurable Number)
  2.1 While Symbols Remaining And Limit Not Exceeded
    (Configurable Limit)
    2.1.1 Find Next Candidate Symbol for Replacement
    2.1.2 Generate Checksum with Candidate Replacement
    2.1.3 If Checksum Matches Expected Checksum
      – Return "found correction"
    2.1.4 Else
      – Recursively Invoke Main Algorithm
      – If It Returns "found correction"
      Return "found correction"
  2.2 End While
3 End For
4 Return "give up"

One of the biggest advantages to this error correction approach is that it does not add processing or storage overhead to the error free operation of a memory system. All overhead will be realized after the conventional memory system detects an error and SELAC begins to locate and correct the error. There is no memory system overhead to SELACs correction approach. There is processor overhead that is highly implementation and configuration dependent to correct errors on a known corrupted data block. The current implementation takes less than five seconds on average to correct two corrupted data values in a single data block of 64 bytes.

4 Experimental Results

SELAC was implemented in the C language on a Hewlett-Packard workstation running the UNIX operating system.

In order to test SELAC's error correcting abilities, we randomly inserted errors on large data files after SELAC had time to analyze a fifty to a hundred uncorrupted data blocks of the file being used for testing. The nature of the inserted errors varies from experiment to experiment.

It is important to understand that the approach to error correction discussed in this paper is not just a single implementation. There are many variables in data format of input files as well as in configuration data of the error correction process itself. So to give a single set of statistics to indicate how well this approach works would not be a very useful information.

Probably the most impactful variant in using this approach is in the nature of the data being input for error simulation. It is observed that text data is very easy to correct, whereas binary data is very difficult to correct. For each type of data used in experimentation, a measure of the amount of information present is given by the formula for *H(S)* which is called the entropy of the zero-memory source [1]. This entropy function is important because by measuring the amount of information present in an average data block, it is possible by subtraction to measure the amount of inherent redundancy in the average data block. If there is a large amount of inherent redundancy, one can expect better results from this error correction approach than if there is not much inherent redundancy. In the following formula for the entropy function, S_i is symbol number *i* and *P* is the probability of that symbol appearing at any given position.

$$H(S) \equiv \sum_S P(S_i) \log \frac{1}{P(S_i)}$$

All of the following results were generated using the same search configuration. However, as the tables indicate, the data configuration varies in such things a checksum type, number of errors present, checksum length, and block size. The reader may see the next section on *General Results* before coming to any hard conclusions about the success or the potential for success of the error correction approach taken in this project.

In Table 2, the first four columns describe the characteristics of data that SELAC is experimenting on. The "% Correct" column shows the percentage of erred blocks that were corrected properly by SELAC. The "% False Correction" column shows the percentage of erred blocks that SELAC provided a correction but that correction was not the right correction. The column labeled "% Given Up On" shows the percentage of erred blocks that SELAC did not produce any correction for.

Table 2. Text Oriented Data Results

Error Count	Chksum Type	Chksum Size bits	Block Size bytes	Correct %	False Correction %	Given Up On %
1	Equal Weight	16	64	96	4	0
1	Prime Weight	16	64	96	4	0
2	Equal Weight	14	64	85	15	0
2	Prime Weight	14	64	85	15	0
2	Equal Weight	16	64	86	14	0
2	Prime Weight	16	64	86	14	0
2	Prime Weight	18	64	86	14	0
2	Prime Weight	16	256	56	38	6
3	Equal Weight	16	64	72	28	0
3	Prime Weight	16	64	72	28	0
3 Bit Burst	Prime Weight	16	64	96	3	1
6 Bit Burst	Prime Weight	16	64	96	4	0
9 Bit Burst	Prime Weight	16	64	92	7.5	0.5

As you can see from the table above, this error correcting approach gives a varying degree of success depending heavily upon the data configuration. One may notice that the performance of the algorithm was good for a single bit error, although it is very easy to beat this percentage with a Single Error Correcting - Double Error Detecting (SEC-DED) code [4]. Furthermore, it is feasible to design an algorithm that would try every single one-bit error possibility and find one that will cause the checksum to match. With the prime weighted checksum, it is typically possible to find a single one-bit corrected block that will satisfy the checksum. With the equally weighted checksum, this trivial correction approach would not be feasible; nor would that be feasible for two or more bit errors. All in all, one has to conclude that this algorithm does not do well in correcting single bit errors when compared to traditional error correction methods.

It is more interesting to consider the 2 bit error examples. This error correction scheme is able to correct such errors with 86 percent accuracy. It did not do nearly as well with 256 data bytes. However, further study is needed to determine if a larger search space and/or a larger checksum could improve these results.

What is really fascinating is the results in correcting the burst type errors. In these experiments, a burst error was simulated by randomly selecting a place in the data block and changing B bits from that location forward to ones. Notice that this may not result in B bit errors. Some of the bits could already have been ones. In the above table, it shows that 9 bit burst errors were corrected with a 92 percent success ratio. These burst errors contain anywhere from 1 to 9 actual bit errors. These burst errors can corrupt one or two data characters. The error correcting algorithm does not know how many errors it is looking for.

The following table shows the results of running this error correcting algorithm on other types of data. Table 2 was all on a text oriented data file with an zero memory entropy function equal to 4.70 which means that theoretically it would take 4.70 bits to represent each data element. Table 3 shows the results of running this error correcting algorithm on other types of data. The column labeled "Entropy" is the entropy function of the data as given in equation 2. The next three columns are the same as in Table 2.

Table 3. Results on Various Types of Data

Data Type	Entropy	% Correct	% False Correction	% Given Up On
Text Oriented	4.70	96	3	1
Loadable Binary File	5.87	45	55	0
Pascal Source	4.84	94	4	2
Proprietary Language Source	4.51	86	12	2
Pascal Listing	3.44	94	5	1
Compressed Text Oriented	7.85	0	100	0

All of the above tests were run with prime number weighted checksums, 16 checksum bits, 64 data bytes, and 6 bit burst errors.

4.1   Generalization of the Results

As has been stated already, one has to be cautious in interpreting the preceding results. By tuning the search configuration, the results can change. The search configuration used for all of the above tests was made to be reasonable for a text oriented file of 64 byte data blocks, however not much effort was made to reach the best performance.

One of the features of the SELAC is that for each error (or set of errors), it figures out where in the search the error corrector would have stumbled upon the actual error if it had searched that far. This information is accumulated regardless of whether the error corrector actually finds the error or not. Further, this information is calculated for the first, second, third, etc., errors in the data block. The error whose location would be tried for error correction first is considered the first error. After correcting that error, the next error whose location would be tried for correction (after recalculating all heuristics) is considered the second error and so on. The next two tables show where the errors would be found. These tables are very useful for tuning the search configuration. In both tables, each column corresponds to a position in the block of a possible error, in the order that they would be considered by SELAC. Each row corresponds to a candidate replacement in the order that they would be considered by SELAC.

Table 4. Location of First Error in Search

Symbol #	Depth -> 0	1	2	3	4	5	6	7	8	9	10	11	12	13	14	15	V
0	39	1	1	.	.	.	.	.	.	.	.	.	.	.	.	.	.
1	12	.	1	.	.	.	.	.	.	.	.	.	.	.	.	.	.
2	12	.	.	.	.	.	.	.	.	.	.	.	.	.	.	.	.
3	5	.	.	.	.	.	.	.	.	.	.	.	.	.	.	.	.
4	2	.	.	.	.	.	.	.	.	.	.	.	.	.	.	.	.
5	4	.	.	.	.	.	.	.	.	.	.	.	.	.	.	.	.
6	1	.	.	.	.	.	.	.	.	.	.	.	.	.	.	.	.
7	1	.	.	.	.	.	.	.	.	.	.	.	.	.	.	.	.
8	.	.	.	.	.	.	.	.	.	.	.	.	.	.	.	.	.
9	3	.	.	.	.	.	.	.	.	.	.	.	.	.	.	.	.
10	1	.	.	.	.	.	.	.	.	.	.	.	.	.	.	.	.
11	1	.	.	.	.	.	.	.	.	.	.	.	.	.	.	.	.
12	3	.	.	.	.	.	.	.	.	.	.	.	.	.	.	.	.
13	1	.	.	.	.	.	.	.	.	.	.	.	.	.	.	.	.
14	1	1	.	.	.	.	.	.	.	.	.	.	.	.	.	.	.
15	.	.	.	.	.	.	.	.	.	.	.	.	.	.	.	.	.
16	1	.	.	.	.	.	.	.	.	.	.	.	.	.	.	.	.
17	1	.	.	.	.	.	.	.	.	.	.	.	.	.	.	.	.
18	.	.	.	.	.	.	.	.	.	.	.	.	.	.	.	.	.
19	1	.	.	.	.	.	.	.	.	.	.	.	.	.	.	.	.
20	.	.	.	.	.	.	.	.	.	.	.	.	.	.	.	.	.
21	1	.	.	.	.	.	.	.	.	.	.	.	.	.	.	.	.
22	2	.	.	.	.	.	.	.	.	.	.	.	.	.	.	.	.
23+	4	.	.	.	.	.	.	.	.	.	.	.	.	.	.	.	.

Table 5. Location of Second Error in Search

Symbol #	0	1	2	3	4	5	6	7	8	9	10	11	12	13	14	15	V
0	24	1	2	3	2	.	.	1	.	.	.	2	.	.	.	.	1
1	5	.	1	.	1	.	.	.	.	.	.	.	1	.	.	.	.
2	3	.	.	.	.	.	.	.	.	.	.	.	.	.	.	.	.
3	1	1	2	.	.	.	.	.	.	.	.	.	.	.	.	.	.
4	1	.	.	.	1	.	.	.	.	.	.	.	.	.	.	.	.
5	.	1	.	.	.	.	.	.	.	.	.	.	.	.	.	.	.
6	2	.	.	.	.	.	.	1	.	.	.	.	.	.	.	.	.
7	.	.	.	.	.	.	.	.	.	.	.	.	.	.	.	.	.
8	.	.	.	.	.	.	.	.	.	.	.	.	.	.	.	.	.
9	.	.	.	.	.	.	.	.	.	.	.	.	.	.	.	.	.
10	.	1	.	.	.	.	.	.	.	.	.	.	.	.	.	.	.
11	1	.	.	.	.	.	.	.	.	.	.	.	.	.	.	.	.
12	.	.	.	.	.	.	.	.	.	.	.	.	.	.	.	.	.
13	1	.	1	.	.	.	.	.	.	.	.	.	.	.	.	.	.
14	.	.	.	.	.	.	.	.	.	.	.	.	.	.	.	.	.
15	1	.	.	.	.	.	.	.	.	.	.	.	.	.	.	.	.
16	.	.	.	.	.	.	.	.	.	.	.	.	.	.	.	.	.
17	.	.	.	.	.	.	.	.	.	.	.	.	.	.	.	.	.
18	.	.	.	.	.	.	.	.	.	.	.	.	.	.	.	.	.
19	.	.	.	.	.	.	.	.	.	.	.	.	.	.	.	.	.
20	.	.	.	.	.	.	.	.	.	.	.	.	.	.	.	.	.
21	.	.	.	.	.	.	.	.	1	.	.	.	.	.	.	.	.
22	.	.	.	.	.	.	.	.	.	.	.	.	.	.	.	.	.
23+	1	.	.	.	.	.	.	.	.	.	.	.	.	.	.	.	.

Notice that the error corrector can very often locate the position of the error on the very first try. And it can find the correct replacement character often on the first try. Almost all errors that are going to be corrected are found in the first few tried locations, and within the first few attempted replacement characters. The last column in the tables show all errors that would be found in the 16th or higher attempted location. The last row shows all errors that would be found in the 23nd or higher attempted symbol for replacement. The error correction was not configured to try all positions in the tables, so some of the errors were not found because the error corrector did not search enough.

It would be tempting to think that by increasing the scope of search one can increase the error correction capabilities just that easily. Obviously, this is going to cost more CPU time, but that is not all. In almost all cases that the search algorithm failed, it was because it found a set of changes that would be a false correction. If the search space is made too large in some places, it is more likely to fail because of a false correction, even though it was on its way to finding the real correction. For example, if you try 22 possible replacements in the location that is most likely to be in error, you may miss a good correction in the second most likely error position because you found a false correction in the first most likely error position. It would be a worthwhile improvement to have the error corrector search a more elaborate sequence through the tables to minimize the probability of a false correction. For example the 1st replacement character in the second most likely error location may be more likely a valid correction than the 20th replacement character in the first most likely error location.

5   Comparison With Existing Error Correction Schemes

A systematic code with the SEC-DED property would require 11 check bits for 64 data bytes[2]. Most of the experiments ran by SELAC ran with 16 check bits. When run with 11 check bits, SELAC succeeded still 96% of the time verses 100% for the SEC-DED method. Of course, the method implemented for this paper has the benefit of being able to correct 66% of all double bit errors with the same amount of check bits on certain types of data.

A systematic code with the Double Error Correcting – Triple Error Detecting (DEC-TED) [4] property would require 21 check bits for 64 data bytes [2]. When run with 21 check bits on double bit errors, SELAC succeeded 86% of the time verses 100% for DEC-TED. This implementation was able to correct 60% of all triple bit errors with the same amount of check bits on certain types of data verses 0% for the DEC-TED approach.

In order to compare quantitatively, let us consider a storage device in which each bit has a 0.01% chance of becoming corrupted. A 64 byte data block would be correct 95.001% of the time. It would have a single error 4.865% of the time. So a SEC-DED scheme could produce the correct data 99.866% of the time. This new error correction scheme could correct (4.865 * 0.96)% of all single bit errors. Counting all the double errors it could also correct, the new error correcting scheme would produce the correct data 99.672% of the time. If this new statistical approach were employed only after the SEC-DED scheme failed, just counting double errors that could be corrected, the combined scheme could produce correct output 99.948% of the time. This is a moderately good improvement under the given assumptions. It should be noted that this analysis is valid only for text oriented data.

The error correction capabilities of SELAC is more impressive for burst errors. For example, consider a 6 bit burst error on the same type of data. Assuming that there is an equal density of ones and zeros in the data, the chance of a burst corrupting only one bit is 9.38%. So the SEC-DED algorithm can correct less than 10% of the errors, whereas the new statistical approach can correct 83% of these errors.

6   Conclusions and Further Direction

In this paper we present statistical algorithms to locate and correct errors in secondary storage mediums. The algorithms were implemented as a program called SELAC and its error locating and correcting capabilities were studied. For multi-bit errors and burst errors, SELAC outperformed SEC-DED, and DEC-TED, whereas for single and double errors, SEC-DED and DEC-TED proved better. Since SELAC does not cost any additional processing time or storage overhead until after an error has occurred, it would be worthwhile sitting on top of existing error correction schemes for better overall coverage.

Error location and correction by SELAC on the loadable binary file was not nearly as successful as the other types of data, as was expected. There is a great deal of structure to such files, however it is only that it is not nearly as obvious to the error correcting algorithm since the structure has little to do with conditional probability of bytes following bytes. It will be interesting to see if it is possible to come up with algorithms that perform error correction on these type of files by making elaborate use of the structure of the machine language in the file.

So far, little effort has been made in tuning the search configuration to improve the coverage on even text-oriented files. As was stated before, effort in obtaining better results is likely to be somewhat fruitful. One simple, but probably very valuable improvement would be to make the error corrector step through the possible corrections in an order that better matches the likely places of the errors. More research is needed to decide when to stop searching for one error and to start looking for two (likewise two to three). The experiments presented in this paper have been run for up to three errors. The results above would probably have been much better if the algorithm focused more on locating the first and the second error and less on the third.

Another important area of investigation is the reliability of the correction especially in critical applications. More research is needed to determine how to evaluate the level of confidence of a candidate correction block so applications can determine if they should reject a correction due to a lack of confidence.

References

[1] Norman Abramson. *Information Theory and Coding*. McGraw-Hill Book Company, New York, 1963.

[2] C.L. Chen and M.Y. Hsiao. Error-correcting codes for semiconductor memory applications: A state-of-the-art review. In *IBM Journal of R&D*, March 1984.

[3] V.S.S. Nair and J.A. Abraham. Real number codes for fault-tolerant matrix operations on processor arrays. In *IEEE Transactions on Computers*, April 1990.

[4] J. Wakerly. *Error-Detecting Codes, Self-Checking Circuits and Applications*, chapter 1. Elsevier North Holland Inc. New York, 1978.

Panel Session: Common Techniques in Fault-Tolerance and Security

Common Techniques Panel:
Common Techniques in Fault-Tolerance and Security

Karl N. Levitt and Steven Cheung
Department of Computer Science
University of California, Davis
Davis, California 95616
U.S.A.
levitt@cs.ucdavis.edu, cheung@css.ucdavis.edu

It is our opinion that the security community should draw on the techniques developed by the fault-tolerance community to a significant extent; presently, there seems to be less opportunity for transfer in the other direction. The fault-tolerance community has given significant attention to the development of systems where reliability is a key requirement which pervades all aspects of the design. Security, although clearly a requirement, is often achieved through retrofitting, e.g. intrusion detection, and firewalls.

To motivate a serious attempt to explore the utility of the fault-tolerance paradigm to security, we propose a system to significantly improve the security of large networks. The system will provide a capability to deal with a variety of attacks on network components, e.g. on user accounts, hosts, and routers. It will provide a capability to detect attacks through real-time detection (e.g. intrusion detection) and diagnosis. In addition, it will provide a capability to halt the spread of attacks (e.g. through dynamically settable firewalls). And, it will provide a capability to recover to a secure state. Thus, this system embodies most of the steps common to reasonably complex, albeit state-of-the-art fault tolerant systems.

The attached table is an initial attempt to provide security counterparts to the most common fault-tolerance terms. A few issues bear some discussions.

Is there a conceptual difference between the threat to security and that to reliability? Catherine Meadows speaks in more detail to this question. The expansionist's view is that security is just another subarea of Dependable Computing, since it deals with just another manifestation

Basics

Fault Tolerance	Security
Faults	Vulnerabilities (e.g. software bugs, improper system setups)
Latent faults	Time bombs, Trojan horses
Intermittent faults	Race conditions
Errors	Exploitation of vulnerabilities (intentionally or accidentally)
Failure	Violation of policies
Random/accidental in nature (except Byzantine faults)	Malicious/intentional in nature
Independent fault assumptions	Correlations among attacks (e.g. trusts among hosts allow attacks to spread)

Evaluation

Fault Tolerance	Security
Reliability	Mean time between break-ins, covert channel capacity
Availability	Denial of service

Techniques

Fault Tolerance	Security
Redundancy, majority voting	Sending packets over multiple communication paths, storing critical files in more than one sites, using multiple servers for authentication
Error detection or correcting codes	Cryptography
Heterogeneity (e.g. N-version programming)	Having heterogeneous hosts and routers which run different communication protocols; cost: standardization of protocol and OS
Error containment	Access control, firewalls, diversity

Diagnosis/ Detection

Fault Tolerance	Security
System diagnosis (e.g. active probing for faults)	ISS, COPS, Netsweep, NSM, IDS, anomaly and misuse detection, auditing, testing or monitoring by site administrators, virus scanners, integrity checking

Response

Fault Tolerance	Security
Reconfiguration	Disconnecting a host from the net, adding rules to the packet filter to block packets from suspicious hosts, routing packets around suspicious routers
Recovery from failure (e.g. recovery blocks)	Recovery from attacks (e.g. reload a clean copy of the routing table to a compromised router, changing passwords).

of faults. However, the exact nature of the threat to security deserves a closer look. The threat to security is usually a human or processes (or programs) that trace their ancestry to humans. Being human in origin, the security threat can adapt as it attempts to thwart the defenses launched against the threat. This situation does not arise in the fault-tolerance world, except that the entire motivation for Byzantine agreement is the hypothesis that faults can have malicious manifestations, and it might be prudent to study this model in connection with security. There seem to be security threats that do not correspond to faults studied by the fault-tolerance community. As Fred Cohen has noted, it is unlikely that a computer virus could accidentally appear in a program. On the other hand, a common malicious program that causes denial of service in some Unix systems could be accidentally introduced: a program that claims all of the system's process table slots by creating processes faster than they are killed.

Can security requirements be expressed probabilistically? Most security requirements are expressed as negative properties, for example, files are not to be released to unauthorized users. Generally, the requirement is not expressed as the probability of unauthorized release of a file should be less than some very small number, mainly because there is no clear approach to determine if such a requirement is realized by the underlying system. A counterexample to the "assertion" approach to security is the information-theoretic approach used to compute the bandwidth of covert channels.

Error Containment: Redundancy is effective only if the likely faults affect a single protected component, e.g., a processing element. Moreover, fault-masking prevents the fault from inducing errors that propagate beyond the component that suffered the fault. There seems to be a related concept in the security domain. If a computer on a network is compromised by an attacker, it should be difficult for him to use this compromised machine as a base to attack other machines. (Of course, the Unix concept of trust among systems in a singly administered domain is a counterexample to this principle, favoring ease of use over security.) Access control mechanisms and firewalls associated with network components can block or at least limit the spread of attacks.

Recovery Blocks, Diagnostic Routines: The technique of recovery blocks rests on the effectiveness of acceptance test in detecting errors, due to hardware or software faults. For the most critical applications, recovery blocks would not be used as it is impossible to make precise statements about the effectiveness of acceptance tests. However, the increasingly promising technique of intrusion detection involves comparing a user's (or a computer's or a network's or entire system's) current activity with expected behavior. If the comparison is with respect to the user's previous activity, then it is *anomaly detection* that is being implemented. If the comparison is with attack profiles, then it is *misuse detection*. In either case, it is the strength of the comparison algorithms that determines the effectiveness of the technique. Moreover, the operating system itself supplies audit logs that constitute the raw data for the comparison. When the system-generated audit trails are at too low a level of abstraction to enable effective detection, it is incumbent on the application programs to generate audit logs appropriate to the system's security requirements. Particular problems are what support the operating system should provide for application-specific auditing and can the audit trails be made tamperproof? Also yet to be seriously considered are the issues of security-related audit trails for server-based operating systems (such as Mach) and for network components (such as routers, name servers, file servers, etc.) Feather et al.'s paper uses security-like techniques, anomaly and misuse detection, to locate faulty components on Ethernets.

The fault-handling cycle: In an advanced fault-tolerant system, the handling of a fault can involve the following steps: error detection, damage assessment, reconfiguration, and recovery. Current attempts at security are much less ambitious, typically relying on prevention or on attack detection that triggers a message to be sent to a security officer; thereafter it is the responsibility of the human security officer to deal with the situation, e.g., to remove an offending user or site, to request additional audit logs for a particular user, and to save audit logs as evidence. We envision that human intervention at this level will not be possible for much longer, particularly when attacks are on a network's infrastructure or have the potential for rapid propagation, such as the Internet worm. The intrusion detection system of the future will be much like the alluded to advanced fault-tolerant systems. Just as a fault-tolerant system might contain fault handling sites (that effect fault masking, fault analysis, reconfiguration, and recovery — all carried out through communication among the sites), the intrusion detection system should have intrusion management components distributed across a large network with a protocol for exchanging information among the sites that impact the network's security.

Redundancy: Yves Deswarte has been considering the problem of dispersing the blocks of a file among sites. The goal is to protect against unauthorized release of files due to a single site being subverted. The similar problem of avoiding unauthorized access to a message in transmission can be solved by the decomposition of the message into blocks that are guaranteed to be dispersed among different routes between message source and destination. Of course, encryption for confidentiality, including threshold schemes, can be used in conjunction with dispersal.

Here is a final question that bears on the analysis of a system with respect to a reliability or security goal.

Can the reliability or security performance of a system be degraded by fault-tolerance or security measures? In the case of fault-tolerance the answer is yes. A fault-tolerant system contains more components than its non-redundant counterpart, and if the redundancy is not carefully managed, the result could be a less reliable system. A similar situation can exist when measures are taken to improve security. Tables holding information on the security state of a system can be used by an attacker to identify vulnerable sites. Also, a secure system might include mechanisms to logically disconnect sites that have been determined to be subverted; an attacker could use these mechanisms to disconnect sites that might expose him or even to disable quickly an entire subnetwork.

Improving Security by Fault Tolerance

Yves Deswarte
LAAS-CNRS & INRIA
7, Avenue du Colonel Roche
31077 Toulouse Cedex, France
deswarte@laas.fr

Security techniques must accompany the current technological evolution: systems are more distributed and mobile, more complex, and more flexible to fit user requirements. Distribution and mobility imply that hardware faults are more frequent, more administrators and operators are involved, and more intrusions are likely to occur. Complexity and flexibility increase the probability of design faults. Secure systems have to deal with all these classes of faults: hardware faults, design faults, intrusions, malicious or careless administrators or operators. Fault tolerance should then help to make systems more secure.

Fault tolerance can be defined as a property of a computing system to perform its tasks correctly even if it is affected by faults. These faults can be either internal faults, i.e., failures of some system components, or external faults such as environment anomalies or erroneous interactions with users, operators or maintenance staff.

According to such definitions, some common practice techniques for security can be viewed as contributing to fault tolerance. For instance, protection mechanisms prevent unauthorized access, but also contribute to the detection and confinement of errors.

Moreover, since the correctness of some components is mandatory if security is to be enforced, it would be desirable to make these components fault tolerant. In particular, a *Trusted Computing Base* is a "single point of failure", both for security and for reliability: if it fails, the TCB would either deny authorized access or enable unauthorized access. To prevent TCB failures, fault tolerance can be envisaged to deal with design faults, hardware faults, and even faulty interactions with trusted users.

However, application of fault tolerance techniques to security has to confront a fundamental contradiction: fault tolerance requires redundancy, and redundancy is detrimental to confidentiality. For example, data replication, while commonly used to tolerate accidental faults, implies that several information copies are present in the system. An intrusion into a part of the system is thus more likely to lead to sensitive information disclosure. Thus the challenge is to add redundancy while preserving confidentiality.

A first class of solutions consists of using coding theory to design error-correcting codes which would also be good ciphers. Threshold schemes [6] are an example of such codes: the information is coded as a set of shadows such that gathering any T shadows (T being the threshold) is sufficient to reconstitute the information, while no information is obtained from less than T shadows. This technique is efficient for the storage of small sensitive data items, but is inadequate for storing large files and for processing numerical data.

For the storage of large files, information dispersal techniques appear to be more efficient. In this case, files are split and stored on several storage units, and it is necessary to gather the information from several storage units to reconstitute the files. Two variants have been proposed. In one of these, Rabin suggests first to cipher the file, then to code it with a specific error-correcting code and to disperse the resulting data on different storage sites [5]. LAAS proposed another variant called Fragmentation-Redundancy-Scattering (FRS) [3]: the file is first ciphered by means of a stream cipher, the resulting data are distributed into fragments, and fragments are replicated and scattered among a set of storage sites. Confidentiality is increades due to the difficulty for an intruder to know in which order the different fragments have to be reassembled, rather than on the strength of the cipher.

For data processing itself, ciphering techniques are inadequate: most numerical operations and comparisons have to be run on deciphered data. LAAS and the University of Newcastle are currently developing an FRS technique for data processing [2]. The application software is developed according to an object model and objects are decomposed into smaller objects recursively until each object either carries no significant information (and then can be considered as a fragment), or cannot be decomposed any more (e.g., a character string). Undecomposable objects containing significant information have to be ciphered. Then redundancy can be added to fragments (e.g., by replication), and the resulting fragments can be scattered.

Fault tolerance techniques can also be applied to deal with the problem of malicious or careless operators and administrators: separation of duty can be viewed as a form of fault tolerance. Majority voting has also been successfully applied at LAAS to tolerate intrusions into security servers, including intrusions by privileged users [1].

Even design faults can be tolerated by means of design diversity [4]. In this case, fault tolerance can be an attractive alternative to formal verification, when the complexity of the system or of its environment is close to or exceeds the current limitations of formal methods.

All these techniques show that fault tolerance can improve security.

References

[1] Yves Deswarte, Laurent Blain, and Jean-Charles Fabre. Intrusion tolerance in distributed systems. In *Proceedings of the 1991 International Symposium on Research in Security and Privacy*, pages 194–201. IEEE, May 1991.

[2] Jean-Charles Fabre, Yves Deswarte, and Brian Randell. Designing secure and reliable applications using fragmentation-redundancy-scattering: an object oriented approach. In *Proceedings of the 1st European Dependable Computing Conference (EDCC-1)*. Springler-Verlag Lecture Notes in Computer Science, October 1994.

[3] Jean-Michel Fray, Yves Deswarte, and David Powell. Intrusion-tolerance using fine-grain fragmentation-scattering. In *Proceedings of the 1986 International Symposium on Security and Privacy*, pages 210–221. IEEE, April 1986.

[4] Mark K. Joseph and Algirdas Avizienis. A fault-tolerance approach to computer viruses. In *Proceedings of the 1988 International Symposium on Security and Privacy*, pages 52–58. IEEE, May 1988.

[5] Michael O. Rabin. Efficient dispersion of information for security, load balancing and fault tolerance. *Journal of the ACM*, 36(2):335–348, April 1989.

[6] Adi Shamir. How to share a secret. *Communications of the ACM*, 22(11):612–613, November 1979.

The Need for a Failure Model for Security

Catherine Meadows
Naval Research Laboratory
Code 5543
Washington, DC 20375
U.S.A.
meadows@itd.nrl.navy.mil

Researchers in fault tolerance have long made use of the notion of a failure model, which describes the different ways a component in a system can fail. For example, a node can fail quietly (that is, it send out no information), it can fail with respect to timing (that is, send out information too late), it can fail arbitrarily, or it can fail maliciously. The intent of the failure model for fault tolerance is to make it possible to develop different types of algorithms that address different kinds of failures.

Security has traditionally followed a different type of approach. When security is modeled, it is assumed that the system is trying to operate in the face of a hostile adversary with unlimited capabilities in certain areas. For example, the modeling of secure operating systems assumes the existence of Trojan Horse code that is able to signal information along covert channels, while the modeling of secure protocols assumes the existence of an intruder who is able to read and modify all traffic, gain control of nodes, and compromise old secret information. In the actual development of secure systems, such assumptions may be relaxed, of course, but, for theoretical models the most stringent assumptions usually apply.

It is my opinion that it is time to start moving away from worst-case assumptions and begin developing a failure model for computer security. The reasons are two-fold: first, that the theories devoted to the more traditional computer security problems are reaching the stage of maturity where such models are necessary, and that is impossible even to consider the less traditional problems without at least the beginning of such a model.

We begin with the issue of maturity. The concentration on worst-case assumptions is characteristic of a developing theory; one's first goal is to test the limits of theory by applying it to the most severe cases possible. It is also usually paradoxically the case that it is easier to develop theories under worst-case assumptions; worst-case assumptions are usually the simplest to develop or formulate. However, as a field matures, the limitations of this approach become more clear. Theories developed for worst-case assumptions may turn out to be impractical to apply. Moreover, it may also turn out that, as we delve more deeply into the problem, that it is always possible to extend our notion of a "worst-case" assumption so that our model fails to handle it. Such has been the case in natural language processing, for example; it is always possible to come up with some kind of bizarre formulation that a given theory cannot handle but is that is nevertheless natural language.

My position is that we are beginning to reach this point in the theory of secure systems. This has been particularly noticeable in the recent developments of theories of information flow. For any theory that has been developed, it has been possible to come up with a type of covert channel that can defeat it. This has culminated in theories that are able to handle such subtle notions as probabilistic channels that convey information by varying the probability that events will occur [1].

Such theories are difficult to apply, both because of the difficulty of calculating the probabilities of events in a realistic system, and because of the amount of variables that must be considered. Perhaps at this point it is time to re-evaluate these information flow theories by developing realistic "failure models" for covert channels and evaluating theories in terms of these models.

The second reason for the need for failure models is the changing emphasis of security research. In the past research in secure computer systems has concentrated on secrecy. In theory, as long as one has sound access controls and has eliminated all covert channels, there is no way that secrecy can be compromised in a secure system, even though all code not entrusted with enforcing the security policy is actively attempting to subvert it. This is not the case with other security attributes such as integrity or denial of service. An access control mechanism can decide which processes modify what data items, and in what order, but whether the data items are modified correctly is the responsibility of the processes themselves. Likewise, an access control mechanism can assist a process in providing a service by preventing other processes from denying it accesses to resources such as memory, but it cannot guarantee that that process will actually provide the service. Thus, if we assume that processes are always actively attempting to subvert the security policy will not result in meaningful model of a secure system. On the other hand, we do not have the resources to guarantee that all processes are completely trustworthy. We will need some notion of different degrees of trust, which can be supplied by a failure model. One needs to be able to characterize the different ways that a process can attempt to subvert the security policy, and to have some means of ranking these different kinds of attacks in order of their likelyhood.

The failure model we are attempting to build should have two attributes. First of all, it should characterize the different attacks that can be made against the components of a secure system. For example, in a network, we might include both passive and active wiretapping in our failure model. Secondly, it should include some way of ranking the the different kinds of attacks in order of their likelyhood.

Much of the work on characterizing failures is spread throughout the literature, and it remains only to pull it together. Indeed, work along these lines has already been done by Landwehr et al. in [3].

The problem of ranking, however, is more of a challenge. In some cases such a ranking is straightforward. For example, the ability to perform active wiretapping subsumes the ability to perform passive wiretapping. Other cases are not so clear. For example, it is generally believed that timing channels are harder to exploit than storage channels. But there exist high capacity timing channels that are easier to exploit than many low capacity storage channels [2].

Moreover, since we are dealing with an intelligent adversary, the likelyhood of a failure can change according to circumstances. For example, as attackers become more experienced and sophisticated, certain attacks may become more likely. On the other hand, as security measures become more sophisticated, certain kinds of failures may become less likely, as attackers turn to other, easier, ways to break into systems. We conclude that any ranking of failures must be nonlinear and allow for overlaps, and such a ranking will probably have to be dynamic. Moreover, the ranking may change, not only from system to system, but during the lifetime of the system itself.

Such a failure model will not be easy to design, and we may have to be satisfied by approximate success. However, it is my belief that even a limited success in this direction will vastly improve our ability to evaluate the success of our attempts to provide security.

References

[1] J. W. Gray III. Toward a Mathematical Foundation for Computer Security. In *Proceedings of the 1991 IEEE Symposium on Security and Privacy*, pages 21–34. IEEE Computer Society Press, May 1991.

[2] W. M. Hu. Reducing Timing Channels with Fuzzy Time. In *Proceedings of the 1991 IEEE Symposium on Security and Privacy*, pages 8–24. IEEE Computer Society Press, May 1991.

[3] C. E. Landwehr, A. R. Bull, J. P. McDermott, and W. Choi. A Taxonomy of Computer Program Security Flaws with Examples. NRL Report 9591, Naval Research Laboratory, November 1993.

Reliability and Security

Peter G. Neumann*
Computer Science Laboratory
SRI International
Menlo Park, California 94025-3493
U.S.A.
neumann@csl.sri.com

Yogi Berra, the great American sage, might have said

It ain't dependable until its dependable.

Considerable commonality exists between reliability and security. Both are weak-link phenomena. Good design involves avoiding single weak links. However, on many occasions, multiple weak links can be, and have been, known to occur — accidentally or intentionally.

From the anecdotal evidence (such as the archives of the Risks Forum), there are clearly similarities among some of the causes of accidental and intentional events. Many accidental cases could alternatively have occurred intentionally. For example, the 1980 ARPAnet collapse and the 1990 AT&T network-wide propagation of network node crashes could each have been triggered intentionally. The serious outages that resulted from a variety of cables being accidentally cut (Chicago, New York, White Plains, etc.) would have had the same results if they been perpetrated intentionally. Cases of accidental electromagnetic interference could have arisen as malicious rather than natural acts. Similarly, some intentional acts could alternatively have occurred accidentally (such as rm *). A hybrid case is provided by the Internet Worm of November 1988, in which a combination of intentional and accidental circumstances greatly increased the magnitude of the consequences.

Certain security measures may be desirable to hinder malicious penetrators, but do relatively little to reduce hardware faults and software mistakes in design and implementation. Certain

---

reliability measures may be desirable to provide hardware fault tolerance, but do not increase security. On the other hand, a combination of good system architectural approaches and sound software engineering practice (with abstraction, layering, encapsulation, etc.) can enhance both security and reliability. It is highly desirable to consider both within a common system framework, and then to show how certain application properties such as human safety depend upon them.

The bottom line is that we must anticipate threats to both reliability and security, and must not dwell on artificial distinctions between accidental and intentional threats. We must anticipate all threats to system and network dependability. Yogi-stically,

> *It ain't secure if it ain't reliable.*
> *It ain't reliable if it ain't secure.*

Fault Tolerance and Security

Brian Randell
Department of Computer Science
University of Newcastle upon Tyne
Newcastle upon Tyne, NE1 7RU, UK
Brian.Randell@newcastle.ac.uk

When some years ago my colleague John Dobson and I presented a paper[2] to the Oakland Conference, suggesting that many fault tolerance techniques that were standard in the reliability world were of equal potential utility in the security field, our paper (which was entitled Building Reliable Secure Systems out of Unreliable Insecure Components) was regarded as quite controversial. Our paper was an outcome of our belated recognition of a conceptual inadequacy of earlier work at Newcastle on the design of multi-level secure distributed systems[5]. This work had avoided the many problems involved in producing and formally validating a general purpose operating system kernel for use as a trusted computing base, by instead isolating separate security domains into physically separate computers, each running standard Unix, and using encrypted inter-computer communications to impose the required security policy. This was done without modifying standard system interfaces, as was parallel work on at Newcastle on building TMR-based highly reliable Unix systems. It was therefore very easy to construct distributed multi-level secure Unix systems out of highly reliable (i.e., triplicated) computers, and to know that the security and the reliability mechanisms could not interfere with each other.

But we did not have means of allowing for reliability problems in the security mechanisms themselves - we implicitly assumed that these mechanisms were by (some unspecified means) designed, implemented and validated in such a way as to be utterly trustworthy. Yet the general design approach we had developed years earlier at Newcastle based on the concept of idealized fault-tolerant components[1] provided an immediate solution - at least in principle - to this problem. This viewpoint led to our realizing that the computer security community was making surprisingly little use of fault tolerance techniques to complement the fault avoidance techniques on which it was placing so much reliance. Yet, given our view of both security and reliability as special aspects of a more general concept of dependability, such combined use of fault tolerance and fault avoidance (and also fault removal) was to us the obvious way to go.

We had fun in our Oakland paper pointing out that work on computer security seemed to have been ignoring the lessons of history - in medieval times no military architect would have thought of protecting the security of a castle with but a single line of defence. Rather there would be defences in depth (moats *and* walls, drawbridges *and* portcullises, etc.). *And* there would be predesigned strategies for coping with breaches of particular defences - in other words, of handling exceptions! We therefore suggested that, for example, instead of relying solely on formal validation of a user authenticator, one might employ N-version programming. A number of independent teams would each be tasked with building authenticator sub-systems, which would then be integrated together using some appropriate sort of voting scheme (perhaps unanimity rather than majority). And as another example we sketched a scheme in which a filing system and its underlying disk management system each had their own individually appropriate security specification, and means of attempting to meet this specification - with the filing system taking responsibility for trying to ensure that any disk-level security breaches that nevertheless occurred did not escalate to user-level.

Somewhat to our surprise, it was later claimed that the use of fault tolerance, necessitating as it does the use of redundancy, was *conceptually* inappropriate to security (see, for example,[4];[6]). For example, it was argued that if there was redundant information (whether code or data), this increased the amount of information which an intruder could attempt to access and perhaps modify, and thus necessarily eased his/her task. The claimed conceptual inappropriateness of fault tolerance for security is the issue that I wish to address here.

Clearly, merely providing redundancy is counter-productive — in fact of both reliability and security. Fault tolerance involves redundancy and consistency checking of this redundancy. The aim is always to try to reduce the likelihood of faults turning into failures failures being defined with respect to some dependability specification. A triply-modular-redundant system, for example, does not consist of just a triplicated component rather this set of components are embedded in a voting scheme architecture. However, a TMR system whose voters are inadequately reliable (either by virtue of the ease with which they can be by-passed, or the inaccuracy of their voting procedures) will not be an improvement on the basic single component. And the base components have to have some minimum level of reliability before their use in a TMR system will be of benefit.

When we consider the use of fault tolerance for purposes of security, the relevant failures and hence faults must be defined with respect to appropriate security specifications, *at each level of the system.* (And when, as is usually the case, reliability *and* security are needed, such specifications must cover both aspects.) Components and architectures which are totally inadequate with regard to their individual security specifications will not in combination provide enhanced system security. For example, if the triplicated user authorization systems authenticate non-users, or refuse to authenticate bona fide users, too frequently, their use in a TMR system will do little to stop intruders getting in, and will increase the likelihood of service being denied to bona fide users, respectively. It is failure to appreciate this point that has led to the claimed special inappropriateness of fault tolerance to security. In fact, rather than being special to security, the underlying issue relates equally to reliability. In both cases, replicated components constitute a greater amount of things that can go wrong, or can suffer damage, and so increase the fault rate they have to have appropriate characteristics and be used in an appropriate architecture for fault tolerance to outweigh this disadvantage (and the additional complexity involved) and result in a decreased failure rate. (The fact that the definition of appropriate specifications and architectures can be a demanding task[4] does not invalidate the conceptual point I am making indeed it is my experience that the conceptual clarity introduced by viewing security and reliability together, as aspects of the single concept of dependability, can assist with this task[3].)

One final point: in principle, conventional fault tolerance design techniques, based on estimated fault frequencies, can enable fault-tolerant systems to be designed that it is known beforehand will achieve specified levels of dependability. The fact that trustworthy quantitative estimates of the likelihood of security faults are rarely available does not invalidate the idea of using fault tolerance for security purposes, any more than it invalidates the use of design fault tolerance, where such estimation is similarly difficult. Rather it throws one back on that good old stand-by engineering judgement but this after all is what is already presumably being employed in any decisions about the adequacy of, say, formal validation activities (activities whose efficacy are equally difficult to estimate quantitatively) to achieve systems that have the required security characteristics.

References

[1] T. Anderson and P.A. Lee, *Fault Tolerance: Principles and Practice*, Prentice Hall, 1981.

[2] J.E. Dobson and B. Randell, "Building Reliable Secure Systems out of Unreliable Insecure Components," *Proc. Conf. on Security and Privacy*, Oakland, California, IEEE, 1986.

[3] J.C. Fabre and B. Randell, "An Object-Oriented View of Fragmented Data Processing for Fault and Intrusion Tolerance in Distributed Systems," *Proc. ESORICS~'92*, Toulouse, 1992.

[4] M. Joseph, "Integration problems in Fault-Tolerant, Secure Computer Design," *Proc.1st IFIP Int. Conf. on Dependable Computing for Critical Applications (DCCA)*, pp. 348-364, Santa Barbara, California, Springer-Verlag, 1989. (Published as Dependable Computing for Critical Applications, ISBN 0-387-82249-6)

[5] J.M. Rushby and B. Randell, "A Distributed Secure System," *Computer*, vol. 16, no. 7, pp.55-67, 1983.

[6] R. Turn and J. Habibi, "On the Interactions of Security and Fault Tolerance," *Proc. 9th National Security Conf.*, pp. 585-594, 1986.

Common Techniques in Fault Tolerance and Security (and Performance!)

Kent D. Wilken
Department of Electrical and Computer Engineering
University of California, Davis
Davis, California 95616-5294
U.S.A.
wilken@ece.ucdavis.edu

The commonality between fault tolerance techniques and security techniques is considered here as it relates to the design of general purpose computer systems. The thesis of this presentation is that although fault tolerance and security are important computer system attributes, the computer industry is driven by cost/performance. It is incumbent upon fault-tolerant-computing and computer-security researchers to seek techniques and mechanisms that have commonality with those used to achieve high performance, so that new fault tolerance or security measures have a minimal impact on cost/performance and are likely to be included in new systems. An example of a traditional technique that provides fault tolerance, security, and high performance is given below, and an emerging approach is shown to be connected to each of these properties. Other promising areas are suggested where connections among these three computer system properties might be established.

First, consider a textbook example that might not commonly be considered as a mechanism that improves cost/performance and provides security and fault tolerance: virtual memory. An important aspect of virtual memory is that it achieves good cost/performance in the memory hierarchy by exploiting locality. However, the page table that virtual memory requires is a convenient location to store access control information, and the virtual-to-real address translation provides a convenient time to validate access privileges for security. Various types of software faults cause a memory location to be accessed that is not allocated. The virtual memory mechanism can detect such accesses to non-allocated pages, generating a 'segmentation fault' on Unix systems when this occurs.

As a second example, a connection is established among techniques and mechanisms that monitor programs to detect hardware faults, to detect computer viruses, and to profile a program's performance to improve compiler optimizations. Several researchers have proposed *signature monitoring* techniques that use a variation of block error detection codes to encode a program at compile time and later check the program during execution, allowing concurrent detection of certain processor and program-memory errors [5]. Joseph and Avizienis [3] and Wilken and Shen [8] established a connection between signature monitoring and computer virus detection based on the notion that signature monitoring can detected an improperly signatured virus during program execution. However, to be effective against the computer virus threat, signatures must be concealed using cryptographic techniques. Although the cryptographic techniques increase the cost of the monitoring mechanism, the increased cost is offset by the added ability to monitor for hardware faults and for computer viruses.

For signature monitoring, the signatures that are embedded in the program reduce performance because during certain clock cycles a signature instruction must executed rather than a normal program instruction. Wilken [6] shows that minimum performance overhead can be achieved by placing justifying signature instructions at locations that create a maximum weight spanning tree in the program control flow graph. The resulting object code satisfies Kirchhoff's second law, namely the signature (voltage) around any program graph cycle must be zero. Interestingly, recent work in program profiling uses a maximum spanning tree to allow a program to be profiled with minimum overhead [2], a result that Knuth shows can be derived from Kirchhoff's first law [4]. The significant cost for signature monitoring is the performance overhead and the memory overhead for the signature instructions. A principle cost for on-line program profiling is the performance overhead and the memory overhead for the instruction that increments each profile counter. However, because these two type of monitoring probes have the same spanning tree location requirements, the cost of each probe could be shared by including a new instruction in the architecture, INCREMENT & SIGNATURE, an instruction that increments the specified destination register, and uses the instruction's immediate field to hold a justifying signature. INCREMENT & SIGNATURE could be included in a typical RISC architecture with very simple modification. The signature could be concealed as proposed in [8], creating a monitoring mechanism that facilitates high performance, fault tolerance, and security.

Fault-tolerant-computing and computer-security researchers should try to discover connections among performance, fault tolerance and security beyond the examples given here. One promising area is using compiler and architecture techniques, which are generally associated with achieving high performance. Recent work in fault tolerant computing uses compiler and architecture techniques to achieve low-cost software and hardware fault detection [7], and to provide low-cost rollback and recovery [1]. Using compiler and architecture techniques to improve security seems to be a promising notion. For example, it might be possible for a secure compiler to insert covert channels in valid software that could be observed by an architected monitor, allowing software used by intruders to be detected.

References

[1] N.J. Alewine, S.-K. Chen, C.-C. Li, W.K. Fuchs, and W.-M. Hwu. Branch recovery with compiler-assisted multiple instruction retry. In *Proc. 22nd FTCS*, pages 66–73. IEEE, 1992.

[2] T. Ball and J. Larus. Optimally profiling and tracing programs. In *Principles of Programming Languages*, pages 59–70, January 1992.

[3] M. Joseph and A. Avizienis. A fault tolerance approach to computer viruses. In *Proc. Symp. on Security and Privacy*, pages 52–58. IEEE, 1988.

[4] D. Knuth. *The Art of Computer Programming: Vol. 1, Fundamental Algorithms*, volume 1. Addison Wesley, 1973.

[5] A. Mahmood and E. McCluskey. Concurrent error detection using watchdog processors - a survey. *IEEE Transactions on Computers*, 37(2):160–174, Feb. 1988.

[6] K. Wilken. An optimal graph-construction approach to placing program signatures for signature monitoring. *IEEE Transactions on Computers*, 42(11):1372–1381, Nov 1993.

[7] K. Wilken and T. Kong. Efficient memory access checking. In *Proc. 23nd FTCS*, pages 566–575. IEEE, 1993.

[8] K. Wilken and J. Shen. Concurrent error detection using signature monitoring and encryption. In *Proc. 1st IFIP Working Conference on Dependable Computing for Critical Applications*, pages 365–384. Springer-Verlag, 1991.

Real-Time Systems

Upper and Lower Bounds on the Number of Faults a System Can Withstand Without Repairs*

Michel Goemans, Nancy Lynch and Isaac Saias
Laboratory for Computer Science
Massachusetts Institute of Technology
Cambridge, Massachusetts 02139
U.S.A.
{goemans,lynch,saias}@theory.lcs.mit.edu

Abstract

We consider the following scheduling problem. A system is composed of n processors drawn from a pool of N. The processors can become faulty while in operation and faulty processors never recover. A report is issued whenever a fault occurs. This report states only the existence of a fault, but does not indicate its location. Based on this report, the scheduler can reconfigure the system and choose another set of n processors. The system operates satisfactorily as long as at most f of the n selected processors are faulty. We exhibit a scheduling strategy allowing the system to operate satisfactorily until approximately $(N/n)f$ faults are reported in the worst case. Our precise bound is tight.

Key words: fault tolerance, maximum matching, redundancy, scheduling.

1   Introduction

Many control systems are subject to failures that can have dramatic effects. One simple way to deal with this problem is to build in some redundancy so that the whole system is able to

*Research supported by research contracts AFOSR-89-0271, ONR-N00014-91-J-1046, NSF-CCR-8915206 and DARPA-N00014-89-J-1988.

function even if parts of it fail. In a general situation, the system's manager has access to some observations allowing it to control the system efficiently. Such observations bring information about the state of the system that might consist of partial fault reports. The available controls might include repairs and/or replacement of faulty processors.

To model the problem, one needs to make assumptions regarding the occurrence of faults. Typically, they are assumed to occur according to some stochastic process. To make the model more tractable, one often considers the process to be memoryless, i.e. faults occur according to some exponential distribution. However, to be more realistic, many complications and variations can be introduced in the stochastic model, and they complicate the time analysis. Examples are: a processor might become faulty at any time or only during specific operations; the fault rate might vary according to the work load; faults might occur independently among the processors or may depend on proximity. The variations seem endless and the results are rarely general enough so as to carry some information or methodology from one model to another.

One way to derive general results, independent of the specific assumptions about the time of occurrence of faults, is to adopt a *logical time* that, instead of following an absolute frame, is incremented only at each occurrence of a fault. Within this framework, we measure the maximal number of faults to be observed until the occurrence of a crash instead of the maximal time of survival of a system until the occurrence of a crash.

As an introduction to this general situation, we make the following assumptions and simplifications:

Redundancy of the system: We assume the existence of a pool $\mathcal{N}$ composed of N identical processors from among which, at every time t, a set S_t of n processors is selected to configure the system. The system works satisfactorily as long as at least $n - f$ processors among the n currently in operation are not faulty. However, the system cannot tolerate more than f faults at any given time: it stops functioning if $f + 1$ processors among these n processors are faulty.

Occurrence of faults, reports and logical time: We consider the situation where only processors currently under use can become faulty and where failures do not occur simultaneously. We furthermore assume that, whenever a processor fails, a report is issued, stating that a failure has occurred, but without specifying the location of the failure. (Reporting additional information might be too expensive or time consuming.) Based on these reports, the scheduler might decide to reconfigure the system whenever such failure is reported. As a result, we restrict our attention to the discrete model, in which time t corresponds to the t-th failure in the system.

Repairs: No repair is being performed.

Deterministic Algorithms: We assume that the scheduler does not use randomness.

Since the universe consists of only N processors, and one processor fails at each time, no scheduling policy can guarantee that the system survives beyond time N. (A better a priori upper bound is $N - n + f + 1$: at this time, only $n - f - 1$ processors are still non-faulty. This does not allow for the required quorum of $n - f$ non-faulty processors.) But some scheduling policies seem to allow the system to survive longer than others. An obviously bad policy is to choose n processors once and for all and never to change them: the system would then collapse at time $f + 1$. This paper investigates the problem of determining the best survival time.

This best survival time is defined from a worst-case point-of-view: a *given* scheduler allows the system to survive (up to a certain time) only if it allows it to survive against *all* possible failure patterns in which one processor fails at each time.

Our informal description so far apparently constrains the faults to occur in on-line fashion: for each t, the t-th fault occurs before the scheduler decides the set S_{t+1} to be used subsequently. However, since we have assumed that no reports about the locations of the faults are available, there is no loss of generality in requiring the sets S_t to be determined a priori. (Of course, in practice, some more precise fault information may be available, and each set S_t would depend on the fault pattern up to time t.) Also, as we have assumed a *deterministic* scheduler, we can assume that the decisions $S_1, \ldots, S_N$ are *revealed* before the occurrence of any fault. We express this by saying that the faults occur in an off-line fashion.

2   The Model

Throughout this paper, we fix a universal set $\mathcal{N}$ of processors, and let N denote its cardinality. We also fix a positive integer n ($n \leq N$) representing the number of processors that are needed at each time period, and a positive integer f representing the number of failures that can be tolerated ($f < n$).

We model the situation described in the introduction as a simple game between two entities, a *scheduler* and an *adversary*. The game consists of only one round, in which the scheduler plays first and the adversary second. The scheduler plays by selecting a sequence of N sets of processors (the *schedule*), each set of size n, and the adversary responds by choosing, from each set selected by the scheduler, a processor to kill. We consider only sequences of size N because the system must collapse by time N, since, by definition, a new processor breaks down at each time period.

Formally, a schedule $\mathcal{S}$ is defined to be a finite sequence, $S_1, \ldots, S_N$, of subsets of $\mathcal{N}$, such that $|S_t| = n$ for all t, $1 \leq t \leq N$. An $\mathcal{S}$-*adversary* $\mathcal{A}$ is defined to be a finite sequence, $s_1, \ldots, s_N$, of elements of $\mathcal{N}$ such that $s_t \in S_t$ for every t.

Let $\mathcal{S}$ be a schedule, and $\mathcal{A}$ an $\mathcal{S}$-adversary. Define the *survival time*, $T(\mathcal{S}, \mathcal{A})$, to be the largest value of t such that, for all $u \leq t$, $|\{s_1, \ldots s_u\} \cap S_u| \leq f$. That is, for all time periods u up to and including time period t, there are no more than f processors in the set S_u that have failed by time u.

We are interested in the minimum survival time for a particular schedule, with respect to arbitrary adversaries. Thus, we define the *minimum survival time* for a schedule, $T(S)$, to be $T(S) \stackrel{\text{def}}{=} \min_A T(S, A)$. In this definition, the minimum is taken over all S-adversaries. An adversary A for which $T(S) = \min_A T(S, A)$ is said to be *optimum* for S. Finally, we are interested in determining the schedule that guarantees the greatest minimum survival time. Thus, we define the *optimum survival time* T_{opt}, to be $\max_S T(S) = \max_S \min_A T(S, A)$. Also define a schedule S to be *optimum* provided that $T(S) = T_{opt}$. Our objectives in this paper are to compute T_{opt} as a function of N, n and f, to exhibit an optimum schedule, and to determine an optimum adversary for each schedule.

3   The Result

Recall that $1 \leq f < n \leq N$ are three fixed integers. Our main result is stated in terms of the following function defined on the set of positive real numbers (see Figure 1):

$$h_{n,f}(k) \stackrel{\text{def}}{=} \left\lfloor \frac{k}{n} \right\rfloor f + \left(k - \left\lfloor \frac{k}{n} \right\rfloor n + f - n \right)^+ ,$$

where $(x)^+ = \max(x, 0)$. In particular, $h_{n,f}(k) = \frac{k}{n} f$ when n divides k.

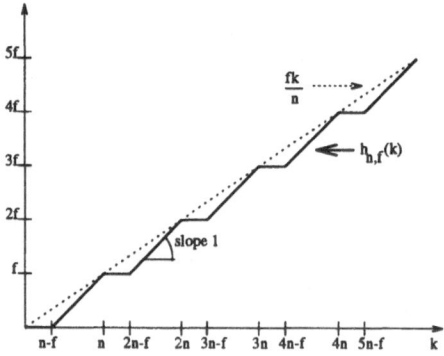

Figure 1: The function $h_{n,f}(k)$

The main result of this paper is:

Theorem 3.1

$$T_{opt} = h_{n,f}(N).$$

We will present our proof in two lemmas proving respectively that T_{opt} is no smaller and no bigger than $h_{n,f}(N)$.

Lemma 3.2

$$T_{opt} \geq h_{n,f}(N).$$

Proof: Consider the schedule $S_{trivial}$ in which the N processors are partitioned into $\lfloor \frac{N}{n} \rfloor$ batches of n processors each and one batch of $p = N - \lfloor \frac{N}{n} \rfloor n$. Each of the first $\lfloor \frac{N}{n} \rfloor$ batches is used f time periods and then set aside. Then, the last batch of processors along with any $n - p$ of the processors set aside is used for $(f + p - n)^+$ time periods. It is easy to see that no adversary can succeed in killing $f + 1$ processors within a batch before this schedule expires. ∎

In order to prove the other direction of Theorem 3.1, we need the following result about the rate of increase of the function $h_{n,f}(k)$.

Lemma 3.3 *For $0 \leq k$ and $0 \leq l \leq n$ we have $h_{n,f}(k) \leq h_{n,f}(k+l) + n - l - f$.*

Proof: Notice first that $h_{n,f}(k) = h_{n,f}(k+n) - f$ for all $k \geq 0$. Moreover, the function h increases at a sublinear rate (see Figure 1) so that, for $p, q \geq 0$, we have $h_{n,f}(p+q) \leq h_{n,f}(p)+q$. Letting $p = k + l$ and $q = n - l$, we obtain

$$h_{n,f}(k) = h_{n,f}(k+n) - f \leq h_{n,f}(k+l) + n - l - f,$$

which proves the lemma. ∎

4 The Upper Bound

In this section we establish the other direction of the main theorem. We begin with some general graph theoretical definitions.

Definition 4.1

- For every vertex v of a graph G, we let $\gamma_G(v)$ denote the set of vertices adjacent to v. We can extend this notation to sets: for all sets C of vertices $\gamma_G(C) \stackrel{\text{def}}{=} \cup_{v \in C} \gamma_G(v)$.

- For every bipartite graph G, $\nu(G)$ denotes the size of a maximum matching of G.

- For every pair of positive integers L, R, a *left totally ordered* bipartite graph of size (L, R) is a bipartite graph with bipartition $\mathcal{L}, \mathcal{R}$, where $\mathcal{L}$ is a totally ordered set of size L and $\mathcal{R}$ is a set of size R. We label $\mathcal{L} = \{a_1, \ldots, a_L\}$ so that, $a_i < a_j$ for every $1 \le i < j \le L$. For every $\mathcal{L}' \subseteq \mathcal{L}$ and $\mathcal{R}' \subseteq \mathcal{R}$, the subgraph induced by $\mathcal{L}'$ and $\mathcal{R}'$ is a left totally ordered bipartite graph with the total order on $\mathcal{L}$ inducing the total order on $\mathcal{L}'$.

- Let G be a left totally ordered bipartite graph of size (L, R). For $t = 1, \ldots, L$, we let $I_t(G)$ denote the left totally ordered subgraph of G induced by the subsets $\{a_1, a_2, \ldots, a_{t-1}\} \subseteq \mathcal{L}$ and $\gamma_G(a_t) \subseteq \mathcal{R}$.

Let us justify quickly the notion of left total order. In this definition, we have in mind that $\mathcal{L}$ represents the labels attached to the different times, and that $\mathcal{R}$ represents the labels attached to the available processors. The times are naturally ordered. The main argument used in the proof is to reduce an existing schedule to a shorter one. In doing so, we in particular select a subsequence of times. Although these times are not necessarily consecutive, they are still naturally ordered. The total order on $\mathcal{L}$ is the precise notion formalizing the ordering structure characterizing time.

Consider a finite schedule $S = S_1, \ldots, S_T$. In graph theoretic terms, it can be represented as a left totally ordered bipartite graph G with bipartition $\mathcal{T} = \{1, 2, \ldots, T\}$ and $\mathcal{N} = \{1, 2, \ldots, N\}$. There is an edge between vertex $t \in \mathcal{T}$ and vertex $i \in \mathcal{N}$ if the processor i is selected at time t. The fact that, for all t, $|S_t| = n$ translates into the fact that vertex $t \in \mathcal{T}$ has degree n. For such a bipartite graph, the game of the adversary consists in selecting one edge incident to each vertex $t \in \mathcal{T}$.

Observe that the adversary can kill the schedule at time t if it has already killed, before time t, f of the n processors used at time t. It then kills another one at time t and the system collapses. In terms of the graph G, there exists an adversary that kills the schedule at time t if and only if the subgraph $I_t(G)$ has a matching of size f, i.e. $\nu(I_t(G)) \ge f$. Therefore, the set $\mathcal{P}$ that we now define represents the set of integers L and R for which there exists a schedule that survives at time L, when R processors are available.

Definition 4.2 Let L and R be two positive integers. $(L, R) \in \mathcal{P}$ iff there exists a left totally ordered bipartite graph G of size (L, R) with bipartition $\mathcal{L}$ and $\mathcal{R}$ satisfying the two following properties:

1. All vertices in $\mathcal{L}$ have degree exactly equal to n,

2. For every $t = 1, \ldots, |\mathcal{L}|$, all matchings in $I_t(G)$ have size at most equal to $f - 1$, i.e. $\nu(I_t(G)) \le f - 1$.

The main tool used in the proof of Theorem 3.1 is the following duality result for the maximum bipartite matching problem, known as Ore's Deficiency Theorem [3]. A simple proof of this theorem and related results can be found in [2].

Theorem 4.1 *Let G be a bipartite graph with bipartition A and B. Then the size $\nu(G)$ of a maximum matching is given by the formula:*

$$\nu(G) = \min_{C \subseteq B} \left[|B - C| + |\gamma_G(C)| \right]. \tag{1}$$

The following lemma is crucial for our proof.

Lemma 4.2 *There are no positive integers L and R such that $(L, R) \in \mathcal{P}$ and such that $L > h_{n,f}(R)$.*

Proof:

Working by contradiction, consider two positive integers L and R such that $(L, R) \in \mathcal{P}$ and $L > h_{n,f}(R)$. We first show the existence of two integers L' and R' such that $L' < L$, $(L', R') \in \mathcal{P}$ and $L' > h_{n,f}(R')$.

Let $\mathcal{L} = \{a_1, a_2, \ldots, a_L\}$ and $\mathcal{R} = \{b_1, b_2, \ldots, b_R\}$ be the bipartition of the graph G whose existence is ensured by the hypothesis $(L, R) \in \mathcal{P}$.

We apply Theorem 4.1 to the graph $I_L(G)$ where we set $A = \{a_1, a_2, \ldots, a_{L-1}\}$ and $B = \gamma_G(a_L)$. Let C denote a subset of B for which the minimum in (1) is attained. (C is possibly empty.) Define $\mathcal{L}' \stackrel{\text{def}}{=} \mathcal{L} - (\{a_L\} \cup \gamma_{I_L(G)}(C))$ and $\mathcal{R}' \stackrel{\text{def}}{=} \mathcal{R} - C$ and let L' and R' denote the cardinalities of $\mathcal{L}'$ and $\mathcal{R}'$. Hence, $L' = L - 1 - |\gamma_{I_L(G)}(C)|$ so that $L' < L$. Consider the bipartite subgraph G' of G induced by the set of vertices $\mathcal{L}' \cup \mathcal{R}'$. In other words, in order to construct G' from G, we remove the set $C \cup \{a_L\}$ of vertices and *all* vertices adjacent to some vertex in C. We have illustrated this construction in Figure 2. In that specific example, $n = 4$, $f = 3$, $L = 6$ and $R = 7$, and $h_{4,3}(7) = 5$. One can show that $C = \{b_5, b_6, b_7\}$ and, as a result, G' is the graph induced by the vertices $\{a_1, a_2, a_3, a_4, b_1, b_2, b_3, b_4\}$. The graph G' has size $(L', R') = (4, 4)$.

We first show that $(L', R') \in \mathcal{P}$. Since the vertices in $\mathcal{L}'$ correspond to the vertices of $\mathcal{L} - \{a_L\}$ not connected to C, their degree in G' is also n. Furthermore, G', being a subgraph of G, inherits property 2 of Definition 4.2. Indeed, assume that there is a vertex $a_{t'}$ in G' such that $I_{t'}(G')$ has a matching of size f. Let t be the label of the corresponding vertex in graph G. Since the total order on $\mathcal{L}'$ is induced by the total order on $\mathcal{L}$, $I_{t'}(G')$ is a subgraph of $I_t(G)$. Therefore, $I_t(G)$ would also have a matching of size f, a contradiction.

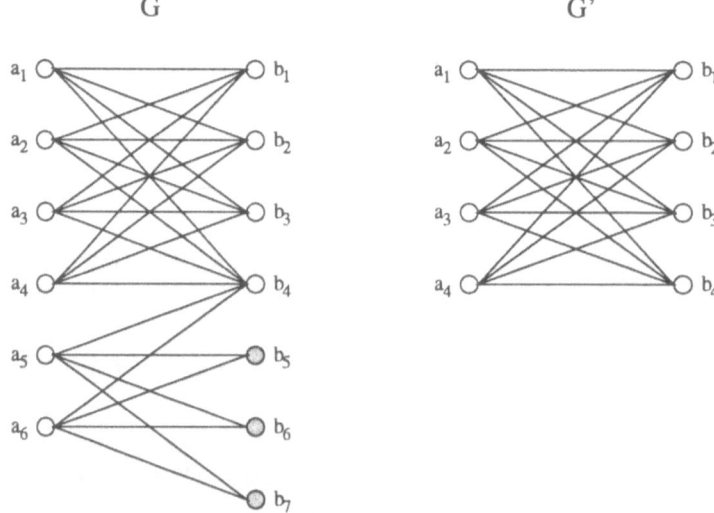

Figure 2: An example of the construction of G' from G. The vertices in C are darkened.

Let us show that $L' > h_{n,f}(R')$. The assumption $(L, R) \in \mathcal{P}$ implies that $f - 1 \geq \nu(I_L(G))$. Using Theorem 4.1 and the fact that $B = \gamma_G(L)$ has cardinality n, this can be rewritten as

$$
\begin{aligned}
f - 1 &\geq \nu(I_L(G)) = |B - C| + |\gamma_{I_L(G)}(C)| \\
&= n - |C| + |\gamma_{I_L(G)}(C)|.
\end{aligned} \tag{2}
$$

Since $C \subseteq B \subseteq \mathcal{R}$, we have that $0 \leq |C| \leq n \leq R$ and, thus, the hypotheses of Lemma 3.3 are satisfied for $k = R - |C|$ and $l = |C|$. Therefore, we derive from the lemma that

$$
h_{n,f}(R') = h_{n,f}(R - |C|) \leq h_{n,f}(R) + n - |C| - f.
$$

Using (2), this implies that

$$
h_{n,f}(R') \leq h_{n,f}(R) - |\gamma_{I_L(G)}(C)| - 1.
$$

By assumption, L is strictly greater than $h_{n,f}(R)$, implying

$$
h_{n,f}(R') < L - 1 - |\gamma_{I_L(G)}(C)|.
$$

But the right-hand-side of this inequality is precisely L', implying that $L' > h_{n,f}(R')$.

We have therefore established that for all integers L and R such that $(L, R) \in \mathcal{P}$ and $L > h_{n,f}(R)$, there exists two integers L' and R' such that $L' < L$, $(L', R') \in \mathcal{P}$ and $L' > h_{n,f}(R')$. Among all such pairs (L, R), we select the pair for which L is minimum. By the result that we just established, we obtain a pair (L', R') such that $(L', R') \in \mathcal{P}$ and $L' < L$. This contradicts the minimality of L.

■

Lemma 4.3
$$
T_{opt} \leq h_{n,f}(N).
$$

Proof: By assumption, $(T_{opt}, N) \in \mathcal{P}$. Hence this result is a direct consequence of Lemma 4.2

■

This Lemma along with Lemma 3.2 proves Theorem 3.1.

In the process of proving Lemma 3.2 we proved that $S_{trivial}$ is an optimum schedule. On the other hand, the interpretation of the problem as a graph problem also demonstrates that the adversary has a polynomial time algorithm for finding an optimum killing sequence for each schedule S. When provided with S, the adversary needs only to compute a polynomial number (actually fewer than N) of maximum bipartite matchings, for which well known polynomial algorithms exist (for the fastest known, see [1]).

5 Future Research

The problem solved in this paper is a first step towards modeling complex resilient systems and there are many interesting extensions. We mention only a few.

An interesting extension is to consider the case of a system built up of processors of different types. For instance consider the case of a system built up of a total of n processors, that is reconfigured at each time period and that needs at least g_1 non-faulty processors of type 1 and at least g_2 non-faulty processors of type 2 in order to function satisfactorily. Assume also that these processors are drawn from a pool $\mathcal{N}_1$ of N_1 processors of type 1 and a pool $\mathcal{N}_2$ of N_2 processors of type 2, that $\mathcal{N}_1 \cap \mathcal{N}_2 = \emptyset$, and that there are no repairs. It is easy to see that the optimum survival time T_{opt} is at least the survival time of every strategy for which the number of processors of type 1 and type 2 is kept constant throughout. Hence:

$$T_{opt} \geq \max_{\{(n_1,n_2); n_1+n_2=n\}} \min \left(h_{n_1,n_1-g_1}(N_1), h_{n_2,n_2-g_2}(N_2) \right).$$

It would be an interesting question whether T_{opt} is exactly equal to this value or very close to it.

Extend the definition of a *scheduler* to represent a randomized scheduling protocol. (Phrased in this context, the result presented in this paper is only about deterministic scheduling protocols.) A scheduler is called *adversary-oblivious* if it decides the schedule independently of the choices $s_1, s_2, \ldots$ made by the adversary. An *off-line* adversary is an adversary that has access to the knowledge of the full schedule $S_1, S_2, \ldots$ before deciding the full sequence $s_1, s_2, \ldots$ Note that, by definition, off-line adversaries make sense only with adversary-oblivious schedulers. By comparison, an *on-line* adversary decides for each time t which processor s_t to kill, without knowing the future schedule: at each time t the adversary decides s_t based on the sole knowledge of $S_1, \ldots, S_t$ and of $s_1, \ldots, s_{t-1}$. In this more general framework, the quantity we want to determine is

$$T_{opt} \stackrel{\text{def}}{=} \max_{S} \min_{A} E\left[T(S, A) \right]. \tag{3}$$

For an adversary-oblivious, randomized scheduler, one can consider two cases based on whether the adversary is on-line or off-line. As is easily seen, if the adversary is off-line, randomness does not help in the design of optimal schedulers: introducing randomness in the schedules cannot increase the survival time if the adversary gets full knowledge of the schedule before committing to any of its choices. As a result, the off-line version corresponds to the situation investigated in this paper.

It would be of interest to study the online version of Problem (3). On-line adversaries model somewhat more accurately practical situations: faults naturally occur in an on-line fashion and the role of the program designer is then to design a scheduler whose expected performance is optimum. Hence, comparing the two versions of Problem 3 would allow to understand how much randomness can help in the design of optimum, adversary-oblivious, schedulers.

For instance, in the case where $N = 4, n = 2$ and $f = 1$, and where on-line adversaries are considered, a case analysis shows that T_{opt} is equal to 9/4 for randomized algorithms. A direct application of Theorem 3.1 shows that that $T_{opt} = 2$ for deterministic algorithms. [4] presents a full treatment of the case $f = 1$. It is shown that, in this case, the optimal randomized strategy is the greedy strategy: "At each time t, select the set S_t uniformly at random among the sets maximizing the probability of survival at time $t + 1$ when the adversary makes its decisions at random."

Going towards even more realistic and complex situations, we can also study the case where the scheduler is provided at each time with some partial information about the fault process.

Acknowledgments

We want to thank Stuart Adams from the Draper Laboratories who suggested the problem to us and George Varghese for helpful suggestions.

References

[1] J.E. Hopcroft and R.M. Karp, "A $n^{5/2}$ algorithm for maximum matching in bipartite graphs", *SIAM J. Computing*, 2, 225–231 (1973).

[2] L. Lovász and M. Plummer, "Matching theory", *North-Holland Mathematical Studies 121*, Annals of Discrete Mathematics, 29, (1986).

[3] O. Ore, "Graphs and matching theorems", *Duke Math. J.*, 22, 625–639 (1955).

[4] I. Saias, "Proving Correctness of Randomized Distributed Algorithms", *Ph.D. Thesis, Massachusetts Institute of Technology*, to be completed June 1994.

Scheduling Fault Recovery Operations for Time-Critical Applications

Sandra Ramos-Thuel and Jay K. Strosnider
Department of Electrical and Computer Engineering
Carnegie Mellon University
Pittsburgh, Pennsylvania 15213
U.S.A.

Abstract

This paper introduces algorithms for scheduling fault recovery operations on systems which must preserve the timing correctness of critical application tasks in the presence of faults. The algorithms are based on methods to reserve time for the processing of recovery tasks at the design stage. This allows recovery tasks to be scheduled with very low run-time overhead, complementing or reducing the need for hardware replication to support dependable operation. Although previous work has advocated the use of reservation methods, there exists no formal methodology for allocating such time. A methodology is developed which facilitates the difficult task of verifying the timing correctness of a desired reservation strategy. In addition, simulation results are presented which give insight into the effectiveness of different reservation strategies in averting timing failures under a variety of transient recovery loads. [1]

1   Introduction

Fault-tolerant, real-time systems require correct, time-constrained results in the presence of faults. Missed deadlines in many high dependability systems can result in significant property damage or loss of human life. Historically, designers relied almost exclusively upon massive replication to achieve their dependability goals. Research suggests that not only is this approach inadequate for dealing with certain fault classes, but also that it is inappropriate for

[1]This research is supported in part by a grant from the Office of Naval Research under contract N00014-92-J-1524, by the Federal Systems Division of IBM Corporation under University Agreement Y-278067, and by AT&T Bell Laboratories under the Cooperative Research Fellowship Program.

many applications with strict space, weight, and cost constraints. A cost-effective solution to this problem is to exploit time redundancy, complementing or reducing the required level of hardware replication for a system's tolerance of transient faults.

Time redundancy can be designed into a system by reserving time for the service of fault recovery procedures, should they occur. These *reservation methods* are intended to assist the scheduler in fulfilling the challenging task of jointly satisfying the timing requirements of fault-free application tasks as well as those of any recovery procedures resulting from the detection of software or transient hardware faults. The strategy for allocating the reserved time is greatly influenced by the underlying scheduling philosophy of the real-time system.

Two scheduling philosophies are distinguished. One uses heuristics to generate a pre-planned schedule of activities to be executed at run-time, such that all task timing requirements are met. Since the function of the scheduler is to perform a table lookup, these real-time systems are referred to as *table-driven or time-driven systems*. Another philosophy advocates the use of priority mechanisms to resolve contention conflicts among real-time tasks at run-time. These priority mechanisms rely on analytical methods to ensure that the timing specifications are met, without the need to store a table explicitly describing the schedule. Real-time systems using priority mechanisms for scheduling are known as *priority-driven systems*.

Table-driven systems are attractive due to their low scheduling overhead. However, the memory requirements may be very large, modifying the tables can be very difficult and tedious, and the run-time flexibility is limited to non-existent [1]. The analytical techniques for evaluating timing correctness in priority-driven systems, on the other hand, facilitate the design of real-time systems with complex timing requirements including task synchronization, transient overloads, and support for fault-tolerant operation.

While priority-driven systems have gained recent popularity, most real-time systems developed to date are table-driven (e.g., MARS [2], MARUTI [3], FTMP [4], and many traditional in-flight control applications). As a result, reservation methods for enhancing the dependability of real-time workloads have only focused on table-driven systems [2, 5, 6, 7]. Such methods usually involve creating backup schedules [6, 8, 9] or embedding time slots in the primary schedule which are explicitly reserved for the service of recovery procedures [10, 11]. The strategies for reserving time are usually *ad hoc*, guided by heuristics that are rarely ever stated explicitly. Even in cases where the strategies are well described, an evaluation of their effectiveness in enhancing a system's ability to deliver timely fault recovery service is lacking. This motivates us to: (a) identify fundamental methods of reserving time for recovery; (b) develop and apply this methodology within the context of priority-driven systems; and (c) evaluate and compare their effectiveness.

In this paper, we propose two algorithms to reserve time for the recovery of periodic real-time application tasks on a uniprocessor. We assume the tasks are scheduled according to a fixed-priority scheduling mechanism, in which the priorities of the tasks are assigned before run-time and remain fixed during operation. Our reservation algorithms are based on the observation that there are only two fundamental ways in which time can be reserved for the recovery of real-time tasks. Reserved time is either *dedicated* to the recovery of some individual tasks, as in the *Private Reservation Algorithm*, or it is *shared* by all tasks, as in the *Communal Reservation Algorithm*. Analytical models for ensuring the timing correctness of each algorithm are developed. A common problem with reservation methods is that the reserved utilization is usually wasted if no faults occur. To measure the maximum waste of processing potential that can occur during fault-free operation, we present an analytical framework and illustrate this cost with sample workloads. Finally, simulation results are presented which give insight into the effectiveness of each algorithm in averting timing failures caused by the scheduler's inability to ensure timely recovery.

The paper is organized as follows. In Section 2, we discuss basic issues related to partitioning the reserved processing time among the tasks in a real-time workload. Our reservation algorithms are presented in Section 3 and simulation results which compare their performance are reported in Section 4. A summary and conclusions are presented in Section 5.

2   The Time Partitioning Problem

The reservation of processing time for recovery requires that two basic design decisions be made. First, one must decide whether the reserved time should be partitioned in a shared or in a dedicated manner. A *dedicated partitioning* approach means that when time is reserved, it is statically bound to the recovery of individual real-time tasks. On the other hand, a *shared partitioning* approach reserves a pool of recovery time which is dynamically allocated to failed application tasks on a contention basis. Clearly, hybrid partitioning schemes are also possible.

Second, one must decide whether the access privileges conferred to each application task for the usage of reserved recovery time will be fair or unfair. *Fairness* implies that all tasks will have an equal share, in principle, of the reserved time. (Note that the issue of fairness applies to both shared and dedicated partitioning strategies.) On the contrary, the allocation is *unfair* if recovery time is explicitly biased in favor of some tasks, at the expense of others. Hence, unfair allocation policies are those which tend to discriminate between tasks by favoring the timely recovery of some tasks over others. The issues of *time partitioning* and *allocation fairness* are fundamental to the design of any reservation strategy.

In this paper we explore both shared and dedicated time partition strategies, while focusing on fair allocation schemes only. However, the same underlying methodology proposed for fair allocation can be used to support unfair allocation schemes.

3   Reservation Algorithms

This section starts by presenting the analytical framework used as a basis to develop and evaluate the timing correctness of our reservation algorithms. The assumptions made in this analysis are discussed and a presentation of the algorithms follows. The section concludes with an evaluation of the utilization costs associated with supporting each algorithm.

3.1   Framework and Assumptions

Consider a real-time system with n periodic tasks, $\tau_1, \ldots, \tau_n$. Each task, τ_i, has a worst-case computation requirement C_i, a period T_i, an initiation time $\phi_i \geq 0$ or offset relative to some time origin, and a deadline D_i, specified relative to the task arrival time and assumed to satisfy $D_i \leq T_i$. The parameters C_i, T_i, ϕ_k, and D_i are known deterministic quantities. We require that these tasks be scheduled according to a fixed priority algorithm, such as the deadline monotonic algorithm, in which tasks with small values of D_i are given relatively high priority [12]. We assume that the periodic tasks are indexed in priority order with τ_1 having highest priority and τ_n having lowest priority. For simplicity, we refer to those levels as $1, \ldots, n$ with 1 indicating highest priority and n the lowest. A periodic task, say τ_i, gives rise to an infinite sequence of *jobs*. The kth such job is ready at time $\phi_i + (k-1)T_i$ and its C_i units of required execution must be completed by time $\phi_i + (k-1)T_i + D_i$ or else a *timing fault* will occur. A task set in which all job deadlines are guaranteed to be met is said to be *schedulable*.

Liu and Layland [13] proved that a task τ_i is guaranteed to be schedulable if the deadline for its *first* job is met when it is initiated at the same time as all higher priority tasks, i.e., $\phi_k = \phi_i$, for $k = 1, \ldots, i-1$. This is because the time between the arrival of a task's job and the completion of its service, referred to as its *response time*, is maximized when it arrives at the same instant at which all tasks of equal and higher priority arrive. The phasing scenario in which the initiation times for all tasks are equal is known as the *critical instant*, which is the worst-case phasing. It follows that a workload is schedulable under a fixed-priority assignment if the deadline for the first job of every task starting at a critical instant is met.

Liu and Layland also developed a sufficient test for the schedulability of a task set in which task deadlines are equal to the periods, $D_i = T_i$. They proved that if the workload utilization is less than 69%, a fixed-priority assignment exists for the tasks which guarantees that the task set is schedulable. Assigning the priorities to the tasks according to the Rate Monotonic (RM) algorithm was shown to be optimal in that no other fixed-priority assignment can guarantee the schedulability of a task set which cannot be scheduled with a RM priority assignment. RM scheduling assigns priorities to tasks according to their periods, such that τ_i has higher priority than τ_j if $T_i < T_j$. Ties are resolved arbitrarily.

Lehoczky, *et al.*, extended these results by deriving a necessary and sufficient schedulability criterion for fixed-priority workloads under critical instant phasing. This criterion dictates that a task τ_i is schedulable iff

$$min_{\{0<t\leq D_i\}} \{W_i(t)/t\} \leq 1, \tag{1}$$

where $W_i(t)$ is the cumulative work that has arrived from priority levels 1 to i in the time interval $[0,t]$ under critical instant phasing and is computed as

$$W_i(t) = \sum_{j=1}^{i} C_j \cdot \lceil t/T_j \rceil. \tag{2}$$

Intuitively, task τ_i is schedulable if $W_i(t)/t \leq 1$, because there exists some time t before the task's deadline D_i for which the elapsed time t is at least as great as the time required to complete all the work that has arrived, including the C_i units for τ_i. It follows that the entire task set is schedulable if the maximum value of $W_i(t)/t$ over the minima computed for each task τ_i, $i = 1, \ldots, n$, is also less than or equal to one, as indicated by

$$max_{\{1\leq i\leq n\}} \ min_{\{0<t\leq D_i\}} \{W_i(t)/t\} \leq 1. \tag{3}$$

This analytical framework allows us to evaluate the schedulability of a task set assuming that no processing time is reserved for recovery. The following sections will extend this framework to explore and evaluate the timing impact caused to a real-time workload when different reservation methods are supported. In particular, two reservation approaches represented by the Private and Communal Reservation algorithms are analyzed. The Private Reservation Algorithm reserves time for the recovery of individual real-time tasks while the Communal Reservation Algorithm reserves a shared pool of recovery time which is distributed among tasks on-line on a contention basis. Our algorithms are developed under the following assumptions:

- *A1:* All operating system overheads such as context switching, preemption, etc., are assumed to be zero and any task can be instantly preempted.

- *A2:* Tasks are ready at the start of their period and do not suspend themselves or synchronize with any other task.

- *A3:* The recovery of any faulty periodic task τ_i involves retrying the faulty task or executing an alternate task. In either case the retry has a known worst-case execution time C_{rec_i}. (To simplify our analysis, we further assume that the retry execution time is proportional to the primary execution time of the task, i.e., $C_{rec_i} = k \times C_i$, where k is a recomputation scale factor greater than zero.)

- *A4:* Each recovery task must be completed by the deadline of the associated periodic task found to be faulty.

3.2   Private Reservation Algorithm

The Private Reservation Algorithm (PRA) attempts to reserve time for the specific recovery of each task τ_i, for $i = 1, \ldots, n$. The objective of this algorithm is to guarantee that each job for task τ_i can be retried x_i times without violating the timing correctness of any other task. The number of times each task can be retried, x_i, is referred to as the number of *retry tickets*. Because the reserved time is dedicated to each task so that there is no on-line contention for recovery time, it is said that x_i retry operations are *unconditionally guaranteed* for each task.

Since we are limiting our attention to fair allocation strategies, it is desired that all tasks be given the same number of retry tickets, i.e., $x_1 = x_2 = \ldots = x_n$. If this is the case, it is said that all tasks have equal recovery privileges. When the utilization of the real-time workload is high, it may be impossible to grant equal recovery privileges to all tasks. In this case, a mechanism for discriminately conferring retry tickets to tasks is needed. These two cases are now addressed in detail.

3.2.1   Equal Recovery Privileges Per Task

To guarantee that each job for task τ_i can be retried x_i times without violating the timing correctness of any other task is equivalent to saying that the processing time to be devoted to τ_i is given by

$$C_i + x_i \times C_{rec_i} = C_i\{ 1 + x_i \times k\}. \tag{4}$$

This implies that the schedulability test for task τ_i must now consider not only the processing requirements of all fault-free higher priority tasks, as in Equation (2), but also the time it would take to service all their retry operations, should they fail. These observations lead to the following theorem.

Theorem 1 *A task τ_i in a real-time workload in which any failed jobs can unconditionally be retried x_i times is schedulable iff*

$$max_{\{1 \le i \le n\}}\ min_{\{0 < t \le D_i\}}\ \{W_i'(t)/t\} \le 1 \tag{5}$$

$$W_i'(t) = \sum_{j=1}^{i} (1 + x_j \times k)C_j \cdot \lceil t/T_j \rceil. \tag{6}$$

Proof: According to Liu and Layland's results, the response time of τ_i is maximized when it arrives at the same instant in which all fault-free tasks of higher priority arrive. If any higher priority task τ_j fails, we wish to unconditionally guarantee that it can be retried up to x_j times without violating the schedulability of τ_i. Hence, if τ_j fails sometime prior to τ_i's completion, the response-time of τ_i will be further increased by the time required to service the x_j retries for τ_j, which amounts to $(x_j \times k \times C_j)$ time units. Hence, the worst-case response time of τ_i, under faulty operation, occurs when *all* tasks of higher and equal priority, including τ_i, fail and exhaust their maximum number of retry tickets.□

Equation (5) gives us a schedulability test for a task set with some guaranteed recovery properties. Hence, given the timing requirements of a task set and some desirable recovery properties, this test tells us if they can simultaneously be satisfied. It would also be useful to answer the reverse question: What are the constraints on a task set for which some desired recovery properties can be guaranteed? This question is answered in the ensuing discussion.

The *breakdown utilization* of a task set is the utilization at which the real-time application tasks fully utilize the processor [13]; that is, the task set is schedulable but any proportional increase in the computation times of all tasks will make it unschedulable. The breakdown utilization of a task set gives us an upper bound on the processing capacity attainable if no processing time is reserved for recovery. Lehoczky, *et al.* [14] introduced a closed-form expression for computing the breakdown utilization of a task set. The breakdown utilization U_{BD} of a task set with utilization $U_W = \sum_{i=1}^{n} C_i/T_i$, is given by

$$U_{BD} = U_W \times \Delta^* \tag{7}$$

$$\Delta^* = [\, max_{\{1 \leq i \leq n\}} \, min_{\{0 < t \leq D_i\}} \, \{W_i(t)/t\} \,]^{-1}$$

Reserving time for recovery will obviously reduce the processing capacity available to the application workload. Hence, if some processing time is reserved, the breakdown utilization is unattainable under fault-free operation. As a result, we need to define upper utilization bounds which demarcate the maximum utilization attainable by a fault-free application workload after a portion of the processing capacity is reserved for recovery. These bounds are referred to as *fault-free utilization bounds*.

We now compute the fault-free utilization bounds given by the PRA for the case in which all tasks have an equal number of retry tickets. Note that for this case, $W_i'(t)$ is further simplified to

$$W_i'(t) = (1 + x \times k) \sum_{j=1}^{i} C_j \cdot \lceil t/T_j \rceil. \tag{8}$$

Similar to the computation of the breakdown utilization given in Equation (7), we can compute the fault-free utilization bound for a given number of retry tickets x, or U_{BD}' as

$$U_{BD}' = U_W \times [max_{\{1 \leq i \leq n\}} min_{\{0 < t \leq D_i\}} \{W_i'(t)/t\}]^{-1}. \tag{9}$$

$$U_{BD}' = \frac{U_{BD}}{(1 + x \times k)}. \tag{10}$$

Task	T_i	C_i	D_i	Utilization
τ_1	4	0.25	4	6.25%
τ_2	6	0.50	6	8.33%
τ_3	10	1.50	10	15.00%
τ_4	14	1.25	14	8.92%

Table 1: Timing Requirements for Example

x	0	1	2	3	4
U'_{BD}	0.856	0.428	0.285	0.214	0.171

Table 2: Fault-Free Utilization Bounds for Various Numbers of Retry Tickets

Hence, for a given recomputation scale factor k and number of retry tickets x, U'_{BD} represents a corresponding fault-free utilization bound. Note that for the special case in which tasks have no retry tickets, x is equal to 0 and no time is reserved for recovery. Hence, the fault-free utilization bound is equal to the task set's breakdown utilization, as expected.

We now illustrate the framework presented so far for the PRA with an example. Consider a set of 4 tasks, τ_1, τ_2, τ_3, and τ_4, with timing requirements shown in Table 1. This task set has a utilization U_W of 38.5% which is below Liu and Layland's least upper utilization bound of 69% so it is schedulable. The breakdown utilization of this task set is found to be 85.6%, using Equation (7). If we assume that each task retry is equivalent to re-executing the entire primary task, then $C_{rec_i} = C_i$ and $k = 1$. Using Equation (10) we compute the fault-free utilization bounds for different numbers of retry tickets (assuming all tasks have an equal number of tickets). As shown in Table 2, each value of U'_{BD} is simply obtained as

$$U'_{BD} = \frac{0.856}{(1+x)}. \qquad (11)$$

Given that the utilization of the task set is known, we can now look at Table 2 to determine how many retry tickets can be given to the tasks without violating schedulability. One retry ticket may be given to each task as long as the utilization does not exceed 42.8%. Moreover, to reserve two retry tickets per task the utilization cannot exceed 28.5%. Hence, for a workload utilization of 38.5%, we find that a maximum of one retry ticket can be given to each task.

Now let us imagine that this same task set is being executed on a processor that runs at $\frac{3}{4}$ the speed, so that the execution times for all tasks are longer. The utilization of the task set then becomes 51.3%. Consulting Table 2 we find that no retry tickets can be given to all the tasks, because the workload utilization exceeds the fault-free utilization bound for the reservation of one retry ticket per task, that is, $U'_{BD} = 42.8\%$. Hence, it is impossible to guarantee equal recovery privileges for *all* tasks. However, it may still be possible to reserve some retry tickets for one or more tasks. At this point, we are inevitably faced with the decision of choosing which of the tasks will have dedicated recovery time. Consequently, some strategy for discriminating among the tasks is needed.

3.2.2 Discriminatory Recovery Privileges

Let us represent a given assignment of retry tickets to tasks by a *ticketing vector*. Thus, a ticketing vector contains the number of retry tickets, x_i, given to each task τ_i, for $i = 1, \ldots, n$. To minimize the difference in recovery privileges among the tasks, we limit our attention to the case in which tasks can hold no more than 1 retry ticket. Therefore, the ticketing vector is represented by $[\, x_1 x_2 \ldots x_n \,]$, where x_i is 1 or 0 depending on whether or not the corresponding task τ_i has a reserved retry ticket.

We now revisit the previous example, for the case in which the workload utilization is 51.3%. There are 15 ways in which one or zero retry tickets can be given to 4 tasks, such that they do not simultaneously hold a ticket (2^n-1, where n=4). These options are represented by 15 distinct ticketing vectors. Of these 15 possible vectors, some may depict combinations that are unschedulable. These are combinations for which the required time to be reserved would cause the deadlines of real-time tasks to be missed. Hence, the only acceptable ticketing vectors are those which represent ticket allocation choices that are schedulable.

Using the schedulability test given in Theorem 1, we find that 12 out of the 15 combinations are schedulable, while 3 are unschedulable. All schedulable combinations, except for the trivial case of no tickets, are summarized in Table 3, ordered by the total number of tasks covered with a ticket. There is one acceptable ticketing vector which covers 3 tasks, 6 which cover 2 tasks, and 4 which cover 1 task. This amounts to 11 schedulable combinations in which at least one of the tasks are covered by a ticket.

This set of 11 vectors can be further reduced by observing that there is a dominance effect on the coverage provided by some of the vectors. For example, the [1101] vector dominates the [1001] vector because it covers all the tasks that the latter covers. By considering this dominance effect, we reduce the set of vectors to 4, namely, [1101], [0011], [0110], and [1010]. Only one of these 4 vectors can be used by the PRA. Heuristics are needed to make this selection.

For instance, one heuristic is to maximize the number of tasks covered, since that intuitively suggests that the recovery properties of the task set are maximized. This heuristic would lead to the selection of the ticketing vector [1101]. Another heuristic is to try to maximize the covered utilization of the workload. This heuristic favors a ticket assignment which maximizes the sum of the utilizations of covered tasks. The ticketing vector chosen by this heuristic would

Covering 3	Covering 2	Covering 1
[1101]	[0011]	[0001]
	[0101]	[0010]
	[0110]	[0100]
	[1001]	[1000]
	[1010]	
	[1100]	

Table 3: Feasible Ticketing Vectors for Example at 51.3%

be [0011], which covers a 47.9% of the 51.3% workload utilization. Note that the utilization covered by vector [1101] is somewhat lower (47.1%). The highlight of this example is to observe that although retry tickets cannot be given to all tasks equally at high utilization levels, heuristics allow us to select a subset of tasks for which dedicated recovery time can be feasibly reserved.

The identification of feasible options requires the exploration of a potentially large combinatorial space of possible ticketing vectors. Therefore, heuristics are needed not only to select the desired ticketing vector, but also to guide the process of searching for feasible options. To address this problem, we implemented two simple algorithms for computing feasible ticketing vectors. One algorithm is based on the heuristic which attempts to maximize the number of tasks covered, while the other algorithm tries to maximize the covered workload utilization. (The algorithms are reported in [15].) Applying these algorithms to the task set in Table 1, we find a number of feasible ticketing vectors with an associated fault-free utilization bound, as indicated in Table 4. This table can now be consulted to select a ticketing vector for the task set when the utilization is above 42.8%. For example, when the utilization is 51.3%, note that one heuristic would yield the vector [1101] while the other would select the vector [0011]. The effectiveness of these two heuristics will depend on the failure characteristics of the different tasks.

3.3   Communal Reservation Algorithm

The Communal Reservation Algorithm (CRA) attempts to reserve a shared pool of recovery time which is dynamically allocated to failed tasks on a contention basis. The objective of this algorithm is to guarantee that a total of X retry operations may be serviced within a given time interval $[t_a, t_b]$, with no specific number of retries guaranteed for any particular task. Stated otherwise, the cumulative number of retry operations serviced for arbitrary tasks in any interval of time $[t_a, t_b]$ must not exceed X. Hence, X denotes the total number of retry tickets that may be assigned to arbitrary tasks. Since retry tickets are now shared, there is no notion of retry tickets per task, x_i, as in the Private Reservation Algorithm (PRA). Therefore, it is said that

Max. # Tickets		Max. Covered Util.	
Vector	U'_{BD}	Vector	U'_{BD}
[0000]	0.856	[0000]	0.856
[1000]	0.734	[0010]	0.642
[1100]	0.616	[0011]	0.531
[1101]	0.513	[0111]	0.467
[1111]	0.428	[1111]	0.428

Table 4: Fault-Free Utilization Bounds for Heuristically Obtained Ticketing Vectors for Example

retry operations are *conditionally guaranteed* for each task, depending on the number of shared retry tickets available.

Three things are needed to fully specify the CRA, namely, the interval of time $[t_a, t_b]$, the number of retry tickets X which can be assigned over the interval, and the contention resolution policy for the shared tickets. We choose an interval of time equal to the duration of the largest task period, T_n, because this gives us a clear way of evaluating the effect of altering the number of retry tickets on the workload schedulability. Having chosen the duration of the interval, we can analytically compute the maximum number of retry tickets which can be assigned subject to guaranteed schedulability. Finally, the contention resolution policy determines how tickets should be given to faulty tasks as they request to be retried. We assume retry tickets are assigned on a first-come first-serve (FCFS) basis.

As in the PRA, we focus on a fair allocation strategy. Thus, it is desired that, in principle, all tasks be given the same opportunity to contend for the shared retry tickets. This means that none of the tasks should be prevented from contending for tickets in the shared pool. All tasks should have equal contention privileges. As before, we observe that when the workload utilization is high, the contention privileges for some tasks may have to be withdrawn in order to maintain schedulability. These two cases are addressed as follows.

3.3.1 Equal Contention Privileges Per Task

In order to guarantee that X retry operations can be serviced without a deadline violation, we must make sure that all the tasks can withstand the worst-case timing impact of such a recovery workload. This means that each task τ_i must be able to tolerate the largest delay which can be caused by the service of X retry operations. This observation leads to the following theorem.

Theorem 2 *A task τ_i in a real-time workload in which any failed jobs can conditionally be retried X times is schedulable iff*

$$max_{\{1 \leq i \leq n\}} \, min_{\{0 < t \leq D_i\}} \{W_i''(t)/t\} \leq 1 \qquad (12)$$

$$W_i''(t) = W_i(t) + X \times k \times max\{C_j \mid j = 1, \ldots, i\} \qquad (13)$$

where $W_i(t)$ is defined in Equation (2) and k is a recomputation scale factor ($k > 0$).

Proof: Assume that there exists a set of X retry operations which when serviced, cause task τ_i to miss its deadline. Let us refer to the cumulative processing time of these X retry operations as W_r. If τ_i misses its deadline, it must be the case that there is no scheduling point $t \leq D_i$ for which the cumulative fault-free periodic work, $W_i(t)$ and the recovery workload W_r can be serviced. Now let us subject the schedulability of τ_i to the constraint that it can withstand a delay equal to the largest value of W_r possible and still find a satisfying solution in its set of scheduling points. Having met this constraint, it is clearly impossible that τ_i misses its deadline.□

Equation (12) gives us a schedulability test for a task set which allocates a maximum of X retry tickets under the CRA. We now wish to compute the fault-free utilization bounds associated with this algorithm.

Recall that the breakdown utilization for a task set can be computed by multiplying the workload utilization U_W by a scale factor Δ^*, shown in Equation (7). Similarly, the fault-free utilization bound for a given number of retry tickets X can be computed by multiplying the workload utilization by a scale factor Δ_X^* given by

$$\Delta_X^* = [max_{\{1 \leq i \leq n\}} min_{\{0 < t \leq D_i\}}\{W_i''(t)/t\}]^{-1}. \qquad (14)$$

Hence, a fault-free utilization bound is equivalent to

$$U_{BD}' = U_W \times \Delta_X^*. \qquad (15)$$

Note that the special case in which there are no retry tickets $X = 0$ and the fault-free utilization bound is equal to the task set's breakdown utilization.

Let us revisit the task set example examined under the PRA and described in Table 1. Using Equations (14) and (15), we compute the fault-free utilization bounds for values of X ranging from 0 to 4. Results are shown in Table 5. Note that these bounds are higher than the comparable values computed for the PRA in Table 2. This illustrates that the cost of reserving recovery time is higher if the partitioning is dedicated rather than shared. However, when the recovery time is dedicated, the retry operations for any given task are unconditionally guaranteed because there is no contention for tickets.

x	0	1	2	3	4
U'_{BD}	0.856	0.654	0.553	0.479	0.416

Table 5: Fault-Free Utilization Bounds for Various Numbers of Retry Tickets

With a utilization of 38.5%, 4 shared retry tickets can be reserved for the task set ($X = 4$). Even if the workload utilization is increased to 51.3%, 2 retry tickets can still be reserved. Now consider that the task set in Table 1 is executed on a processor that is half as fast, so that the utilization is doubled to 77%. At this utilization we can no longer grant all tasks equal contention privileges because it exceeds the fault-free utilization bound for $X = 1$, or 65.4%. Once again we are faced with the need to provide discriminatory treatment to some tasks in order to guarantee that schedulability is maintained.

3.3.2 Discriminatory Contention Privileges

Unlike the treatment of this case seen in the PRA, in the CRA there is no need for heuristics to determine how to discriminate between the tasks. Discrimination is strictly dictated by schedulability constraints. The contention privileges for a task are simply withdrawn if its retry will violate schedulability of the task set.

Why is this case inherently different from the PRA? In the PRA any task holding a retry ticket is guaranteed one retry operation. Hence, if for instance, 2 out of 4 tasks hold a retry ticket, up to two recovery operations can co-exist. However, if the CRA has one retry ticket, then one and only one recovery operation can exist at any given point in time; task retries are mutually exclusive. Consequently, at high loads the question to be answered is not whether or not a certain task τ_i should be *assigned* a retry ticket. Rather, the question is whether or not a task τ_i should be allowed to *contend* for the retry ticket. Since contention privileges solely depend on schedulability constraints, no heuristics are required.

Similar to the ticketing vector, we now define a *contention vector* to describe contention privileges for the tasks. The contention vector consists of n binary-valued elements of the form x_i which denote whether or not τ_i is allowed to contend for the ticket. A value of $x_i = 0$ indicates that τ_1 cannot contend for the ticket and a value of 1 indicates that it can.

Each task τ_i has an associated fault-free utilization bound above which its retry would cause a deadline to be missed. Let us refer to this bound as U'_{BD_i}. This bound is computed by setting all the values in the contention vector to 0, except the value corresponding to τ_i, or x_i, which is set to 1. After setting the contention vector in this manner, we compute U'_{BD_i} as $U_W \cdot \Delta_i^\star$, where

$$\Delta_i^\star = [max_{\{1 \leq j \leq n\}} min_{\{0 < t \leq D_j\}}\{W'_j(t)/t\}]^{-1}. \tag{16}$$

Contention Vector	U'_{BD_i}	i
[0000]	0.856	None
[1000]	0.811	1
[1100]	0.770	2
[1101]	0.674	4
[1111]	0.654	3

Table 6: Fault-Free Utilization Bounds for the Mutually Exclusive Retries of Tasks in Table 1

$$W'_j(t) = W_j(t) + k \cdot C_j \cdot x_j \qquad (17)$$

If we compute U'_{BD_i} for each one of the tasks of Table 1, assuming a value of unity for the scale k, we get 81.1% for τ_1, 77% for τ_2, 65.4% for τ_3, and 67.4% for τ_4, as shown in Table 6. If the workload utilization is below the bound computed for a given task, the task can be retried without violating the schedulability of the task set. Note that the utilization bounds show us where changes occur in the contention vector. For instance, if the workload utilization is 66% then tasks τ_1, τ_2, and τ_4 may contend for the retry ticket but if it is 68% only τ_1 and τ_2 may contend for the ticket. By the same token, if the utilization exceeds 81.1%, none of the tasks may be retried. In general, a similar set of fault-free utilization bounds may be computed for arbitrary task sets.

3.4   Comparative Utilization Costs

The cost of supporting the Communal and Private Reservation algorithms is a reduction in the processing capacity available to the fault-free workload. To provide some insight into the relative costs of these two algorithms, we chose three real-time application workloads for study, namely, a task set for an Inertial Navigation System (INS), an Avionics task set, and a Multimedia task set. These task sets are representative of common real-time applications and were introduced in [16], [17], and [18], respectively.

Figure 1 illustrates reservation costs associated with the three task sets when the number of retry tickets is varied from 0 to 10 and the recomputation scale factor $k = 1$. The vertical axis shows the fault-free utilization bounds obtained by using Equations (14), (15), and (10). The upper curves show the fault-free utilization bounds for the CRA while the bottom curves correspond to the PRA. For example, Figure 1.a shows that the INS task set requires that 50% of the utilization be reserved by the PRA to guarantee one retry per task. On the other hand, only 10% of the utilization is needed by the CRA to handle a single retry of any of the INS tasks. The points in which $x = 0$ depict the breakdown utilizations for these task sets, i.e.,

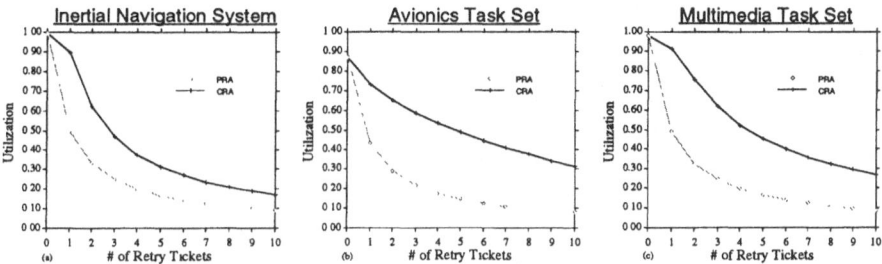

Figure 1: Illustrating Costs of Reservation algorithms: (a) INS ; (b) Avionics ; (c) Multimedia Task Set

when no time is reserved for recovery. It is shown that the INS and Multimedia task sets have a breakdown utilization approaching 100%, while the Avionics task set has a breakdown utilization of about 87%.

These graphs show that the relative cost of sharing reserved recovery time is much lower than dedicating time to the recovery of individual tasks. This cost difference is particularly dramatic for the case in which the number of retry tickets is low, such as 1 or 2. A common assumption made by designers is that the probability of tasks failing twice in succession is negligible. Thus, reservation methods usually allocate enough time so that any given task may be retried at most once. Since the number of retry tickets is usually no greater than 1, the CRA is expected to entail a significantly lower cost than the PRA in practice.

4 Performance Evaluation

To evaluate and compare the effectiveness of the Private and Communal Reservation algorithms, we simulate the injection of a series of faults into the Inertial Navigation System (INS) task set. The INS task set consists of 6 periodic tasks, with an average processing load of 88% under fault-free operation.

We assume that faults are detected at the completion of a task's execution, consistent with the Recovery Block model introduced by Randell [19]. The detection of faults causes a stream of recovery tasks to be generated, each requiring to be serviced prior to the deadline of the periodic task that was found faulty. If the scheduler can guarantee that a recovery task will be serviced before its deadline, the request is said to be *accepted*. Otherwise, the request is *rejected* or denied service, because there is no value in servicing a recovery task that cannot be completed by its deadline.

In the absence of data for modeling software faults during the operational phase, we adopted the conventional assumption that the times between faults in a single program are exponentially distributed [20, 21]. Mean fault interarrival times were chosen so that the expected size of the generated transient recovery loads varied between loads close to 0%, to loads causing the processor to experience a joint periodic and recovery processing load around 100%. Each recovery load has a characteristic retry rate, or mean number of task retries requested per second, as will be shown later on.

The effectiveness of each reservation algorithm was measured by its probability of successfully scheduling recovery requests, using a metric referred to as recovery coverage. We define *recovery coverage* as the conditional probability that the deadline for a recovery task is guaranteed, given that a periodic application task has failed and issued a recovery request. This definition of coverage is analogous to that given by Bouricius, *et al.* for the recovery from permanent hardware faults [22]. It is also similar to the widespread coverage metric used to gauge the effectiveness of error detection mechanisms. The coverage provided by a reservation algorithm is empirically computed as the percentage of recovery requests accepted for service relative to the total number of recovery requests issued to the scheduler. Therefore,

$$Coverage = \frac{(\# \; req. \; accepted)}{(\# \; req. \; issued)} = \frac{N_{accepted}}{N_{issued}}. \tag{18}$$

Coverage values range from 0.0 to 1.0, and it is desired that values be close to unity.

Experiments were conducted in the following manner. First, processing time for recovery was reserved by the PRA and the CRA. Recall that the PRA represents reserved time by a ticketing vector while the CRA does so by a contention vector and a number of total retry tickets to be shared by the tasks. Then a fault injection profile is created based on a desired set of injection parameters, such as a mean retry rate with an exponential distribution. This profile is used to generate the recovery workload which the scheduler is requested to schedule at run-time in addition to servicing all periodic tasks. The simulator generates a *recovery log* describing all scheduling decisions concerning the acceptance or rejection of each recovery request. The recovery log is analyzed to compute a coverage estimate.

Let us start by discussing how the CRA and PRA reserve processing time in the INS task set. Figure 1.a showed the fault-free utilization bounds for the INS task set as the number of retry tickets reserved by the PRA and the CRA increases from 0 to 10. Since the fault-free utilization bound for 1 shared retry ticket is above 88%, the CRA can reserve 1 retry ticket to be allocated to any one failed task at run-time. Hence, one recovery task can be serviced during any interval of time of duration equal to the largest task period. All tasks have equal contention privileges so they can all contend for this retry ticket. On the contrary, the fault-free utilization bound for 1 retry ticket per task is about 50% for the PRA, which is below the 88% task set utilization. Hence, the PRA cannot reserve 1 retry ticket per task without violating the schedulability of the task set; retry privileges can be given to some, but not all, of the tasks.

Using the heuristic of maximizing covered utilization, we compute the fault-free utilization

Ticketing Vector	U'_{BD_i}	i
[000000]	0.994	None
[000001]	0.967	6
[000101]	0.946	4
[000111]	0.854	5
[010111]	0.774	2
[011111]	0.777	3
[111111]	0.497	1

Table 7: Fault-Free Utilization Bounds for the INS task set under the PRA

bounds, U'_{BD_i} for a set of ticketing vectors in which a subset of the tasks holds a ticket. These bounds are shown in Table 7. The rightmost column indicates the priority of the task associated with the fault-free utilization bound in the row. For instance, the bound for τ_5 is 85.4%, so τ_5 cannot hold a retry ticket if the utilization of the task set exceeds 85.4%. Since the utilization of the task set is 88%, the PRA finds that the most protection it can provide to the tasks is given by the ticketing vector [000101]. Therefore, tasks τ_4 and τ_6 hold 1 retry ticket, while the other tasks hold none.

A comparison of the coverage provided by the PRA and the CRA to the INS task set is shown in Figure 2, with mean retry rates fluctuating between 80 and 8000 retries per second. The upper horizontal axis shows the mean recovery load resulting from each burst of task retries, expressed as a percentage of processor utilization. For the retry rates shown, the recovery load goes up to 15.37%, for a joint load of 103.37%.

The coverage provided by the CRA is highest when the recovery load is low. The leftmost point on the graph corresponds to a recovery load of about 0.11%. As the retry rate increases, the performance of the CRA drops significantly. When the recovery load reaches 15.37%, the coverage provided by CRA has dropped to 0.107. The explanation for this behavior is that the CRA is highly sensitive to the degree of contention for the shared retry tickets. If the demand for retry tickets never exceeds the maximum supply of tickets, all retry requests get accepted and coverage is 1.0. However, if the demand tends to exceed the maximum supply of tickets, the scheduler is usually saturated. This means that the number of retry operations contending for tickets surpasses the maximum number of tickets available on average. Consequently, beyond the saturation point, an increase in the retry rate causes a proportional increase in the rejection rate.

The performance curve for the PRA differs from that of the CRA in that it is not a smooth, monotonically decreasing function of the retry rate. The coverage provided by the PRA fluctuates around some constant value which is relatively insensitive to variations in the retry rate. We observed that this behavior is typical of algorithms which provide highly biased

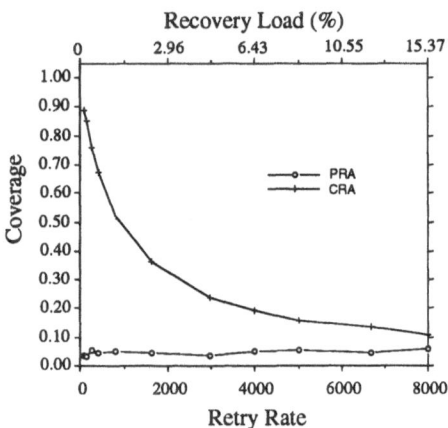

Figure 2: Coverage results for the INS task set (Periodic load: 88%)

coverage for the tasks. The coverage is biased if one or more tasks have no protection against timing failures while others have a coverage close to 1.0. Since the PRA ticketing vector is [000101], only tasks τ_4 and τ_6 have reserved processing time for recovery. Every recovery request issued by tasks τ_1, τ_2, τ_3 and τ_5 is rejected because the PRA cannot reserve time for their recovery, so their coverage is 0.0. Since tasks τ_4 and τ_6 each have 1 retry ticket per job, their retry requests are always accepted as long as a job does not fail more than once in succession. Our simulation studies indicate that multiple job failures are unlikely to occur except for situations in which the retry rates are very high. Therefore, the coverage provided to τ_4 and τ_6 is close to 1.0.

To further explore the performance differences between the CRA and PRA, we measured the coverage provided by each algorithm when the periodic processing load is decreased to 40% and 67%, as shown in Figure 3. Recall that such a decrease in processing load can result from executing the task set on a faster processor.

Referring to Figure 1.a, if the periodic load is 40%, the PRA can reserve 1 retry ticket per task, so the ticketing vector is [111111]. For this load, the CRA can reserve 3 retry tickets which can be requested by any one of the tasks. If the periodic load is 67%, the PRA can no longer reserve retry tickets for all the tasks and the ticketing vector becomes [011111]. At this load, the CRA can still grant equal contention privileges to all the tasks but the number of shared retry tickets is reduced to 1.

As in Figure 2, the coverage provided by the CRA is highest when the recovery load is low and coverage drops significantly as the recovery load is increased. Even a recovery load of only 2.60% when the periodic load is as low as 40% causes the coverage of the CRA to drop from

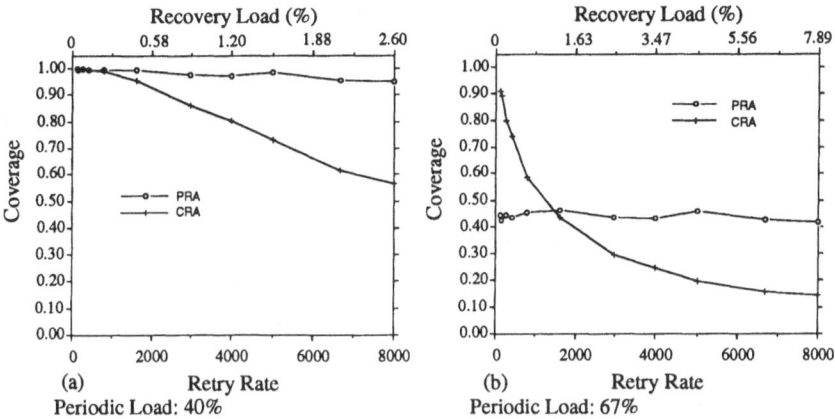

Figure 3: INS: Sensitivity of Coverage to Variations in Periodic Load

1.0 to 0.57 (Figure 3.a). This verifies once again the sensitivity of the CRA algorithm to the degree of contention for retry tickets.

The most dramatic performance variation with respect to changes in periodic load is observed for the PRA. The PRA provides a coverage close to 1.0 when the periodic load is 40%, it drops to about 0.47 when the periodic load is 67% (Figure 3.b), then collapses to about 0.05 when the load reaches 88% (Figure 2). This dramatic change in coverage is due to the fact that the PRA tries to reserve relatively large portions of processing time for the dedicated recovery of each individual task. When it fails to do so because of periodic loading constraints, it sacrifices the number of tasks for which time is reserved. This introduces a strong coverage bias, as discussed earlier.

If we refer to the amount of time available for the recovery of a single task τ, as the depth of an algorithm, and we refer to the total number of tasks with access to recovery time as the breadth of an algorithm, it is fair to say that: the CRA strives for breadth, rather than depth, while the PRA strives for depth, rather than breadth. Specifically, the CRA reserves retry tickets that can be contended for by all the tasks but the total supply of retry tickets may be relatively low. On the other hand, the PRA reserves retry tickets which are bound to individual tasks such that the supply for any given task is usually adequate, but not all the tasks are likely to get tickets. Thus, the advantage of the PRA is that it eliminates contention and provides excellent coverage for the protected tasks but it lacks breadth.

The opposing tradeoffs made by the CRA and PRA algorithms are responsible for the performance crossover seen in the graphs of Figure 3 and Figure 2. At a low periodic load of 40%, the PRA is capable of reserving retry tickets for all the tasks, yielding an excellent coverage

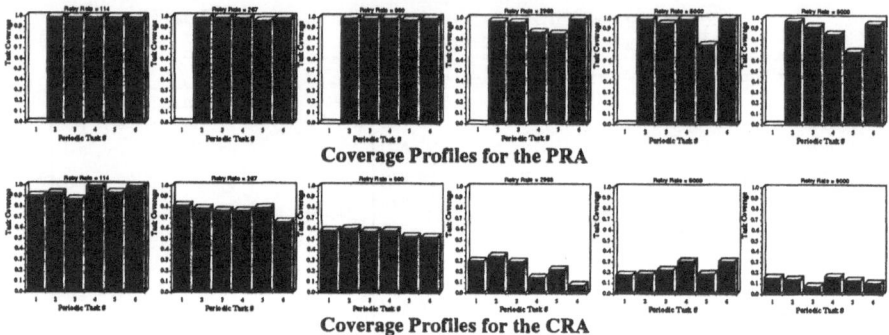

Coverage Profiles for the PRA

Coverage Profiles for the CRA

Figure 4: Coverage Profile for the INS task set (periodic load = 67%)

for the entire workload. The coverage provided by the CRA inevitably degrades due to its sensitivity to contention. At a medium load of 67%, the CRA outperforms the PRA as long as the retry rate is low. As the retry rate is increased, the biased coverage provided by the PRA outweighs the benefits of sharing recovery time among the tasks. Finally, at a high load of 88%, the overall protection provided by the PRA is very poor so the sharing of a single retry ticket done by the CRA is more efficient.

To further illustrate performance differences among the CRA and PRA, we include a coverage profile for the INS task set in Figure 4, for the case in which the periodic load is 67%. This profile illustrates the percentage of recovery requests serviced for each one of the 6 periodic tasks in the set. The top row of histograms is associated with the PRA while the bottom row corresponds to the CRA. Retry rates increase from left to right. These coverage profiles visually highlight the qualitative tradeoff of breadth versus depth suggested earlier. Note how the PRA provides a relatively high coverage for tasks τ_2, τ_3, τ_4, τ_5, and τ_6, at the expense of zero coverage for τ_1. The coverage provided by the CRA, on the other hand, is fairly well balanced among all the tasks, but the average coverage decreases as the retry rate is increased. It is clear that CRA provides better coverage at a low retry rate while PRA provides better coverage at a high retry rate.

5   Summary and Conclusions

Two algorithms were introduced to reserve time for the recovery of time-critical application tasks in fixed-priority systems. The *Private Reservation Algorithm* reserves time for the recovery of individual tasks while the *Communal Reservation Algorithm* reserves a shared pool of recovery time which is allocated to tasks on-line on a contention basis. Analytical formulations were derived to compute the amount of processor utilization that must be reserved by each algorithm to provide a desired level of guaranteed recovery to a given application workload. Using a metric known as *recovery coverage*, the effectiveness of these shared versus dedicated reservation algorithms was studied via simulation. Coverage measures the effectiveness of an algorithm in averting timing failures due to the scheduler's inability to deliver timely recovery service. Results show that dedicated methods provide high coverage over a wide range of recovery loads if the application load is low enough to allow recovery time to be reserved for all application tasks. Shared methods, on the other hand, are preferable when the application load is medium to high, ensuring a better utilization of the reserved time. However, the coverage provided by sharing is high only at low recovery loads and is very susceptible to increases in recovery load. To circumvent these shortcomings, our research is investigating on-line methods to dynamically allocate time for recovery. In addition, we are characterizing and evaluating the performance and implementation overheads associated with reservation versus dynamic allocation methods. These are the topics of a subsequent paper.

References

[1] P. Hood and V. Grover. Designing real-time systems in ada. Technical Report 1123-1, SofTech Inc., 460 Totten Pold Road, Waltham, MA 022540-9197, January 1986.

[2] Kopetz et.al. Distributed fault-tolerant real-time systems: The mars approach. *IEEE Micro*, 9(1):25–40, February 1989.

[3] V. Nirkhe and W. Pugh. A partial evaluator for the maruti hard real-time system. In *Real-Time Systems Symposium*, pages 64–73, Dec. 1991.

[4] T.B. Smith III. *The Fault-Tolerant Multiprocessor Computer*. Moyes Publications, 1986.

[5] Daniel Mosse. *A Framework for the Development and Deployment of Fault-Tolerant Applications in Real-Time Environments*. PhD thesis, University of Maryland, College Park, MD., 1992.

[6] A.L. Liestman and R.H. Campbell. A fault-tolerant scheduling problem. *IEEE Transactions on Software Engineering*, SE-12(11), November 1986.

[7] A. Wei, K. Hiraishi, R. Cheng, and R. Campbell. Application of the fault-tolerant deadline mechanism to a satellite on-board computer system. In *1980 International Symposium on Fault-Tolerant Computing*, pages 107–109, June 1980.

[8] H. Hecht. Fault-tolerant software for real-time applications. *ACM Computing Surveys*, 8(4):391–407, December 1976.

[9] C.M. Krishna and K.G. Shin. On scheduling tasks with a quick recovery from failure. *IEEE Transactions on Computers*, C-35(5):448–455, May 1986.

[10] S. Balaji, L. Jenkins, L.M. Patnaik, and P.S.Goel. Workload redistribution for fault-tolerance in a hard real-time distributed computing system. In *1989 International Symposium on Fault-Tolerant Computing*, pages 366–373, Chicago, Illinois, June 1989.

[11] R.H. Campbell, K.H. Horton, and G.C. Belford. Simulations of a fault-tolerant deadline mechanism. In *1979 International Symposium on Fault-Tolerant Computing*, pages 95–101, Madison, Wisconsin, June 1979.

[12] J. Y.-T. Leung and J. Whitehead. On the complexity of fixed-priority scheduling of periodic real-time tasks. *Performance Evaluation*, 2:237–250, 1982.

[13] C.L. Liu and J.W. Layland. Scheduling algorithms for multiprogramming in a hard-real-time environment. *Journal of the Association for Computing Machinery*, 20(1):46–61, January 1973.

[14] John Lehoczky, Lui Sha, and Ye Ding. The rate-monotonic scheduling algorithm: Exact characterization and average case behavior. In *Real-Time Systems Symposium*, pages 166–171, 1989.

[15] Sandra Ramos Thuel. *Enhancing Fault Tolerance of Real-Time Systems through Time Redundancy*. PhD thesis, Carnegie Mellon University, May 1993.

[16] K. Fowler. Inertial navigation system simulator: Top-level design. Technical Report CMU/SEI-89-TR-38, Software Engineering Institute, January 1989.

[17] D. Locke, D. Vogel, and T.J. Mesler. Building a predictable avionics platform in ada: A case study. In *Real-Time Systems Symposium*, pages 181–189, Dec. 1991.

[18] S. Sathaye, D. Katcher, and J. Strosnider. Fixed priority scheduling with limited priority levels. Technical Report CMU-CDS-92-7, Carnegie Mellon University, August 1992.

[19] B. Randell. System structure for software fault tolerance. *IEEE Transactions on Software Engineering*, pages 220–232, June 1975.

[20] Daniel P. Siewiorek and Robert S. Swarz. *Reliable Computer Systems*. Digital Press, 1992.

[21] J.D. Musa. A theory of software reliability and its application. *IEEE Transactions on Software Engineering*, pages 312–327, September 1975.

[22] W.G. Bouricius. Reliability modeling for fault-tolerant computers. *IEEE Transactions on Computers*, C-20:1306–1311, Nov. 1971.

Evaluation of
Dependability Aspects

Effects of Physical Injection of Transient Faults on Control Flow and Evaluation of Some Software-Implemented Error Detection Techniques

Ghassem Miremadi and Jan Torin
Department of Computer Engineering
Chalmers University of Technology
S-412 96 Göteborg, Sweden

Abstract

An approach for assessing the impact of physical injection of transient faults on control flow behaviour is described and evaluated. The fault injection is based on two complementary methods using heavy-ion radiation and power supply disturbances. A total of 6,000 transient faults was injected into the target microprocessor, an MC6809E 8-bit CPU, running three different benchmark programs. In the evaluation, the control flow errors were distinguished from those that had no effect on the correct flow of control, vis. the control flow OKs. The errors that led to wrong results are separated from those that had no effect on the correct results. The errors that had no effect on either the correct flow or the correct result are specified. Three error detection techniques, namely two software-based techniques and one watchdog timer, were combined and used in the test in order to characterize the detected and undetected errors. It was found that more than 87% of all errors and 93% of the control flow errors could be detected.

Key Words: Experimental Evaluation, Fault-Injection, Transient Faults, Control Flow Checking, Error Detection, Fault-Tolerance, Failure Analysis.

1 Introduction

The prevention of system failure in critical computer applications is of decisive importance, owing to the costly and hazardous consequences. It has been shown, however, that **transient faults** are the major source of system failures in such applications [2], [3], [10], [28]. One way to attack the problem of system failures in order to achieve dependable computing systems is through the utilization of **fault tolerance**. However, a fault tolerant system requires mechanisms for the detection of errors or faults, which are expected to be tolerated. To design error detection mechanisms requires sufficient knowledge of error classes and their effects on system behaviour. Furthermore, a study of error effects aids to characterize the system's susceptibility to faults. An effective approach to reaching these goals is through the use of **fault injection techniques**. However, to obtain confident results, a fault injection method must provide a fault model which is equally as representative as actual faults found in real operations.

A number of studies concerning the impact of transient faults on computer systems have been performed by different researchers during recent years. An investigation of the effect of heavy-ion induced single event upsets is given in [12], which described the usefulness of heavy-ion radiation for the validation of error-handling mechanisms, as well as the variation of error behaviour on the external buses. An extended study based on heavy-ion radiation and power supply disturbances as fault injection methods is also reported in [14]. In [15], gate-level faults are injected in a gate-level simulation of a processor for studying the error manifestation, and the error propagation at the program level. The study in [4] describes a simulation environment for analysis of the impact of transient faults in a microprocessor-based jet-engine controller used in the Boeing 747 and 757 aircrafts. A simulation-based fault injection is presented in [23], where various aspects of the error behaviour of a 32-bit pipelined RISC are studied.

In this study, two complementary fault injection methods, i.e., **heavy-ion radiation** and **power supply disturbances**, are used to characterize some aspects of fault effects that have not been studied in the above research. In the characterization, the control flow errors were distinguished from those that had no effect on the correct flow of control, viz. the control flow OKs. The errors that led to wrong results are separated from those that had no effect on the correct results. The errors that had no effect on either the correct flow or the correct result are specified.

Three error detection techniques, i.e., one **watchdog timer** (WDT) and two software-based techniques, called **Block Signature Self-Checking** (BSSC) and **Error Capturing Instructions** (ECIs), are combined and used in the test in order to characterize the detected and undetected errors. A primary evaluation of the techniques has previously been presented in [20]. The present evaluation of the techniques extends the previous one in two ways. One, improvements are made on the BSSC mechanism which resulted in a significant increase in the error detection coverage. Two, the previous evaluation of the combination of the techniques was preliminary and gave insufficient results; a complete evaluation is performed in the present study. The BSSC and ECI methods are based on inserting redundant machine instructions into the main memory to detect control flow errors. Most of the proposed techniques for detecting control flow errors use a form of extra hardware, such as a watchdog processor [6], [15], [16], [17], [22], [24], [25], [26], [27], [29], [30], or other forms of hardware [9], [11], [32]. Hence, the BSSC and ECI techniques, which use only extra software, are expected to be economical error detection techniques in some industrial microprocessor-based systems. The BSSC technique is an adaptation and simplification of ideas presented by other authors, such as [15], [27]. The basic concept for the ECI technique was first described by Halse and Preece [8]. An evaluation of this technique has previously been presented in [31]. The watchdog timer mechanism was combined with the BSSC and ECI mechanisms in order to enhance the coverage, so that some errors that went undetected by the BSSC and ECI mechanisms could be detected.

The microprocessor tested was an MC6809E 8-bit microprocessor. This simple microprocessor was selected for two reasons. One, the desired results can be obtained faster and also are easier to analyze. Two, this processor has no built-in error detector, which is necessary to be able to evaluate the error detection mechanisms.

The organization of this paper is as follows: Section 2 describes the two fault injection methods that are used in the test. The error detection mechanisms are described in Section 3. Section 4 describes the organization of the experiment as well as the three benchmark programs used in the test. The results concerning the effects of the faults on the control flow behaviour are presented in Section 5. The evaluation results of the error detection mechanisms are given in Section 6. Finally, a summary and conclusions are given in Section 7.

2 Fault Injection Methods

2.1 Heavy-Ion Radiation

Bombarding integrated circuits with heavy-ion radiation from a ^{252}Californium source can cause single event upsets in the circuits. A single event upset is a soft error which is introduced when an ionizing heavy ion penetrates into the depletion region of a reversed biased pn-junction. The heavy ion affects only stored information by changing bit values from "zero" to "one" or vice versa. In most cases, a heavy ion affects only single bits; the occurrence of multiple bit errors is rare. No hardware is damaged by the ions. The changed bit alters the valid behaviour of the circuit as soon as the bit is used. The valid behaviour of the circuit can easily be recovered by resetting it. The single event upsets are assumed to occur randomly both in time and in position within a circuit [18]. Heavy ions are attenuated in air, which makes it necessary to irradiate circuits in vacuum. The lid of the circuit must be removed in order to enable the heavy-ion bombardment. A detailed description of the use of heavy-ion radiation for fault injection experiments is presented in [13].

2.2 Power Supply Disturbances

The power supply disturbances consisted of short voltage drops inserted at the power supply pin of the MC6809E. In a voltage drop, the power supply was sunk from +5V to +0.8V. The propagation delay time, i.e., the hold time at +0.8V, was 50ns. The fall time, from +5V to 0.8V, was 20ns while the rise time, from 0.8V to +5V, was 30ns. This type of voltage drop causes many different kinds of errors in the MC6809E. The voltage drops were inserted by means of a MOS power transistor placed between the power supply and the Vcc pin of the MC6809E. As each voltage drop did not cause errors in the MC6809E, the voltage drops were repeated every 100 ms until an error occurred. The power supply disturbances were injected at random points in time during the program execution. Detailed information on the power supply disturbances is given in [7].

3 Error Detection Mechanisms

The basic concept of the Block Signature Self-Checking technique (BSSC) has previously been reported in [20]. However, two main modifications are made to this technique in this experiment.

The first modification deals with the size of a signature, in which only one byte is used for a signature instead of two bytes as used in the previous one. The other modification makes it possible to check the control flow of a program at both the entry and exit points of the program's blocks. In the previous evaluation of the BSSC, the control flow was checked only at the exit point of the blocks. A description of the modified BSSC is given in the following.

The principal task of the BSSC technique is to check the correct flow of control between different blocks of an application program. To do this, the machine code is partitioned into blocks consisting only of non-branching instructions, called **basic blocks** [1]. Hence, the only legal entry to any basic block is via its first instruction and the only legal exit from it is via its last instruction. All other entries into and exits from the block are illegal. To check the correct entrance and departure in a basic block, the basic block is modified as shown in Figure 3-1. The modified block now consists of four parts: a **call instruction**, immediately before the basic block, to a routine called *entry*; the **basic block** itself; another **call instruction**, immediately after the basic block, to a routine called *exit*; and a **signature** embedded at the end of the modified block. The value of the signature in the basic block corresponds to the number of bytes appearing in that basic block plus the size of a call instruction. The inserted call instruction at the beginning of the block calls the entry-routine, which checks whether the execution of the previous basic block was successfully completed. This is done by checking whether the value of a static variable matches a unique *KEY* value. In normal operation, this *KEY* should have been saved in the static buffer by the exit-routine called in the previous basic block. If this is the case, the first address (m+1) of the basic block is stored in the static buffer and the normal execution continues at the beginning of the basic block. Otherwise, a control flow error is presented. The inserted call instruction at the end of the block calls the exit-routine, which checks that a correct entrance to the current basic block is performed. This is done by adding the current value (m+1) in the static buffer to the embedded signature and comparing the result with the address of the signature. If they are equal, a correct entrance to the block has been performed. Any mismatch check is an indication of a control flow error. In correct operation, the exit-routine stores the unique *KEY* in the static buffer. The exit-routine also modifies the return address on the stack for the purpose of skipping the signature, upon which the correct execution can continue immediately after the signature. To implement the BSSC technique, a program called a **postprocessor** has been developed, which inserts the BSSC mechanisms into the object code generated by a compiler/assembler.

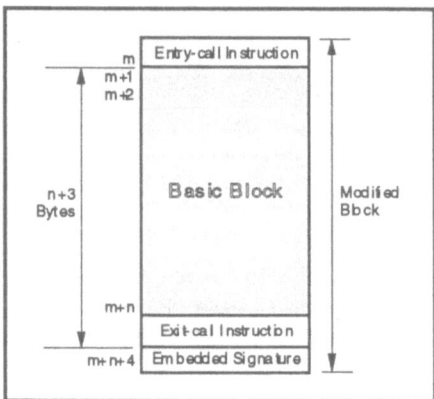

Figure 3-1   Modification of a Basic Block

This modified BSSC technique has three advantages as compared with the previous version of the BSSC. One, the coverage is enhanced because of the added control flow checking at the entry point. Two, the memory overhead is reduced owing to the use of a one-byte signature instead of a two-byte signature. Three, the program code has become position-independent, because the signature is not dependent on the memory addresses. One drawback, however, is that the execution time overhead is increased by 4%.

The BSSC technique is used to detect control flow errors in the program area. However, when the normal execution departs from its correct flow, it may also continue at a random address outside the program area. Hence, it is necessary to combine the BSSC technique with another technique capable of detecting control flow errors outside the program area. One such technique is the Error Capturing Instructions (ECIs), which consists of certain instructions in the instruction repertoire. In the experiment, the ECI mechanisms were inserted throughout the data and unused areas to detect control flow errors occurring in these areas. They are never executed during normal operation and hence their execution is an indication that an illegal jump to these areas has occurred. Examples of instructions which can be used to implement the ECI technique are: unconditional branch instructions, call instructions, jump instructions, software interrupt instructions (SWI) and no-operation instruction (NOP). They can be used to initiate an error handling routine or to initiate an infinite loop in order to prevent erroneous execution. Undefined operation codes in a microprocessor can also be used to implement the ECI technique if the microprocessor has an undefined operation codes detector in its design. In this experiment, the SWI instruction of the MC6809E was used to implement the ECI mechanism. To insert the ECIs throughout the data area, the data segment of a program must be partitioned into blocks. The ECIs can then be inserted between the data blocks. Each ECI mechanism between two data blocks consisted of a block of three consecutive SWI instructions. In the unused area, the entire area must be filled with the ECI mechanisms.

A watchdog timer mechanism (WDT) was also combined with the BSSC and ECI mechanisms in order to determine the kinds of errors that could be detected by the WDT, but which had gone undetected by the BSSC and ECI mechanisms. This is an attempt to enhance the error detection coverage. The programmable timer available in the logic analyzer used in the test was employed as the watchdog timer mechanism in the experiment. In the normal operation of the target microprocessor, an application program was re-executed in an infinite loop. Hence, a particular address of the program appeared periodically on the address bus within a specific time limit. This address was used to reset the watchdog timer. If the time limit had been reached and the address had still not appeared on the address bus, then the watchdog timer triggered the system that an error had occurred. The execution time of an application program was used as the time limit for that program.

4 Experiment Organization

4.1 Experimental System

The experimental system consists of four main parts: the target system under test, a logic analyzer (DAS 9200), a host computer and a controlling microcomputer that controls the experiment. Figure 4-1 provides the basic schematic organization of the experimental system.

The target integrated circuit tested was an MC6809E microprocessor. The microprocessor, together with a board and a clock generator, constituted the target microcomputer system. The board consisted of 48 kilobytes EPROM, 16 kilobytes RWM and some other devices. To determine the occurrence of errors, the target system was provided with another MC6809E microprocessor as a reference. Note that faults were injected only into the target microprocessor. The reference microprocessor was supplied with the same clock generator as the target microprocessor and thus the two microprocessors were operated in synchrony. The external data bus of the target system was connected to the reference microprocessor via a one-directional buffer. This means that the reference microprocessor was allowed to read the content of the data bus. Determination of the occurrence of an error was as follows. All bus signals from the target microprocessor were continually compared with corresponding bus signals from the reference microprocessor during system operation. The occurrence of an error could easily be ascertained as soon as any difference was seen in the comparison between the target and the reference microprocessors. When a difference was seen in the comparison, in other words when the error was manifested, the logic analyzer was triggered (trigg 1). The execution behaviour of the target microprocessor was then recorded by the logic analyzer. A total of 300 bus cycles was recorded for each error, 30 cycles before and 270 cycles after manifestation of the error. After this triggering (trigg 1), another trigger signal (trigg 2) triggered the logic analyzer again. Three events could have generated the second triggering: either the error was detected by a software mechanism, the watchdog timer was reset or time had run out. A total of 60 additional cycles were recorded by the logic analyzer in the second triggering. These cycles were used to determine whether an error which had not been previously detected within 270 cycles in the first recording could be detected later. Hence, a total of 360 bus

c   y   c   l   e   s   w   a   s

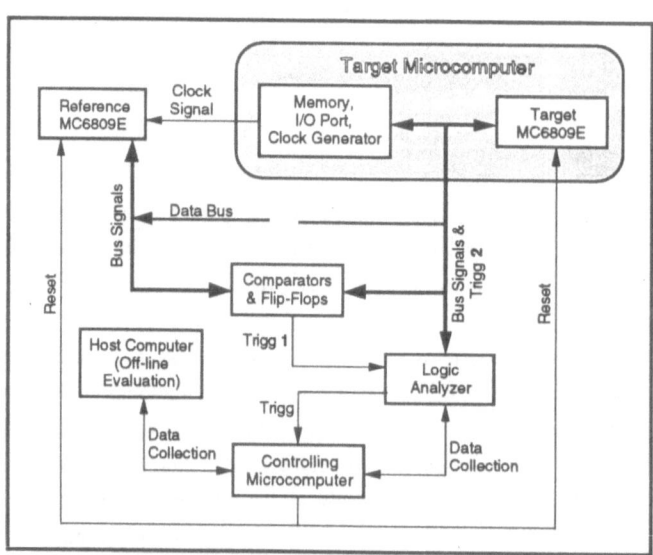

Figure 4-1 Schematic Organization of the Experiment

recorded for each error. After the occurrence of the second trigger signal, the logic analyzer triggered the controlling microcomputer, which transferred the recorded error data from the logic analyzer to the host computer for off-line evaluation. The above recording procedure was repeated for each error.

4.2 Workloads and Overhead

Three workloads, all written in C programming language, were used in the test: a linked list (LL), an integer matrix manipulation (MM) and a quicksort (QS). The task of the linked list program was to search an integer key in a linked list, removing the record associated with that key and then inserting the record into another linked list, according to the ascending order of the key values. The key value of each record is a 16-bit integer. The matrix manipulation program transposed a 5×5 matrix A of 16-bit integers and multiplied the transposed matrix A^T with the source matrix, i.e. $A^T \times A$. The quicksort program sorted an array of pointers to the twenty-two data records, according to the ascending order of the key values in the data records. Here also the key value is a 16-bit integer. The operations of these three workloads should together represent the various types of operations in real application programs.

The memory space of the MC6809E CPU consists of 64 kilobytes. However, the sizes of the programs are very low compared with the memory size, as they occupy only two percent of the memory space. This poses two serious problems. One, error rates may be dependent on memory locations at which the programs will be stored. Two, as the programs are small, they cannot be representative of large programs used in real applications. To minimize these limitations, the memory space was consecutively filled with sufficient copies of the same program. Each program copy owned its private data segment in the data area. About 47.5 kilobytes were required for storing the program codes of all copies, whereas 15 kilobytes were sufficient for the data segments. To protect the program codes against injected faults, EPROM was used for storing the program codes. Figure 4-2 shows the memory map configuration.

The execution procedure of all workloads comprised two phases. In the first phase, the data area was initiated with the ordinary data items. In the second phase, the task of the workloads was performed. A small main program called the program copies randomly until all copies were called. This procedure was repeated in an infinite loop until an error manifested itself.

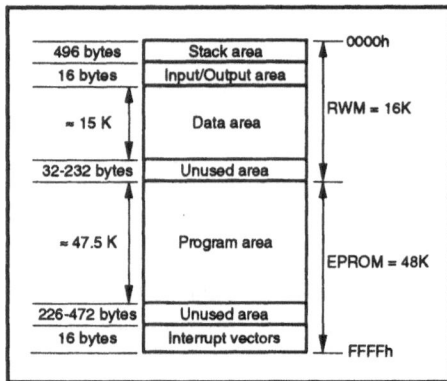

Figure 4-2 Memory Map

Table 4-1 Characteristics of the Workloads

Workloads	L L	M M	Q S
Source Program Size (bytes)	561	418	418
Total Basic Blocks (#)	40	24	28
Basic Blocks With BSSC (#)	19	11	14
Program Size Overhead (%)	23.7	18.4	23.4
Execution Time Overhead (%)	99.9	83.4	115.4
Data Size Overhead (%)	4.6	9.1	5.9
Branch Instructions (%)	26.4	17.9	25.1
Non-Branch Instructions (%)	73.6	82.1	74.9
Used Instructions of the Instruction Set (%)	17.4	17.0	22.2

All program codes were provided with the BSSC mechanism. Each basic block requires seven extra bytes for providing it with a BSSC mechanism. However, providing all basic blocks with the BSSC mechanism results in high overhead in terms of program size and execution time. This is a consequence of the occurrence of many basic blocks which consisted only of a few instructions. In order to reduce the overhead, only basic blocks containing more than five instructions were provided with the BSSC mechanism. For example, only 11 of the total of 24 basic blocks in the matrix manipulation program were provided with the BSSC mechanism. The number of basic blocks provided with the BSSC mechanism for each workload is given in Table 4-1. The table also shows the program size overhead and the execution time overhead for each workload. The program size overhead was between 18% and 24%, depending on the individual workload, while the execution time overhead varied between 83% and 115%.

The sizes of the entry and exit routines used for the BSSC mechanism were 14 and 23 bytes, respectively. Their execution times were 38 and 52 μs [*], respectively. These times, together with the execution time of the call instructions to the entry and exit routines, give a total of 106 cycles of execution time overhead for each BSSC mechanism.

The data size overhead caused by the insertion of ECI mechanisms in the data segments varied between 4% and 9%. The data overhead for each program is shown in Table 4-1. More information about the programs is also given in the table. This includes the percentage of branching and non-branching instructions used in each program, as well as the percentage of instructions that are used from the total instructions in the instruction set.

5 Control Flow Behaviour

This section describes different outcomes of the control flow behaviour. The results are based on a total of 6,000 errors , i.e. 1,000 errors for each of the three workloads and each of the two fault injection methods. The confidence limits given in the results are based on 95%.

---

[*] 1 μs = 1 bus cycle

The control flow behaviour after manifestation of errors is divided into two main classes: Errors that resulted in **control flow errors** and errors that had no effect on the correct control flow, i.e., **control flow OKs**. The results for each workload and for each method of fault injection are shown in Table 5-1. The rate of control flow errors caused by the heavy-ion radiation was about 8% lower than that caused by the power supply disturbances. The control flow errors behaved in different ways. Most errors, i.e. 86%, caused a severe control flow error which

Table 5-1 Different Classes of Control Flow Errors and Control Flow OKs

		Control Flow Errors					Control Flow OKs			
			Undetected					Undetected		
Workloads		Detec-ted	Illegal Reset	Result Wrong	Result OK	Total	Detec-ted	Result Wrong	Result OK	Total
LL	HIR	78.5±2.5	1.5±0.8	1.8±0.8	4.6±1.3	86.4±2.1	2.5±1.0	8.7±1.7	2.4±0.9	13.6±2.1
LL	PSD	96.0±1.2	0.7±0.5	0.4±0.4	0.1±0.1	97.2±1.0	0.5±0.4	1.5±0.8	0.8±0.6	2.8±1.0
MM	HIR	78.8±2.5	0.5±0.4	5.3±1.4	3.6±1.2	88.1±2.0	0.9±0.6	8.0±1.7	2.9±1.0	11.9±2.0
MM	PSD	92.9±1.6	0.1±0.1	2.7±1.0	0.2±0.2	95.9±1.2	0.0±0.0	3.4±1.1	0.7±0.5	4.1±1.2
QS	HIR	81.0±2.4	3.1±1.1	2.5±0.9	5.2±1.4	91.8±1.7	2.2±0.9	1.7±0.8	4.3±1.3	8.2±1.7
QS	PSD	90.2±1.8	6.6±1.5	0.4±0.4	0.5±0.4	97.7±0.9	0.1±0.1	1.4±0.7	0.8±0.6	2.3±0.9
Total	HIR	79.4±1.4	1.7±0.5	3.2±0.6	4.5±0.7	88.8±1.1	1.9±0.5	6.1±0.9	3.2±0.6	11.2±1.1
Total	PSD	93.0±0.9	2.4±0.6	1.2±0.4	0.3±0.2	96.9±0.6	0.2±0.2	2.1±0.5	0.8±0.3	3.1±0.6
Total		86.2±0.9	2.1±0.4	2.2±0.5	2.4±0.4	92.8±0.7	1.0±0.3	4.1±0.5	2.0±0.4	7.2±0.7

was detected by one of the error detection mechanisms, i.e., by the BSSC, the ECI or the WDT. Approximately 6.7% of all control flow errors went undetected by the error detection mechanisms. More than 2% of such control flow errors were related to the errors which caused an illegal reset to the processor. Another type of undetected control flow error caused a deviation of only a few bytes from the correct control flow, upon which the program's execution was resynchronized; Appendix A shows an example of such occurrence. The effect of these errors on the result produced was different. In 2.2% of such errors, the result produced was wrong, while 2.4% of such errors had no effect on the computation, so that the result delivered was correct.

In the class of control flow OKs, 1% of errors was detected by the WDT. An example of the errors in this case is an error that affected the condition code register in which a segment of the program code was unexpectedly re-executed, upon which the time limit of the WDT ran out. About 4% of errors having no effect on the control flow occurred during the data manipulation, upon which the delivered result was wrong. Examples of such a class of errors are those which affected the data bus, or the address bus when it contained an address of the data area. This was also observed for some errors when the address bus contained an address of the stack area. In 2% of control flow OKs, the errors had no effect on either the correct flow or the correct result. The effect of these errors was different. Some affected the data bus during a read or write cycle; however, a correct result was produced. Another kind of error which had no effect on the correct flow or correct result did affect the signals of the processor not used in the experiment, such as the BS and the BUSY signals. Finally, a number of errors affected the AVMA signal; however, no effect was observed on the correct flow and the correct result.

The figure of the 'result OK' in the quicksort program was higher than the corresponding figure in the other programs. This may be owing to the operation in the quicksort program, as the sorting operation was performed according to a recursive manner. So, if an error caused a data element to be incorrectly sorted, it could be corrected in the next recursive calling of the quicksort routine.

A fairly large number of the control flow errors, about 32%, were detected by the WDT. The execution behaviour of these errors is classified into several outcomes and is described below. Nearly 12% of such errors permanently forced the processor into an illegal state, called the **Illegal Processor State** (IPS). Upon this occurrence, the processor performed an abnormal operation which was not interpretable. In approximately 5% of the errors, the processor ended by performing **Increasing Consecutive Reading** (ICR). That is, the processor began to read the content of the memory locations in consecutive increasing order without doing any processing. One interesting reason for most of the ICR outcome was the unexpected execution of an illegal op-code, which forced the processor into this state. In this case, upon the occurrence of an error, an illegal branch was made to a random location of the memory that contained such an illegal op-code.

It was found in the experiment that the execution of the illegal op-codes with the hexadecimal values 14, 15 and CD forced the processor into this state. For a few errors, a similar outcome had appeared, however, as **Decreasing Consecutive Writing** (DCW), i.e., the processor began to write on the memory locations in a consecutive decreasing order. In more than 4% of the errors, the processor went permanently into a state called **Dummy Access State**[*] (DAS) [21]. This is a state indicating that the processor will not use the address bus. Nearly 2% of the errors forced the processor into the "SYNC" state, in which the processor enters a synchronizing state and waits for an interrupt signal from an external device. There were two ways in which the processor was forced into the SYNC state. Sometimes the processor erroneously fetched the content of a memory location that was the operation code of the SYNC instruction (SYNC op-code=13). Alternatively, the processor was forced directly by the errors into the SYNC state. In all of the above cases, the normal processor state was recovered by resetting it. Finally, the remainder of the errors, about 10%, could be detected by the watchdog timer for other reasons. These errors are classed as **Other Time Out** (OTO). A number of such errors affected the address bus when it contained an address of the stack area. In these errors, however no effect was observed in the correct flow within 270 cycles, a control flow error should have been occurred after 270 cycles, because the time limit of the WDT ran out. Another kind of error in the OTO class caused an erroneous branch to other points of the program. Consequently, a segment of the program code was unexpectedly re-executed and the time limit of the WDT expired. For some errors, the processor made an illegal jump to the stack area and continued to execute erroneously in that area. This was also observed for some errors which made an illegal jump to the data area. Finally, a number of the errors in the OTO class entered an infinite illegal loop which was detected by the WDT.

Another class of control flow errors, accounting for about 6.5%, comprised those that caused an illegal hardware or software interrupt. The execution behaviour of such errors is classified as **interrupt outcome**. It comprises four hardware interrupts, namely the RESET, the NMI, the IRQ and the FIRQ signals, and also two software interrupts, namely the SWI2 and the SWI3 instructions. Most of the illegal interrupts, i.e. 5.1%, consisted of the illegal RESET to the processor. One interesting reason for most of the illegal RESETs was the execution of an illegal op-code with the hexadecimal value 3E. These errors constituted 4.6% of the errors, while only 0.5% of errors caused an illegal RESET for the other reasons. The remaining illegal interrupts, i.e. 1.4%, consisted of the other types of interrupt.

---

[*] In dummy access, the processor will output address $FFFF_{16}$, R/W=1 and BS=0.

6 Evaluation of the Error Detection Mechanisms

The error detection mechanisms, i.e., the BSSC, the ECI and the WDT, are evaluated in this section. Table 6-1 illustrates the error detection coverage for each workload. More than 87% of the collected errors were detected by the mechanisms. The BSSC mechanism had the best coverage, which was 42.9%. This is a clear improvement in the coverage compared with the results in the previous version of the BSSC, in which the coverage was only 15%. As shown in the table, the coverage of the BSSC mechanism varies between 37% and 50%, depending on the individual workload. It therefore demonstrates the dependency of the coverage figure on the type of workload used. The inserted ECI mechanisms in the data and the unused areas detected about 10% of the errors, approximately half in each area. More than 32% of the errors which

Table 6-1 Statistics for the Coverage

Workloads		Total Coverage	Total BSSC	Total ECI	Total WDT	HW / SW Interrupts
L L	HIR	81.0 ± 2.4	38.5 ± 3.0	10.2 ± 1.9	29.1 ± 2.8	3.2 ± 1.0
L L	PSD	96.6 ± 1.1	47.8 ± 3.1	7.0 ± 1.6	41.3 ± 3.1	0.5 ± 0.4
MM	HIR	79.8 ± 2.5	39.5 ± 3.0	11.6 ± 2.0	24.4 ± 2.7	4.3 ± 1.3
MM	PSD	93.0 ± 1.6	44.1 ± 3.1	10.0 ± 1.9	38.4 ± 3.0	0.5 ± 0.4
Q S	HIR	83.3 ± 2.3	50.0 ± 3.1	12.8 ± 2.1	17.7 ± 2.4	2.8 ± 1.0
Q S	PSD	90.3 ± 1.8	37.6 ± 3.0	9.5 ± 1.8	43.2 ± 3.1	0.0 ± 0.0
Total	HIR	81.4 ± 1.4	42.7 ± 1.8	11.5 ± 1.1	23.7 ± 1.5	3.4 ± 0.6
Total	PSD	93.3 ± 0.9	43.2 ± 1.8	8.8 ± 1.0	41.0 ± 1.8	0.4 ± 0.2
Total		87.3 ± 0.8	42.9 ± 1.3	10.2 ± 0.8	32.4 ± 1.2	1.9 ± 0.3

went undetected by the BSSC and ECI mechanisms could be detected by the WDT. In Section 5, we discussed the errors which caused illegal interrupt. Of the seven interrupts available in the MC6809E, five were not used in the experiment. Hence, the vector addresses of all unused interrupts were provided with the address to a detection routine. This routine detected 1.9% of the errors when an illegal interrupt occurred. It was found that the coverage in the power supply disturbances was 12% higher than the coverage in the heavy-ion radiation. An important issue when evaluating the coverage figure is the efficiency of the error detection mechanisms detecting control flow errors. The results show that more than 93% of the control flow errors could be detected by the mechanisms.

A summary of the detected and undetected errors is given in Table 6-2. The amount of undetected errors was approximately 12.7%. A fairly large proportion of such errors, about 4.4%, did not affect the computation, so that a correct result was produced. On the other hand, approximately 6.2% of such errors did affect the computation, upon which the result produced was wrong. As was mentioned in Section 5, about 5.1% of the errors caused an illegal RESET to the processor. Of these errors, 3% could be detected by the BSSC mechanism, while the remaining 2.1% errors went undetected. In fact, in some industrial applications, a reset to the processor is sufficient for recovery. Hence, the errors which caused illegal RESET and also the errors that did not affect the correct result can be included in the covered errors. So, adding the errors which caused illegal RESET and the errors that did not affect the correct result to the errors detected by the error detection mechanisms, the real coverage is about 93.8%.

Table 6-2 Statistics for the Coverage, Result OK, Result Wrong and Illegal Reset

Total Detected Errors (%)		Total Undetected Errors (%)				
		Result OK		Result Wrong		Illegal Reset
CFE	CFOK	CFE	CFOK	CFE	CFOK	CFE
86.2 ± 0.9	1.1 ± 0.3	2.4 ± 0.4	2.0 ± 0.4	2.1 ± 0.4	4.1 ± 0.5	2.1 ± 0.4
87.3 ± 0.8		4.4 ± 0.5		6.2 ± 0.6		2.1 ± 0.4

Table 6-3   Statistics for the Detection Latencies

Latencies (cycles)	Workloads								
	L L		M M		Q S		Total		Total
	HIR	PSD	HIR	PSD	HIR	PSD	HIR	PSD	HIR & PSD
Mean Latency	69.3	106.2	76.3	139.3	68.8	82.8	71.5	111.3	90.2
Median Latency	62.0	109.0	63.0	144.0	63.0	80.0	62.0	110.0	84.0
WDT Latency	20000	20000	43585	43585	62635	62635	42073	42073	42073

The statistics for the error detection latencies are shown in Table 6-3. The mean and median latencies given in the table are based on the errors detected within 270 cycles by the BSSC and ECI mechanisms. These errors constituted approximately 95% of all errors detected by the BSSC and ECI mechanisms, while the remaining 5% errors were detected after 270 cycles. The mean and median latencies for the errors detected within 270 cycles were, on average, 90 and 84 cycles, respectively. Latencies for each workload are shown in the table. On average, the mean and median latencies for the errors generated by the heavy-ion radiation were, respectively, 60% and 80% shorter than the latencies for the errors generated by the power supply disturbances. The execution time of a workload was used as the time limit of the watchdog timer for that workload. For example, the time limit for the watchdog timer in the LL workload was 40,000 cycles. However, as it was not possible to measure the exact watchdog timer latency, a reasonable value could be achieved by dividing the time limit by two. Hence, 20,000 cycles are used as the watchdog timer latency for the LL workload. The watchdog timer latency for each workload is also shown in the table. On average, the error detection latency of the watchdog timer was approximated as 42,000 cycles. One important advantage of the BSSC and ECI mechanisms is their very short detection latencies as compared with the detection latencies of the watchdog timer.

7 Summary and Conclusions

The results of the physical fault injection showed that nearly 93% of all errors led to control flow errors, while the remainder of the errors, 7%, had no effect on the correct flow. It was found that the rate of control flow errors caused by the power supply disturbances was 8% higher than the rate of control flow errors caused by the heavy-ion radiation. In 2.4% of the errors the result delivered was correct, although a control flow error was presented. More than 4% of the errors which had no effect on the correct control flow affected the computation, in which the results delivered were wrong. In 2% of the errors, no effect was observed on either the correct flow or the correct result. In about 22% of the errors which resulted in control flow errors, the execution of the workloads was stopped and the processor entered into other states.

More than 87% of all errors and 93% of the control flow errors were detected by the error detection mechanisms. The BSSC mechanism had the best coverage, which was 42.9%. The inserted ECI mechanisms in the data and the unused areas detected about 10% of the errors, approximately half in each area. More than 32% of the errors which went undetected by the BSSC and ECI mechanisms could be detected by the WDT. Approximately 95% of the errors detected by the BSSC and ECI mechanisms were detected within 270 cycles. The mean and median latencies for these errors were, on average, 90 and 84 cycles, respectively. On average, the error detection latency of the errors detected by the watchdog timer was about 42,000 cycles.

An earlier experiment showed that the rates of errors caused by the heavy-ion radiation was lower for the programs written in assembly language than for those written in C language [19]. Different coverage figures were also obtained for different programs. The current and earlier results indicate that workloads, compilers and fault injection methods all have a serious influence on the coverage, the latency and the error rates.

Directions for future work will be to validate the results presented in this paper. So, the two fault injection methods used in this experiment need to be complemented with other fault injection methods, using the same workloads and experimental environment, for example with a software-implemented fault injection method. Also, the experiment needs to be repeated with another microprocessor to obtain new results in order to find whether the new results agree with the results in this paper. Finally, a study of the compilers and the codes of the benchmark programs may reveal how the compilers and the codes of the benchmark programs are correlated to the variation of the coverage, the latency and the error rates.

References

[1]   Aho, A., R Sethi, and J. Ullman, "Compilers: Principles, Techniques, and Tools," *Addison-Wesley*, 1986.

[2]   Ball, H., and F. Hardy, "Effects and Detection of Intermittent Failures in Digital Systems," *1969 FJCC, AFIPS Conference Proceedings*, Vol. 35, pp. 329-335, 1969.

[3]   Castillo, X, S. R. McConnel, and D. P. Siewiorek, "Derivation and Calibration of a Transient Error Reliability Model," *IEEE Transactions on Computers*, Vol. C-31, No. 7, pp. 658-671, July 1982.

[4]   Choi, C., R. Iyer, R. Saleh, and V. Carreno, "A Fault Behaviour Model for an Avionic Microprocessor: A Case Study," *Proceedings of 1st IFIP Working Conference on Dependable Computing for Critical Applications*, pp. 71-77, Santa Barbara, California, Aug. 1989

[5]   Czeck, E. W., D. P. Siewiorek, "Effect of Transient Gate-level Faults on Program Behaviour," *Digest of Papers, 20th Annual International Symposium on Fault-Tolerant Computing (FTCS-20)*, pp. 236-243, Newcastle, UK, June 1990.

[6]   Eifert, J. B., and J. P. Shen, "Processor Monitoring Using Asynchronous Signatured Instruction Streams," *Digest of Papers, 14th Annual International Conference on Fault-Tolerant Computing (FTCS-14)*, pp. 394-399, Kissimmee, Florida, June 1984.

[7]   Gunneflo, U., "The Effects of Power Supply Disturbances on the MC6809E Microprocessor," *Technical Report No. 89, Department of Computer Engineering, Chalmers University of Technology*, Göteborg, Sweden, 1990.

[8]   Halse, R. G., and C. Preece, "Erroneous Execution and Recovery in Microprocessor Systems," *Software and Microsystems*, Vol. 4, No. 3, pp. 63-70, June 1985.

[9]   Iyengar, V. S., and L. L. Kinney, "Concurrent Fault Detection in Microprogrammed Control Units," *IEEE Transactions on Computers*, Vol. C-34, No. 9, pp. 810-821, Sept. 1985.

[10] Iyer, R. K., and D. J. Rossetti, "A Measurement-Based Model for Workload Dependence of CPU Errors," *IEEE Transactions on Computers*, Vol. C-35, No. 6, pp. 511-519, June 1986.

[11] Kane, J. R., and S. S. Yau, "Concurrent Software Fault Detection," *IEEE Transactions on Software Engineering*, Vol. SE-1, No. 1, pp. 87-99, March 1975.

[12] Karlsson, J., U. Gunneflo, and J. Torin, "The Effects of Heavy-Ion Induced Single Event Upsets in the MC6809E Microprocessor," *Proceedings of 4th International GI/ITG/GMA Conference on Fault-Tolerant Computing Systems*, pp. 296-307, Baden-Baden, Sept. 1989.

[13] Karlsson, J., U. Gunneflo, and J. Torin, "Use of Heavy-ion Radiation From 252Californium for Fault Injection Experiments," *Proceedings of 1st IFIP Working Conference on Dependable Computing for Critical Applications*, pp. 79-84, Santa Barbara, California, Aug. 1989.

[14] Karlsson, J., U. Gunneflo, P. Lidén, and J. Torin, "Two Fault Injection Techniques for Test of Fault Handling Mechanisms," *Digest of Papers, IEEE 1991 International Test Conference*, pp. 140-149, Nashville, TN, Nov. 1991.

[15] Lu, D. J., "Watchdog Processors and Structural Integrity Checking," *IEEE Transactions on Computers*, Vol. C-31, No. 7, pp. 681-685, July 1982.

[16] Mahmood, A., and E. J. McCluskey, "Watchdog Processors: Error Coverage and Overhead," *Digest of Papers, 15th Annual International Symposium on Fault-Tolerant Computing (FTCS-15)*, pp. 214-219, Ann Arbor, Michigan, June 1985.

[17] Mahmood, A., and E. J. McCluskey, "Concurrent Error Detection Using Watchdog Processors- A survey," *IEEE Transactions on Computers*, Vol. 37, No. 2, pp. 160-174, Feb. 1988.

[18] Messenger, G. C., and S. A. Milton, "The Effects of Radiation on Electronics System," *Van Nostrand Reinhold Company Inc.*, 1986.

[19] Miremadi, G., "Software Techniques for On-line Error Detection in Microcomputer-based Systems," *Technical Report No. 102L, Department of Computer Engineering, Chalmers University of Technology*, Göteborg, Sweden, 1990.

[20] Miremadi, G., Karlsson, J., U. Gunneflo, and J. Torin, "Two Software Techniques for On-line Error Detection," *Digest of Papers, 22nd Annual International Symposium on Fault-Tolerant Computing (FTCS-22)*, pp. 328-335, Boston, July 1992.

[21] "8-bit Microprocessors Data Manual 1983," *Motorola Semiconductors*, Switzerland, Sept. 1983.

[22] Namjoo, M., "Techniques for Concurrent Testing of VLSI Processor Operation," *Digest of Papers, IEEE 1982 International Test Conference*, pp. 461-468, Philadelphia, Nov. 1982.

[23] Rimén, M., and J. Ohlsson, "A Study of the Error Behaviour of a 32-bit RISC Subjected to Simulation Transient Fault Injection," *Digest of Papers, IEEE 1992 International Test Conference*, pp. 696-704, Baltimore, Nov. 1992.

[24] Saxena, N. R., and E. J. McCluskey, "Control Flow Checking Using Watchdog Assists and Extended-Precision Checksums," *Digest of Papers, 19th Annual International Symposium on Fault-Tolerant Computing (FTCS-19)*, pp. 428-435, Chicago, June 1989.

[25] Schmid, M. E., R. L. Trapp, A. E. Davidoff, and M. Masson, "Upset Exposure by Means of Abstraction Verification," *Digest of Papers, 12th Annual International Symposium on Fault-Tolerant Computing (FTCS-12)*, pp. 237-244, Santa Monica, June 1982.

[26] Schuette, M. A., and J. P. Shen, "Processor Control Flow Monitoring Using Signatured Instruction Streams," *IEEE Transactions on Computers*, Vol. C-36, No. 3, pp. 264-276, March 1987.

[27] Shen, J. P., and M. A. Schuette, "On-Line Self-Monitoring Using Signatured Instruction Streams," *Digest of Papers, 1983 IEEE International Test Conference*, pp. 275-282, Philadelphia, 1983.

[28] Siewiorek, D. P., V. Kini, H. Mashbrun, S. McConnel and M. Tsao, "A Case Study of C.mmp, Cm*, and C.vmp: Part I — Experiences with Fault Tolerance in Multiprocessor Systems," *Proceeding of the IEEE*, Vol. 66, No. 10, pp. 1178-1199, October 1978.

[29] Sridhar, T., and S. M. Thatte, "Concurrent Checking of Program Flow in VLSI Processors," *Digest of Papers, IEEE 1982 International Test Conference*, pp. 191-199, Philadelphia, Nov. 1982.

[30] Wilken, K., and J. P. Shen, "Continuous Signature Monitoring: Low-Cost Concurrent-Detection of Processor Control Errors," *IEEE Transactions on Computer-Aided Design*, Vol. 9, No. 6, pp. 629-641, June 1990.

[31] Wingate, G. A. S., and C. Preece, "Performance Evaluation of a new Design-Tool for Microprocessor Transient Fault Recovery," *Microprocessing and Microprogramming*, Vol. 27, pp. 801-808, 1989.

[32] Yau, S. S., and FU-C. Chen, "An Approach to Concurrent Control Flow Checking," *IEEE Transactions on Software Engineering*, Vol. SE-6, No. 2, pp. 126-137, March 1980.

Appendix A *Program Behaviour When the Error Resulted in Resynchronization of the Program's Execution*

Cycle	Addr	Data	Control	Assembly	Code	
12	cf61	30		leax	D,x	
13	cf62	8b				
14	cf63	ec				
15	cf64	5c	!avma			
16	ffff	e0	!avma			
17	ffff	e0	!avma			
18	ffff	e0	!avma			
19	ffff	e0				
20	cf63	ec		ldd	-$4,u	
21	cf64	5c				
22	cf65	58	!avma			
23	ffff	e0				
24	1c4	0	busy			
25	1c5	1				
26	cf61	30	← Trigg 1	leax	D,x	← **Error manifestation**

Target CPU	cf61	30		ba = 0,	bs = 0,	lic = 0,	avma = 1,	busy = 0,	r/w = 1
Reference CPU	cf65	30		ba = 0,	bs = 0,	lic = 0,	avma = 1,	busy = 0,	r/w = 1

Address bus affected, control flow error, PC affected, 4 bytes backward illegal jump.

27	cf62	8b				
28	cf63	ec				
29	cf64	5c	!avma			
30	ffff	e0	!avma			
31	ffff	e0	!avma			
32	ffff	e0	!avma			
33	ffff	e0				
34	cf63	ec		ldd	-$4,u	← Program resynchronization.
35	cf64	5c				
36	cf65	58	!avma			
37	ffff	e0				
38	1c4	0	busy			
39	1c5	1				
40	cf65	58		aslb		
41	cf66	49				

System-Level Reliability and Sensitivity Analyses for Three Fault-Tolerant System Architectures

Joanne Bechta Dugan* and Michael R. Lyu†
Department of Electrical Engineering
University of Virginia
Charlottesville, Virginia 22903-2442
jbd@Virginia.edu
†Bellcore
Morristown, New Jersey 07692-1910
lyu@bellcore.com

Abstract

This paper discusses the modeling and analysis of three major fault-tolerant software system architectures: DRB (Distributed Recovery Blocks), NVP (N-Version Programming) and NSCP (N Self-Checking Programming). In the system-level reliability modeling domain, fault tree analysis techniques and Markov modeling techniques are combined to incorporate transient and permanent hardware faults as well as unrelated and related software faults. These models are parameterized by a real-world fault-tolerant flight control computer application for evaluations and comparisons. In particular, a series of sensitivity analysis is performed to explore the critical components in each fault-tolerant architecture and display their quantitative impacts to the overall system reliability.

1   Introduction

Since the first computer was invented some forty years ago, human beings have been depending more and more on computers in their daily lives. When the requirements for and dependencies on computers increase, the crises of computer failures also increase. The impact of hardware and software failures range from inconvenience (e.g., malfunctions of home appliances), economic loss (e.g., interceptions of banking systems) to life-threatening (e.g., failures of flight systems). Needless to say, reliability of computer systems becomes the major concern for our society for the 1990's and beyond. Consequently, computer systems that are used for critical applications are designed to tolerate both software and hardware faults by executing multiple software versions on redundant hardware. Many such examples exist in the aerospace industry [23, 7, 21], nuclear power industry [15, 2, 22], and ground transportation industry [6].

The system architectures incorporating both hardware and software fault tolerance are explored in three typical approaches. The Distributed Recovery Blocks (DRB) scheme [9] combines both distributed processing and Recovery Block (RB) [16] concepts to provide a unified approach to tolerating both hardware and software faults. Architectural considerations for the support of N-Version Programming (NVP) [1] were addressed in [10], in which the FTP-AP system is described. The FTP-AP system achieves hardware and software design diversity by attaching application processors (AP) to the byzantine resilient hard core Fault Tolerant Processor (FTP). N Self-Checking Programming (NSCP) [12] uses diverse hardware and software in self-checking groups to detect hardware and software induced errors. The NSCP concept forms the basis of the flight control system used on the Airbus A310 and A320 aircraft [3].

Sophisticated techniques exist for the separate analysis of fault tolerant hardware [5, 8] and software [11, 18, 19], and a few authors have considered their combined analysis [11, 20, 13]. This paper uses a combination of fault tree and Markov modeling as a framework for the analysis of hardware and software fault tolerant systems. The overall system model is a Markov model in which the states of the Markov chain represent the evolution of the hardware configuration as permanent faults occur. A fault tree model for each state in the Markov chain captures the effects of software bugs and transient hardware faults. This hierarchical approach simplifies the development, solution and understanding of the modeling process. In performing each model, the parameter values are derived from the analysis of data collected from an experimental NVP implementation [14]. A number of sensitivity analyses are conducted to study the quantitative behavior of the system reliability with respect to the parameter values.

2   Modeling Methodology

2.1   Assumptions

Task computation. The computation being performed is a task (or set of tasks) which is repeated periodically. A set of sensor inputs is gathered and analyzed and a set of actuations are produced. Each repetition of a task is independent. The goal of the analysis is the probability that a task will succeed in producing an acceptable output.

Software failure probability. Software faults exist in the code, despite rigorous testing. A fault is activated by some random input and produces an erroneous result. Each compu-

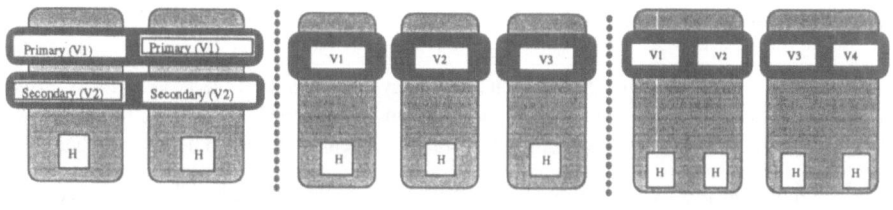

a) Distributed Recovery Block   b) N-version programming   c) N self-checking programming

Figure 1: Structure of a) DRB, b) NVP and c)NSCP

tation of a task receives a different set of inputs which are independent. Thus, a software task has a fixed probability of producing an error for a given task execution.

Constant hardware failure rates. The arrival (activation) rate of *permanent* physical faults is constant and will be denoted by λ.

Transient hardware faults. Transient hardware faults are modeled separately from permanent hardware faults. A transient hardware fault is assumed to upset the software running on the processor and produce an erroneous result which is indistinguishable from an input-activated software error. We assume that the lifetime of transient hardware faults is shorter than a task computation, and thus assign a fixed probability to the occurrence of a transient hardware fault during a single computation.

Related software faults. A related software fault in two different variants produce similar erroneous results on the same input. The two erroneous results match, which will be undetected if the results are compared to each other.

For the comparisons drawn from this study, we assume that the systems are unmaintained. Repairability and maintainability could certainly be included in the Markov model; we have chosen not to include them to make the comparisons clearer. More interesting task computation processes could be considered within this modeling framework as well.

Figure 1 shows the hardware and software error confinement areas [12] associated with the three architectures being considered in this paper. The systems are defined by the number of software variants, the number of hardware replications, and the decision algorithm. The error confinement area covers the region of the system affected by faults in that component.

2.2   System reliability model

A reliability model of an integrated fault tolerant system must include at least three different factors: computation errors, system structure and coverage modeling. In this paper we concentrate on the first two, as coverage modeling has been addressed in detail elsewhere [4].

The computation process is assumed to consist of a single software task that is executed repeatedly, such as would be found in a process control system. The software component performing the task is designed to be fault tolerant. A single task iteration consists of a task execution on a particular set of input values read from sensors. The output is the desired actuation to control the external system. During a single task iteration, several types of events can interfere with the computation. The particular set of inputs could activate a software fault in one or more of the software versions and/or the decider. Also, a hardware transient fault could upset the computation but not cause permanent hardware damage. The combinations of software faults and hardware transients that can cause an erroneous output for a single computation is modeled with a fault tree. The solution of the fault tree yields the probability that a single task iteration produces an erroneous output. We note that in the more general case where more than one task is performed, the analyses of each task can be combined accordingly.

The longer-term system behavior is affected by permanent faults and component repair which require system reconfiguration to a different mode of operation. The system structure is modeled by a Markov chain, where the Markov states and transitions model the long term behavior of the system as hardware and software components are reconfigured in and out of the system. Each state in the Markov chain represents a particular configuration of hardware and software components and thus a different level of redundancy. The fault and error recovery process is captured in the coverage parameters used in the Markov chain [4].

The short-term behavior of the computation process and the long-term behavior of the system structure are combined as follows. For each state in the Markov chain, there is a different combination of hardware transients and software faults that can cause a computation error, and thus a different probability that an unacceptable result is produced.

The fault tree model solution produces, for each state i in the Markov model, the probability q_i that an output error occurs during a single task computation while the state is in state i. The Markov model solution produces $P_i(t)$, the probability that the system is in state i at time t. The overall model combines these two measures to produce $Q(t)$, the probability that an unacceptable result is produced at time t.

$$Q(t) = \sum_{i=1}^{n} q_i P_i(t)$$

We assume that the system is unable to produce an acceptable result while in the failure state, thus ($q_{fail} = 1$).

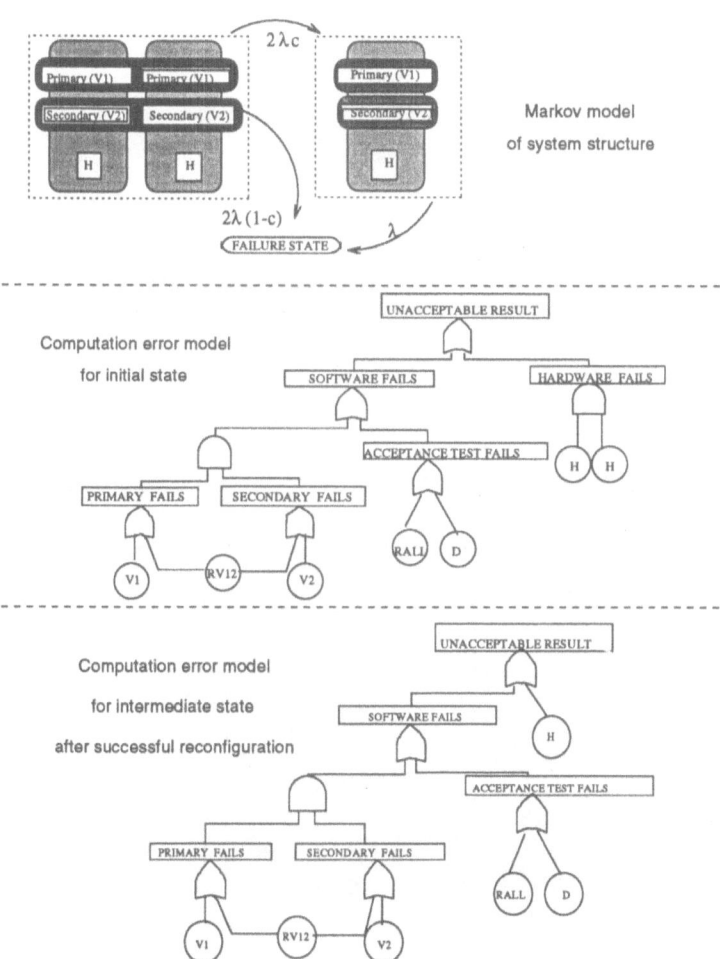

Figure 2: Reliability model of DRB.

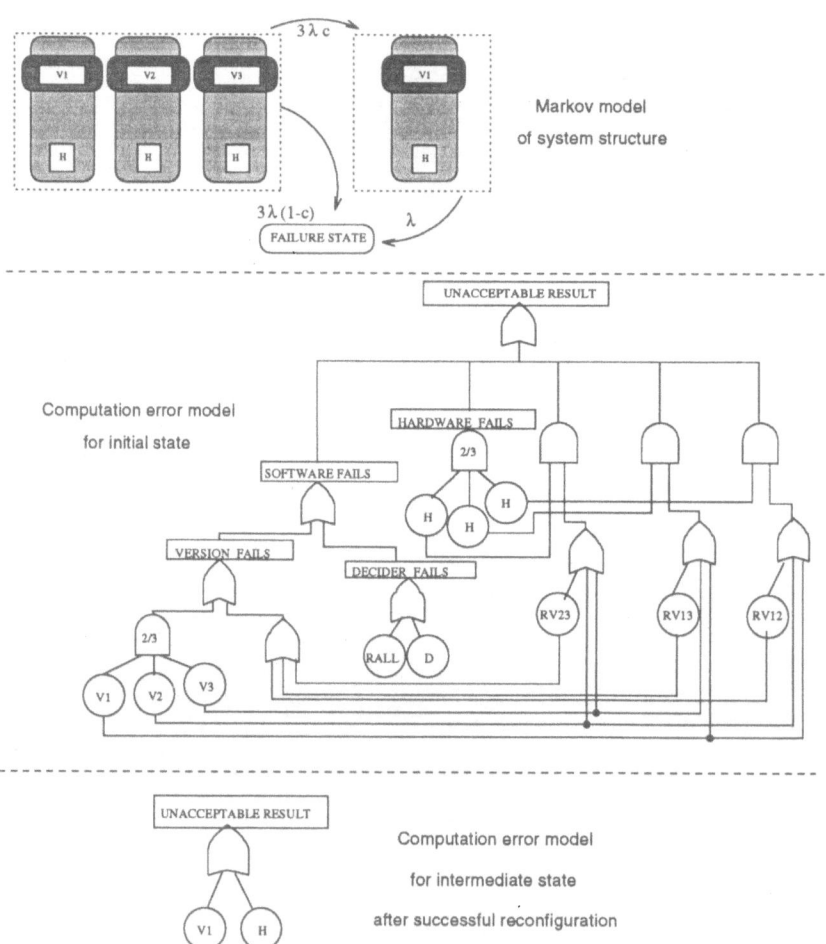

Figure 3: Reliability model of NVP.

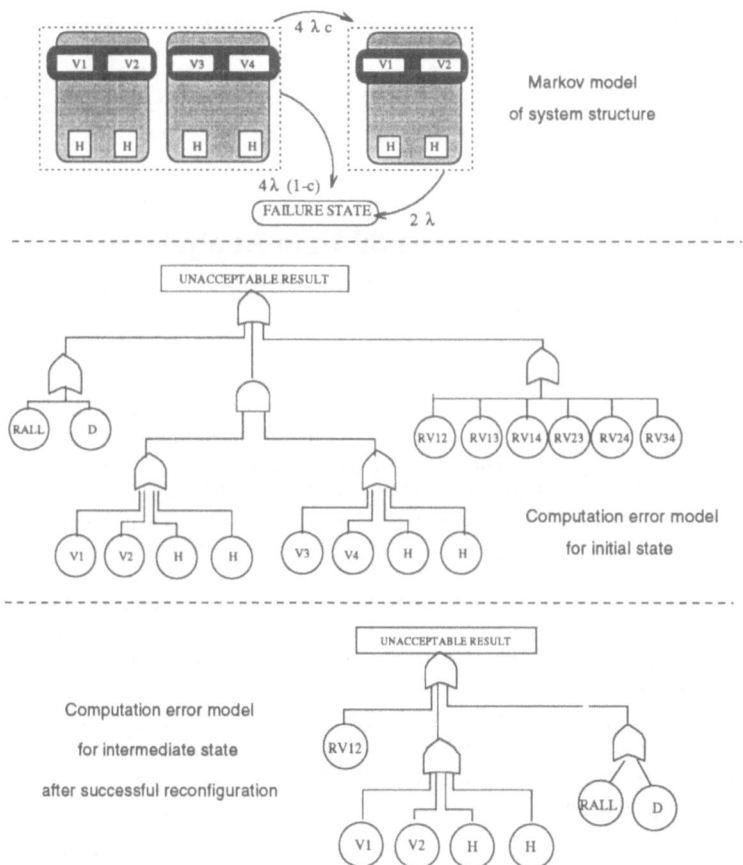

Figure 4: Reliability model of NSCP.

The three-part reliability models used for analysis of DRB, NVP and NSCP are shown in figures 2, 3 and 4, respectively. Each model consists of a three state Markov chain and two fault tree models. The Markov chain shows the evolution of the system structure as permanent physical faults are activated and handled. The fault tree models show the combinations of events which can upset a single task iteration in the full-up (initial) state and the intermediate state of the Markov model. In the Markov model, λ is the rate at which permanent hardware faults are activated and c is the probability that the system can automatically recover from a hardware fault. The basic events in the fault tree represent unrelated software failures (labeled Vi), related software failures between two versions (labeled Vij), related software failures in all versions ($RALL$), decider failures (D) and hardware transients (H).

3   Experimental Data Analysis

3.1   Description of experiment

The parameter values for the models in this paper were determined using actual data derived from an experimental implementation of a real-world automatic (i.e., computerized) airplane landing system, or so-called "autopilot." The software systems of this project were developed and programmed by 15 programming teams at the University of Iowa and the Rockwell/Collins Avionics Division. A total of 40 students (33 from ECE and CS departments at the University of Iowa, 7 from the Rockwell International) participated in this project to independently design, code, and test the computerized airplane landing system, as described in the Lyu-He study [14].

The application used in the Lyu-He study is part of a specification used by some aerospace companies for the automatic (computer-controlled) landing of commercial airliners. The specification can be used to develop the software of a flight control computer (FCC) for a real aircraft, given that it is adjusted to the performance parameters of a specific aircraft. All algorithms and control laws are specified by diagrams which have been certified by the Federal Aviation Administration (FAA). The *pitch control* part of the auto-landing problem, i.e., the control of the vertical motion of the aircraft, was selected for the project in order to fit the 14-week software development time.

By the end of the software development phase, 12 of the 15 programs passed the acceptance test successfully and were engaged in operational testing for further evaluations. The average size of these programs were 1564 lines of uncommented code, or 2558 lines when comments were included. The average fault density of the program versions which passed AT1 (the first step in the Acceptance Test) was 0.48 faults per thousand lines of uncommented code. The fault density for the final versions was 0.05 faults per thousand lines of uncommented code.

The operational environment for the application was conceived as airplane/autopilot interacting in a simulated environment. During the operational phase, 1000 flight simulations were con-

Version Id	Number of failures	Prob. by case	Prob. by time
β	510	0.51	0.000096574
γ	0	0.0	0.0
ϵ	0	0.0	0.0
ζ	0	0.0	0.0
η	1	0.001	0.000000189
θ	360	0.36	0.000068169
κ	0	0.0	0.0
λ	730	0.73	0.000138233
μ	140	0.14	0.000026510
ν	0	0.0	0.0
ξ	0	0.0	0.0
o	0	0.0	0.0
Average	145.1	0.1451	0.000027472

Table 1: Characteristics of accepted programs

ducted. Each flight simulation was characterized by the following five initial values regarding the landing position of an airplane: (1) initial altitude (about 1500 feet); (2) initial distance (about 52800 feet); (3) initial nose up relative to velocity (range from 0 to 10 degrees); (4) initial pitch attitude (range from -15 to 15 degrees); and (5) vertical velocity for the wind turbulence (0 to 10 ft/sec). One simulation consisted of about 5280 iterations of lane command computations (50 milliseconds each) for a total landing time of approximately 264 seconds. For a conservative estimation of software failures in the system, we took the program versions which passed the AT1 for study. The reason behind this was that had the Acceptance Test not included an extra test case after AT1, more faults would have remained in the program versions.

3.2 Failure data analysis

Table 1 shows the software failures encountered in each single version. We examine two levels of granularity in defining software execution errors and correlated errors: "by case" or "by time." The first level was defined based on test cases (1000 in total). If a version failed at any time in a test case, it was considered failed for the whole case. If two or more versions failed in the same test case (no matter at the same time or not), they were said to have coincident errors for that test case. The second level of granularity was defined based on execution time frames (5,280,920 in total). Errors were counted only at the time frame upon which they manifested themselves, and coincident errors were defined to be the multiple program versions failing at the same time in the same test case (with or without the same variables and values).

The accepted programs were then arranged in configurations of 2, 3 and 4 programs, and the error characteristics of each of the configurations is shown in tables 2, 3 and 4. Both the by-case

Category	BY CASE		BY TIME	
	Number of cases	Frequency	Number of cases	Frequency
1 - no errors	53150	0.8053	348259290	0.999192
2 - single error	11160	0.1691	281200	0.000807
3 - two coincident errors	1690	0.0256	230	0.000001
Total	66000	1.0000	1161802400	1.000000

Table 2: Error characteristics for two-version configurations

Category	BY CASE		BY TIME	
	Number of cases	Frequency	Number of cases	Frequency
1 - no errors	163370	0.7426	1160743690	0.999089
2 - single error	51930	0.2360	1056010	0.000909
3 - two coincident errors	4440	0.0202	2700	0.000002
4 - three coincident errors	260	0.0012	0	0.0
Total	220000	1.0000	1161802400	1.000000

Table 3: Error characteristics for three-version configurations

and by-time error detection methods were used. These characteristics were used to determine parameter values for the software failure models of DRB, NVP and NSCP.

Table 5 summarizes the parameters used for the software parameters of the system models. These parameters are derived from a single experimental implementation and so may not be generally applicable. Similar analysis of other experimental data will help to establish a set of reasonable parameters that can be used in models that are developed during the design phase of a fault tolerant system.

Category	BY CASE		BY TIME	
	Number of cases	Frequency	Number of cases	Frequency
1 - no errors	322010	0.65052	2611305000	0.998948
2 - single error	152900	0.30889	2719200	0.001040
3 - two coincident errors	16350	0.03303	31200	0.000012
4 - three coincident errors	3700	0.00747	0	0.0
5 - four coincident errors	40	0.00008	0	0.0
Total	495000	1.0000	2614055400	1.000000

Table 4: Error characteristics for four-version configurations

DRB model	NVP model	NSCP model
BY CASE DATA		
$P_V = 0.095$ $P_{RV} = 0.0167$	$P_V = 0.0958$ $P_{RV} = 0$ $P_{RALL} = 0.0003$	$P_V = 0.106$ $P_{RV} = 0$ $P_{RALL} = 0$
Predicted failure probability (perfect decider, no HW faults)		
0.0265	0.0262	0.0403
Observed failure probability (from the data)		
0.0256	0.0214	0.0406
Probability of decider failure used for system analysis		
0.001	0.0001	0.0001
BY TIME DATA		
$P_V = 0.0004$ $P_{RV} = 8.4 \times 10^{-7}$	$P_V = 0.0003$ $P_{RV} = 6 \times 10^{-7}$ $P_{RALL} = 0$	$P_V = 0.00026$ $P_{RV} = 0$ $P_{RALL} = 1.2 \times 10^{-5}$
Predicted failure probability (perfect decider, no HW faults)		
1×10^{-6}	2.07×10^{-6}	1.23×10^{-5}
Observed failure probability (from the data)		
1×10^{-6}	2.3×10^{-6}	1×10^{-5}
Probability of decider failure used for system analysis		
1×10^{-7}	1×10^{-7}	1×10^{-7}

Table 5: Summary of nominal software parameters used for system analysis

3.3 Hardware parameters

Typical permanent failure rates for processors range in the 10^{-5} *per hour* range, with transients perhaps an order of magnitude larger. Thus we will use $\lambda_p = 10^{-5}$ per hour for the Markov model.

In the by-case scenario, a typical test case contained 5280 time frames, each time frame being 50 ms., so a typical computation executed for 264 seconds. Assuming that hardware transients occur at a rate $\lambda_t = (10^{-4}/3600)$ *per second*, we see that the probability that a hardware transient occurs during a typical test case is

$$1 - e^{-\lambda_t \times 264 \; seconds} = 7.333 \times 10^{-6} \qquad (1)$$

We conservatively assume that a hardware transient that occurs anywhere during the execution of a task disrupts the entire computation running on the host.

For the by-time data, the probability that a transient occurs during a time frame is

$$1 - e^{-\lambda_t \times 0.05 \; seconds} = 1.4 \times 10^{-9} \qquad (2)$$

If we further assume that the lifetime of a transient fault is one second, then a transient can affect as many as 20 time frames. We thus take the probability of a transient to be 20 times the value calculated in equation 2, or 2.8×10^{-8}.

Finally, for both the by-case and by-time scenarios, we assume a fairly typical value for the coverage parameter in the Markov model, $c = 0.999$.

4   Reliability and Sensitivity Analysis

4.1   Reliability analysis

Figure 5 compares the predicted behavior of the three systems. Under both the by-case and by-time scenarios, the recovery block system is most able to produce a correct result, followed by NVP. NSCP is the least reliable of the three. It is noted, however, that the analysis performed in this paper is based on a *reliability* aspect (i.e., whether the system can deliver an acceptable result) rather than on a *safety* aspect (i.e., whether the system can deliver an acceptable result *or* conduct a safety shutdown after detecting an unacceptable condition). NSCP is expected to obtain a much better improvement with respect to the safety analysis. Of course, these comparisons are dependent on the experimental data used and assumptions made. More experimental data and analysis are needed to enable a more conclusive comparison.

Figure 6 gives a closer look at the comparisons between the NVP and DRB systems during the first 200 hours. The by-case data shows a crossover point where NVP is initially more reliable

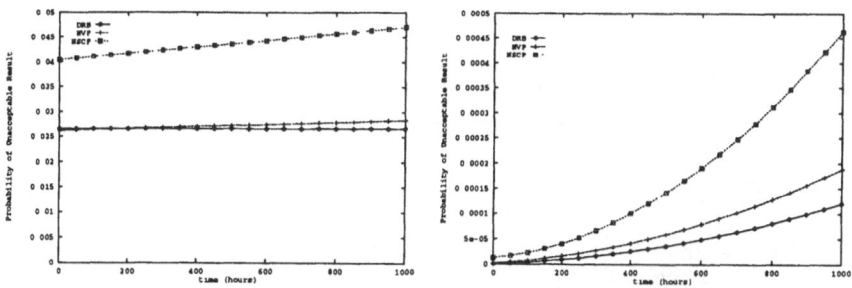

Figure 5: Predicted reliability, by-case data (left) and by-time data (right)

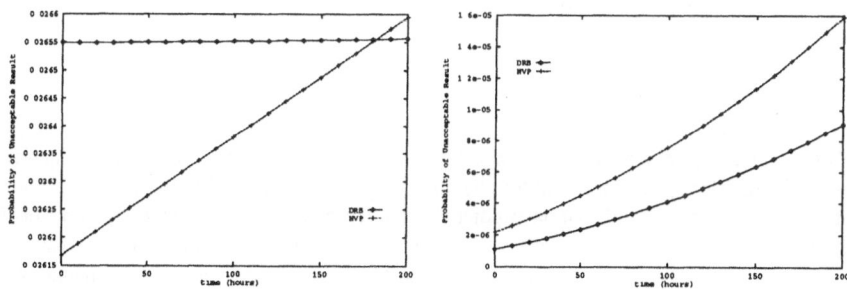

Figure 6: Predicted reliability, by-case data (left) and by-time data (right)

	By CASE Data		By TIME Data	
Parameter	Result	Percent Change	Result	Percent Change
Nominal	0.0265		1.10×10^{-6}	
$P_V + 10\%$	0.0284	7%	1.13×10^{-6}	2.8%
$P_{RV} + 10\%$	0.0282	6.2%	1.18×10^{-6}	7.6%
$P_D + 10\%$	0.0266	1.9%	1.11×10^{-6}	0.9%

Table 6: Sensitivity to parameter change for DRB model

	By CASE Data		By TIME Data	
Parameter	Result	Percent Change	Result	Percent Change
Nominal	0.02617		2.17×10^{-6}	
$P_V + 10\%$	0.03137	19.9%	2.23×10^{-6}	2.6%
$P_{RV} + 10\%$			2.35×10^{-6}	8.3%
$P_{RALL} + 10\%$	0.0262	0.1%		
$P_D + 10\%$	0.02618	0.04%	2.18×10^{-6}	0.5%

Table 7: Sensitivity to parameter change for NVP model

but is later less reliable than DRB. Using the by-time data, there is no crossover point, but the estimates are so small that the differences may not be statistically significant.

For all three systems the probability of producing an unacceptable result is initially much lower with the by-time data than with the by-case data. This analysis dramatizes the potential improvement associated with frequent comparisons (each time frame rather than each test case). The probability of producing an unacceptable result increases with time, as expected, but at 1000 hours is still far below even the initial by-case probability.

4.2 Sensitivity Analysis

To see which parameters are the strongest determinant of the system reliability, we increased each of the failure probabilities in turn by 10 percent and observed the effect on the predicted unreliability. The sensitivity of the predictions to a ten-percent change in input parameters for the DRB model is shown in table 6. It can be seen that the DRB model is most sensitive to a change in the probability of an unrelated fault for the by-case data, and to a change in the probability of a related fault for the by-time data.

Table 7 shows, the change in the predicted unreliability (at $t = 0$) when each of the NVP nominal parameters is increased. For the by-case data, a ten percent increase in the probability of an unrelated software fault results in a twenty percent increase in the probability of an unacceptable result. A ten-percent increase in the probability of a related or decider fault activation has an almost negligible effect on the unreliability. For the by-time data, the proability of a related

	By CASE Data		By TIME Data	
Parameter	Result	Percent Change	Result	Percent Change
Nominal	0.04041		1.237×10^{-5}	
$P_V + 10\%$	0.04833	19.6%	1.243×10^{-5}	0.5%
$P_{RALL} + 10\%$			1.357×10^{-5}	9.7%
$P_D + 10\%$	0.04042	0.02%	1.238×10^{-5}	0.08%

Table 8: Sensitivity to parameter change for NSCP model

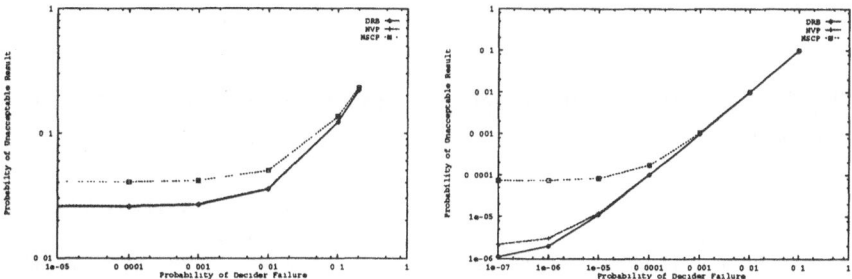

Figure 7: Effect of equal decider failure probabilities, by-case data (left) and by-time data (right)

fault has the largest impact on the probability of an unacceptable result. This is similar to the DRB model.

The sensitivity of the NSCP model to the nominal parameters is shown in table 8. The fault tree models and the sensitivity analysis show that NSCP is vulnerable to related faults, whether they involve versions in the same error confinement area or not.

5 Decider Failure Probability

The probability of a decider failure may be an important input parameter to the comparative analysis of the NVP and DRB systems. In this section we vary the decider failure probability in an attempt to demonstrate its importance. Figure 7 shows, for the by-case and by-time parameterizations, the unreliability of the three systems as the probability of decider failure is varied. For this analysis, we set the probability of failure for the decider to the same value for all three models, and show the probability of an unacceptable result at time $t = 0$.

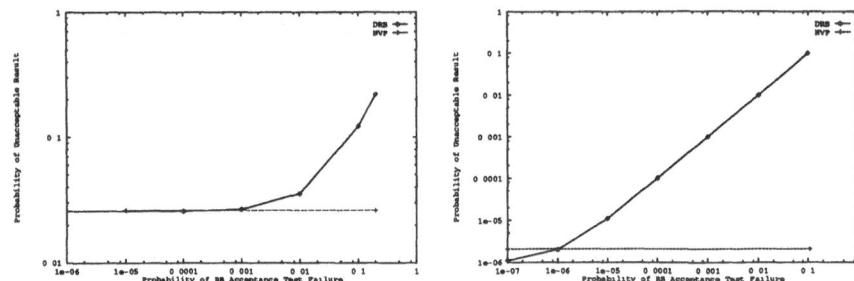

Figure 8: Effect of varying acceptance test failure probability, by-case data(left) and by-time data (right)

For the parameters derived from the experimental data, it seems that DRB and NVP are nearly equally reliable, if both have the same probability of decider failure. However, it is not reasonable for this application to assume equally reliability deciders for both DRB and NVP. The decider for the DRB system is an acceptance test, while that for the NVP is a simple voter and NSCP a simple comparator. For this application, it seems likely that an acceptance test will be more complicated than a majority voter. The increased complexity is likely to lead to a decrease in reliability, with a corresponding impact on the reliability of the system. In fact, reliability of DRB will collapse if the acceptance test in DRB is as complex and unreliable as its primary or secondary software versions. For example, if the probability of failure in acceptance test (P_D) is close to P_V, which is 0.095 by case or 0.0004 by time, then both Figure 7 indicates that DRB will initially perform the worst comparing with NVP and NSCP.

Figure 8 shows how the comparison between DRB and NVP is affected by a variation in the probability of failure for the acceptance test. The parameters for the NVP analysis were held constant, and the parameters (other than the probability of acceptance test failure) for the DRB model were also held constant. Figure 8 shows that the acceptance test for a recovery block system must be very reliable for it to be comparable in reliability to a similar NVP system.

6   Conclusions

We have proposed a system-level modeling approach to study the reliability behavior of three types of fault-tolerant architectures: DRB, NVP and NSCP. Using a recent fault-tolerant software project data, we parameterized the models and displayed the resulting system (un)reliability. The comparisons of the three fault-tolerant architectures were done not only from directly applying the estimated parameters, but from varying the baseline parameters as a sensitivity analysis. Several interesting results were obtained:

1. A drastic improvement of reliability could be observed if a finer and more frequent error detection mechanism could be performed by the decider for each architecture.

2. From the by-case data, varying the probability of an unrelated software fault had the major impact to the system reliability, while from the by-time data, varying the probability of a related fault had the largest impact. This could be due to the fact that the by-time data compares results in a finer granularity level, and is thus more sensitive to related faults among program versions.

3. In comparing the three different architectures, DRB performed better than NVP which in turn was better than NSCP. DRB also enjoyed the feature of relative insensitivity to time in its reliability function. DRB might perform worse than NVP to begin with, but in the long run it could become better.

4. The acceptance test in DRB had to be very reliable for (3) to remain true. If the acceptance test in DRB is as unreliable as its application versions, DRB loses its advantage to NVP and NSCP.

5. NSCP did not seem to perform very well in the reliability analysis. However, it is expected to gain more improvement and close the gap to the other two architectures if a safety analysis is performed.

Needless to say, more data points are wanted for the validation of our models and for more evidences of the advantages and disadvantages of the three fault-tolerant system architectures.

Acknowledgements

This work was partially funded by NASA AMES Research Center under grant number NCA2-617. The authors are grateful to Yu-Tao He and Stacy Doyle for their assistance. The models presented in this paper were solved using SHARPE [17].

References

[1] Algirdas Avižienis. The N-version approach to fault-tolerant software. *IEEE Transactions on Software Engineering*, SE-11(12):1491–1501, December 1985.

[2] P.G. Bishop, D.G. Esp, M. Barnes, P. Humphreys, G. Dahl, and J. Lahti. PODS - a project of diverse software. *IEEE Transactions on Software Engineering*, SE-12(9):929–940, September 1986.

[3] D. Briere and P. Traverse. Airbus A320/A330/A340 electrical flight controls: A family of fault-tolerant systems. In *Proceedings of the 23rd Symposium on Fault Tolerant Computing*, pages 616–623, 1993.

[4] Joanne Bechta Dugan and K. S. Trivedi. Coverage modeling for dependability analysis of fault-tolerant systems. *IEEE Transactions on Computers*, 38(6):775–787, 1989.

[5] Robert Geist and Kishor Trivedi. Reliability estimation of fault-tolerant systems: Tools and techniques. *IEEE Computer*, pages 52–61, July 1990.

[6] Gunnar Hagelin. ERICSSON safety system for railway control. In U. Voges, editor, *Software Diversity in Computerized Control Systems*, pages 11–21. Springer-Verlag, 1988.

[7] A. D. Hills. Digital fly-by-wire experience. In *Proceedings AGARD Lecture Series*, number 143, October 1985.

[8] Allen M. Johnson and Miroslaw Malek. Survey of software tools for evaluating reliability availability, and serviceability. *ACM Computing Surveys*, 20(4):227–269, December 1988.

[9] K.H. Kim and Howard O. Welch. Distributed execution of recovery blocks: An approach for uniform treatment of hardware and software faults in real-time applications. *IEEE Transactions on Computers*, 38(5):626–636, May 1989.

[10] Jaynarayan H. Lala and Linda S. Alger. Hardware and software fault tolerance: A unified architectural approach. In *Proc. IEEE Int. Symp. on Fault-Tolerant Computing, FTCS-18*, pages 240–245, June 1988.

[11] Jean-Claude Laprie. Dependability evaluation of software systems in operation. *IEEE Transactions on Software Engineering*, SE-10(6):701–714, November 1984.

[12] Jean-Claude Laprie, Jean Arlat, Christian Beounes, and Karama Kanoun. Definition and Analysis of Hardware- and Software- Fault-Tolerant Architectures. *IEEE Computer*, pages 39–51, July 1990.

[13] Jean-Claude Laprie and Karama Kanoun. X-ware reliability and availability modeling. *IEEE Transactions on Software Engineering*, pages 130–147, February, 1992.

[14] Michael R. Lyu and Yu-Tao He. Improving the N-version programming process through the evolution of a design paradigm. *IEEE Transactions on Reliability*, June 1993.

[15] C. V. Ramamoorthy, Y. Mok, F. Bastani, G. Chin, , and K. Suzuki. Application of a methodology for the development and validation of reliable process control software. *IEEE Transactions on Software Engineering*, SE-7(6):537–555, November 1981.

[16] Brian Randell. System structure for software fault tolerance. *IEEE Transactions on Software Engineering*, SE-1(2):220–232, June 1975.

[17] R. Sahner and K. S. Trivedi. Reliability modeling using SHARPE. *IEEE Transactions on Reliability*, R-36(2):186–193, June 1987.

[18] R. Keith Scott, James W. Gault, and David F. McAllister. Fault-tolerant software reliability modeling. *IEEE Transactions on Software Engineering*, SE-13(5):582–592, May 1987.

[19] Kang G. Shin and Yann-Hang Lee. Evaluation of error recovery blocks used for co-operating processes. *IEEE Transactions on Software Engineering*, SE-10(6):692–700, November 1984.

[20] George. E. Stark. Dependability evaluation of integrated hardware/software systems. *IEEE Transactions on Reliability*, pages 440–444, October 1987.

[21] Pascal Traverse. Airbus and ATR system architecture and specification. In U. Voges, editor, *Software Diversity in Computerized Control Systems*, pages 95–104. Springer-Verlag, June 1986.

[22] Udo Voges. Use of diversity in experimental reactor safety systems. In U. Voges, editor, *Software Diversity in Computerized Control Systems*, pages 29–49. Springer-Verlag, 1988.

[23] L. J. Yount. Architectural solutions to safety problems of digital flight-critical systems for commercial transports. In *Proceedings AIAA/IEEE Digital Avionics Systems Conference*, pages 1–8, December 1984.

Improving Availability Bounds using the Failure Distance Concept

Juan A. Carrasco
Departament d'Enginyería Electrònica
Universitat Politécnica de Catalunya (UPC)
Diagonal 647, plta. 9, 08028-Barcelona, Spain
carrasco@eel.upc.es

Abstract

Continuous-time Markov chains are commonly used for dependability modeling of repairable fault-tolerant computer systems. Realistic models of non-trivial fault-tolerant systems easily have very large state spaces. An attractive approach which has been proposed to deal with the largeness problem is the use of pruning-based methods which provide error bounds. Using results from Courtois and Semal, a method for bounding the steady-state availability has been recently developed by Muntz, de Souza e Silva, and Goyal. This paper presents a new method based on a different approach which exploits the concept of failure distance to better bound the behavior out of the non-generated state space. The proposed method yields tighter bounds. Numerical analysis shows that the improvement is typically significant.

1 Introduction

Modeling plays an important role in the design, analysis and management of fault-tolerant computer systems. These systems are characterised by exhibiting an stochastic behavior and, accordingly, probabilistic measures are used for their quantitative assessment. Many systems are seen by their users as providing service or not. For these systems, dependability measures such as the availability and the reliability are appropriate. The steady-state availability is a useful measure for repairable systems when the long-term behavior is of interest. In some cases, this measure can be computed using combinatorial techniques [1] or closed-product solution queuing networks [6]. However, in general, the dependencies introduced by lack of coverage, failure propagation, operational configurations and maintenance are such that general-purpose, state level model solution techniques are required. Continuous-time Markov chains (CTMC's) are often used to analyse systems with these dependencies and a number of dependability/performability tools based on these models have been developed in the past (see [11] for a recent review).

Numerical analysis of CTMC dependability models is hampered by the exponential growth of the number of states with the structural complexity of the system. Systems with moderate number of components easily yield CTMC's with millions of states and more. This problem has been attacked in three directions: a) hierarchical model solution [16], b) state lumping techniques [12], and c) pruning techniques. Only the last of them has general applicability. Recently, pruning-based solution methods providing error bounds have been developed for several dependability measures. Bounds for the reliability have been obtained in [2]. A method to bound the steady-state availability has been proposed in [15]. This method has been further developed in [14], [17]. The error bounds offered by these methods are qualitatively superior to the accuracy assessment offered by simulation methods recently developed to attack the largeness problem [10], [3], whose reliability depends on how well the variance is estimated. This makes of great interest the development of efficient bounding techniques.

This paper presents a new method to obtain steady-state unavailability bounds using CTMC models which, typically, gives significantly smaller bands than the method proposed in [15]. Our method uses an upper bound exploiting the fact that, very often, the system is operational a large portion of the time the model is out of the generated state space. The upper bound is developed using the failure distance concept. The rest of the paper is organised as follows. Section 2 describes the availability models under consideration, reviews the method proposed in [15] and, using a regenerative perspective, argues the potential looseness of the steady-state unavailability upper bound given by the method. Section 3 presents the theoretical developments yielding the upper bound used in our method. Section 4 illustrates with examples the reduction in the steady-state unavailability band which our method can achieve and discusses the computational overheads of our method in relation to the method proposed in [15]. Section 5 concludes the paper.

2   Preliminaries

The type of models addressed in this paper are those which result from conceptualising a fault-tolerant system as made up of components which fail and are repaired with constant rates. The system is operational or down as determined by a coherent structure function [1] on the unfailed/failed state of the components. This basically means that repairs cannot take down an operational system and that failures cannot bring operational a down system. A failure of a component can be propagated to other components. In addition, each component can be failed in a finite number of modes. Failure and repair rates and failure propagation can depend on the state of the system. Let $X = \{X(t); t \geq 0\}$ be the CTMC modeling the system, Ω its state space and o the (only) state in which all components of the system are unfailed. We assume that repair transitions involve only one component and that at least a repair transition exists from any state $\neq o$. It follows from the hypotheses that X is finite and irreducible. It is assumed that a high-level description of the model is available from which it is possible to identify the bags of components (we allow component types with instances) which can be failed simultaneously in a single event. Those bags are called *failure events.* E will denote the set of failure events of the model and E_i the set of failure events including i components. It is also assumed that links to the high-level description of the model exist allowing to determine during generation of the CTMC the failure event associated to a failure transition and the component affected by a repair transition. Using this information, it is possible to compute the bag of failed components $F(x)$ in each generated state x.

Let D be the subset of down states and let p_i, $i \in \Omega$ the steady-state probability distribution of X, the steady-state unavailability is defined as $UA = \sum_{i \in D} p_i$. UA is a special case of the more general steady-state reward rate measure $R = \sum_{i \in \Omega} r(i) p_i$, where $r(i), i \in \Omega$ is an arbitrary reward rate structure imposed on X.

Since repair rates are usually several orders of magnitude higher than failure rates, X is highly skewed, i.e., it has a probability distribution concentrated in a small portion of the state space (the states with few failed components). Thus, in general, good approximations for R can be computed using only a small portion of Ω. However, assessing the accuracy of the solution is a difficult problem. The method proposed in [15] was the first to obtain tight bounds in the context of availability modeling. The method can be used to bound any steady-state reward rate measure R (see [17]). Let G be the generated portion of Ω, U the non-generated portion, and S the subset of G through which X can enter G (from U). As in [15], assume that G contains all states with up to a given number K of failed components. The method can be described in terms of the CTMC's X'_i, $i \in S$ which (conceptually) can be obtained from X as shown in Figure 1. First, X_i is obtained from X by redirecting to i all transitions from U to G (S). Second, U is replaced by the states $u_{K+1}, \ldots, u_N$, where each u_k accounts for the subset U_k of U including all the states with k failed components. Failure transitions from G to states in U with k failed components are directed to state u_k. Each state u_k, $k < N$ has transitions to states u_{k+j} with rates $f_j(k)$ chosen to be upper bounds for the sum of the failure transitions rates from any state with k failed components to U_{k+j}. Each state u_k, $k > K + 1$ has also a repair transition to u_{k-1} with a rate $g(k)$ chosen to be a lower bound for the sum of the repair transition rates from any state with k failed components. A similar transition is also introduced from state u_{K+1} to state i. For $f_i(k)$ we can take $\sum_{e \in E_i} \lambda_{ub}(e)$, where $\lambda_{ub}(e)$ is an upper bound to the rate of the failure event e. For $g(k)$ we can take the slowest repair rate of the model. Let $|r|_{lb}$ and $|r|_{ub}$ be, respectively, lower and upper bounds for the reward rate in any state of X, the bounds for R are obtained using the following recipe:

1. for each state $i \in S$ find the steady-state distribution of X'_i and, assigning to the states in G the same reward rate as in X and to the states $u_{K+1}, \ldots, u_N$ a reward rate $|r|_{lb}$ ($|r|_{ub}$),

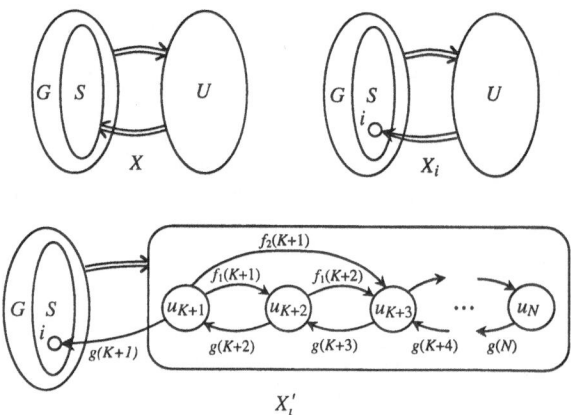

Figure 1: Construction of the CTMC's X_i' used by the bounding method proposed in [Mun89].

    compute the resulting steady-state reward rate $|R_i|_{lb}$ $(|R_i|_{ub})$,

2. $|R|_{lb} = \min_{i \in S} |R_i|_{lb}$, $|R|_{ub} = \max_{i \in S} |R_i|_{ub}$.

Typically S will include all states in G with K failed components and the number of models X_i' to be solved can be large. The computational cost of the method can be reduced at the expense of some looseness of the bounds by the state duplication technique proposed in [15]. In this technique, duplicates of all states in G with more than F failed components are (conceptually) added to U to account for the visits to these states after the number of failed components is made larger than K and before the number of failed components is made equal to F. This state duplication technique can be thought as a redefinition of the CTMC X to which the bounding method is applied. The resulting CTMC's X_i' have the structure depicted in Figure 1, except that the aggregate states u_i will run now from u_{F+1} to u_N and S will only include the states with F failed components which can be reached through repair transitions. Taking F small enough reduces arbitrarely the number of CTMC's X_i' to be solved. A final remark is that G does not have necessarily to include all states with up to a given number of failed components (see [17]). However, it does have to include all states with F or fewer failed components.

The reviewed bounding method was justified in [15] using the exact aggregation theorem for ergodic Markov chains and bounds on conditional steady-state distributions in subsets of Markov chains [7], [8]. Here, it will be discussed using a regenerative perspective. The motivation is to support theoretically our bounding method and ease the comparison between the method proposed in [15] and ours. Let C_i and T_i be, respectively, the expected reward and the expected time in X_i between consecutive jumps from U to i (regeneration points). Let R_i be the steady-state reward rate of X_i. Then, by regenerative theory, $R_i = C_i/T_i$. In addition, using semi-regenerative process theory [5] it is possible to obtain the following result:

Theorem 1 *Let $X = \{X(t); t \geq 0\}$ be a finite irreducible CTMC with state space Ω and reward rate structure $r(i)$, $i \in \Omega$. Let $\Omega = G \cup U$ be a non-trivial partition of Ω ($G, U \neq 0$) and let S be the subset of G through which X can enter G from U. Let R be the steady-state reward rate of X. Let X_i, $i \in S$, be the CTMC obtained from X by redirecting to i the transitions from U to S, assume $X_i(0) = i$, and let R_i be the steady-state reward rate of X_i. Then, $\min_{i \in S} R_i \leq R \leq \max_{i \in S} R_i$.*

Theorem 1 has immediate application to the CTMC's X under consideration. The condition $X_i(0) = i$ is in general required because X_i could contain several closed sets. However, for the CTMC's X considered here, X_i is irreducible, and the steady-state reward rate R_i is independent on the initial distribution of X_i. An sketch of the proof of the theorem is given in the Appendix. The complete proof can be found in [4]. Using Theorem 1, the correctness of the bounds for R computed in the recipe follows from the correctness of the bounds $|R_i|_{lb}$ and $|R_i|_{ub}$ for R_i computed in the first step.

Let $C_{G,i}$ and $C_{U,i}$ denote, respectively, the contributions of the states in G and U to C_i and assume a similar notation for the contributions of the states in G and U to T_i. Then, we have $C_i = C_{G,i} + C_{U,i}$, $T_i = T_{G,i} + T_{U,i}$, and:

$$R_i = \frac{C_{G,i} + C_{U,i}}{T_{G,i} + T_{U,i}}.$$

Consider now the regenerative behavior of X_i' defined by the times at which X_i' hits i from u_{F+1} (analogous to the regenerative behavior considered for X_i). As it will be shown later, the mean time in the states $u_{F+1}, \ldots, u_N$ between regenerations upper bounds $T_{U,i}$, so we can properly call it $|T_{U,i}|_{ub}$. Notice that, since X_i and X_i' enter G through the same state and are identical in G, the mean reward and time in G between regenerations are identical for X_i and X_i'. Then, the lower and upper bounds for R_i computed in the first step of the recipe can be written as:

$$|R_i|_{lb} = \frac{C_{G,i} + |r|_{lb}|T_{U,i}|_{ub}}{T_{G,i} + |T_{U,i}|_{ub}}, \tag{1}$$

$$|R_i|_{ub} = \frac{C_{G,i} + |r|_{ub}|T_{U,i}|_{ub}}{T_{G,i} + |T_{U,i}|_{ub}}. \tag{2}$$

The correctness of these bounds can be justified as follows. Let $g_{lb}(x) = (C_{G,i} + |r|_{lb}x)/(T_{G,i} + x)$, $g_{ub}(x) = (C_{G,i} + |r|_{ub}x)/(T_{G,i} + x)$. Their first derivatives are $dg_{lb}/dx = (|r|_{lb}T_{G,i} -$

$C_{G,i})/(T_{G,i}+x)^2$, $dg_{ub}/dx = (|r|_{ub}T_{G,i}-C_{G,i})/(T_{G,i}+x)^2$. Using $|r|_{lb}T_{G,i} \leq C_{G,i} \leq |r|_{ub}T_{G,i}$, we have $dg_{lb}/dx \leq 0$, $dg_{ub}/dx \geq 0$. Then, since $|r|_{lb}T_{U,i} \leq C_{U,i} \leq |r|_{ub}T_{U,i}$:

$$
\begin{aligned}
R_i &= \frac{C_{G,i} + C_{U,i}}{T_{G,i} + T_{U,i}} \geq \frac{C_{G,i} + |r|_{lb}T_{U,i}}{T_{G,i} + T_{U,i}} = g_{lb}(T_{U,i}) \\
&\geq g_{lb}(|T_{U,i}|_{ub}) = \frac{C_{G,i} + |r|_{lb}|T_{U,i}|_{ub}}{T_{G,i} + |T_{U,i}|_{ub}} = |R_i|_{lb} ,
\end{aligned}
$$

$$
\begin{aligned}
R_i &= \frac{C_{G,i} + C_{U,i}}{T_{G,i} + T_{U,i}} \leq \frac{C_{G,i} + |r|_{ub}T_{U,i}}{T_{G,i} + T_{U,i}} = g_{ub}(T_{U,i}) \\
&\leq g_{ub}(|T_{U,i}|_{ub}) = \frac{C_{G,i} + |r|_{ub}|T_{U,i}|_{ub}}{T_{G,i} + |T_{U,i}|_{ub}} = |R_i|_{ub} .
\end{aligned}
$$

For the particular case of the steady-state unavailability $|r|_{lb} = 0$, $|r|_{ub} = 1$ and the bounds (1), (2) can be written as:

$$
|UA_i|_{lb} = \frac{C_{G,i}}{T_{G,i} + |T_{U,i}|_{ub}} , \tag{3}
$$

$$
|UA_i|_{ub} = \frac{C_{G,i} + |T_{U,i}|_{ub}}{T_{G,i} + |T_{U,i}|_{ub}} . \tag{4}
$$

The examples given in [15] indicate that $|UA|_{ub}$ tends to be much looser than $|UA|_{lb}$. An intuitive explanation for this is the following. Since X_i tends to be highly skewed, typically $|T_{U,i}|_{ub} \ll T_{G,i}$. Since $UA_i = (C_{G,i} + C_{U,i})/(T_{G,i} + T_{U,i})$, the tightness of $|UA_i|_{lb}$ (3) and $|UA_i|_{ub}$ (4) depend mainly on the closeness of $C_{G,i}$ and $C_{G,i} + |T_{U,i}|_{ub}$ to $C_{G,i} + C_{U,i}$. Down states tend to be sparse and, typically, $C_{U,i} \ll T_{U,i}$. Then, $C_{G,i} + |T_{U,i}|_{ub}$ tends to be less closer to $C_{G,i} + C_{U,i}$ than $C_{G,i}$, making $|UA_i|_{ub}$ significantly looser than $|UA_i|_{lb}$.

3   Proposed bounding approach

3.1   Setup

Our method differs from the method given in [15] in the use of tighter upper bounds for UA_i, $i \in S$. We start considering the more general steady-state reward rate measure R and showing how a different upper bound for R_i, $|R_i|'_{ub}$, can be established using an upper bound for $C_{U,i}$. First, $C_{U,i} \leq |r|_{ub}T_{U,i}$ implies

$$
R_i = \frac{C_{G,i} + C_{U,i}}{T_{G,i} + T_{U,i}} \leq \frac{C_{G,i} + C_{U,i}}{T_{G,i} + C_{U,i}/|r|_{ub}} = h(C_{U,i}) ,
$$

with $h(x) = (C_{G,\imath} + x)/(T_{G,\imath} + x/|r|_{ub})$. In addition, $dh/dx = (T_{G,\imath} - C_{G,\imath}/|r|_{ub})/(T_{G,\imath} + x/|r|_{ub})^2 \geq 0$, since $C_{G,\imath} \leq |r|_{ub}T_{G,\imath}$. Thus, $h(x)$ is monotonically increasing and

$$R_\imath \leq h(|C_{U,\imath}|_{ub}) = \frac{C_{G,\imath} + |C_{U,\imath}|_{ub}}{T_{G,\imath} + |C_{U,\imath}|_{ub}/|r|_{ub}} = |R_\imath|'_{ub} \,. \tag{5}$$

Regarding the tightness of $|R_\imath|_{ub}$ and $|R_\imath|'_{ub}$, we have the following result:

Theorem 2 *Assume $C_{G,\imath} < |r|_{ub}T_{G,\imath}$. Then, $|R_\imath|'_{ub} < |R_i|_{ub}$ if and only if $|C_{U,\imath}|_{ub} < |r|_{ub}|T_{U,\imath}|_{ub}$.*

Proof: Consider again the function $h(x) = (C_{G,\imath} + x)/(T_{G,\imath} + x/|r|_{ub})$. For $C_{G,\imath} < |r|_{ub}T_{G,\imath}$, $dh/dx = (T_{G,\imath} - C_{G,\imath}/|r|_{ub})/(T_{G,\imath} + x/|r|_{ub})^2 > 0$. This implies that $h(x)$ is strictly monotonically increasing and, since (5) $|R_\imath|'_{ub} = h(|C_{U,\imath}|_{ub})$ and (2) $|R_\imath|_{ub} = h(|r|_{ub}|T_{U,\imath}|_{ub})$, the result follows. $\bigcirc$

For the steady-state unavailability ($|r|_{ub} = 1$) $|R_\imath|'_{ub}$ (5) is reduced to:

$$|UA_\imath|'_{ub} = \frac{C_{G,\imath} + |C_{U,\imath}|_{ub}}{T_{G,\imath} + |C_{U,i}|_{ub}} \,, \tag{6}$$

where C has the meaning of "mean down time". Also, the fact that the state o is operational and, therefore, has reward rate 0 ensures $C_{G,\imath} < T_{G,\imath}$. Then, Theorem 2 establishes that $|UA_\imath|'_{ub} < |UA_\imath|_{ub}$ if and only if $|C_{U,\imath}|_{ub} < |T_{U,\imath}|_{ub}$.

The bounds $|UA_\imath|_{lb}$ and $|UA_\imath|_{ub}$ are computed in [15] from the steady-state solution of X'_i. In our bounding method, $C_{G,i}$, $T_{G,\imath}$, $|T_{U,\imath}|_{ub}$ and $|C_{U,\imath}|_{ub}$ are computed independently and then combined using (3), (6) to obtain $|UA_\imath|_{lb}$ and $|UA_\imath|'_{ub}$.

3.2 Computation of $T_{G,i}$, $C_{G,i}$ and $|T_{U,i}|_{ub}$

In the following $\tau(v, Z)$ will denote the mean time to absorption in the state or subset of states v of the transient CTMC Z with given initial distribution. Let $\mathbf{A}$ be the restriction of the transition rate matrix of Z to its transient states, $\mathbf{q}$ the column vector giving the initial probability distribution of Z, and $\boldsymbol{\tau}$ the solution of $\mathbf{A}\boldsymbol{\tau} = -\mathbf{q}$. As it is well-known, $\tau(i, Z) = \tau_\imath$.

$T_{G,\imath}$ and $C_{G,i}$ can be computed solving the transient CTMC Y_G^i with initial state i tracking X from i to exit of G:

$$T_{G,i} = \sum_{j \in G} \tau(j, Y_G^i) \,,$$

$$C_{G,\imath} = \sum_{j \in G \cap D} \tau(j, Y_G^\imath) \,.$$

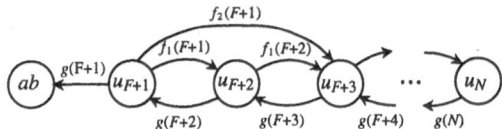

Figure 2: Transient CTMC Y used to derive the bounds $T(k)$ and $C(k)$.

The upper bound $|T_{U,i}|_{ub}$ can be computed using the transient CTMC Y depicted in Figure 2. The boundness of $|T_{U,i}|_{ub}$ will be justified using exact aggregation results for transient CTMC's [9] and the following lemma (see [4] for the proof), closely related to the mean holding time lemma of [15]:

Lemma 1 *Let a transient CTMC Y with the structure depicted in Figure 2 and consider another transient CTMC, Y', with the same structure and such that $f_i'(k) \leq f_i(k)$ and $g'(k) \geq g(k)$. Also assume that Y and Y' have the same initial distribution. Then, $\tau(u_i, Y') \leq \tau(u_i, Y)$, $i = F+1, \ldots, N$.*

Let T_U^s be the mean time spent by X during a visit to U conditioned to entry through state s. T_U^s is the mean time to absorption of the transient CTMC Y_U^s with initial state s tracking X from s to exit of U. Let $Y_U^{s'}$ be the result of the exact aggregation in Y_U^s of the subsets U_k, $k = F+1, \ldots, N$. $Y_U^{s'}$ has the structure of Y and initial state $u_{|F(s)|}$. From exact aggregation results for transient CTMC's [9], $\tau(u_k, Y_U^{s'}) = \tau(U_k, Y_U^s)$ and the transition rates of $Y_U^{s'}$ are convex linear combinations of the transition rates of Y_U^s. More specifically, $\lambda_{u_k,u_l}' = \sum_{j \in U_k} w_j^{k,l} \lambda_{j,U_l}$ and $\lambda_{u_{F+1},ab}' = \sum_{j \in U_{F+1}} w_j^{F+1,ab} \lambda_{j,ab}$, with $w_j \geq 0$, $\sum_j w_j^{k,l} = 1$, $\sum_j w_j^{F+1,ab} = 1$. Consider the "failure" transition rates of $Y_U^{s'}$, $\lambda_{u_k,u_{k+i}}'$, $k \geq F+1$, $i > 0$. Since $f_i(k)$ upper bounds $\lambda_{j,U_{k+i}}$, $j \in U_k$:

$$\lambda_{u_k,u_{k+i}}' = \sum_{j \in U_k} w_j^{k,k+i} \lambda_{j,U_{k+i}} \leq \max_{j \in U_k} \lambda_{j,U_{k+i}} \leq f_i(k) .$$

Using $g(k) \leq \lambda_{j,U_{k-1}}$, $j \in U_k$, $k > F+1$ and $g(F+1) \leq \lambda_{j,ab}$, $j \in U_{F+1}$, it can be similarly shown that $\lambda_{u_k,u_{k-1}}' \geq g(k)$, $k > F+1$ and $\lambda_{u_{F+1},ab}' \geq g(F+1)$. In summary, the transition rates of Y and $Y_U^{s'}$ satisfy the conditions of Lemma 1. Denoting by Y^k the transient CTMC Y with initial state u_k and by $T(k)$ the mean time to absorption of Y^k, and using Lemma 1:

$$T_U^s = \sum_{j=F+1}^N \tau(U_j, Y_U^s) = \sum_{j=F+1}^N \tau(u_j, Y_U^{s'}) \leq \sum_{j=F+1}^N \tau(u_j, Y^{|F(s)|}) = T(|F(s)|) .$$

Let ϕ_s^i be the conditional entry probability distribution of X_t in U through state s. ϕ_s^i can be computed from the mean times to absorption of Y_G^i as

$$\phi_s^i = \sum_{j \in G} \tau(j, Y_G^i) \lambda_{j,s} . \tag{7}$$

Let π_k^i be the probability that X_i enters U through U_k. We have:

$$\pi_k^i = \sum_{s \in U_k} \phi_s^i .$$ (8)

Then, $|T_{U,i}|_{ub}$ can be computed as:

$$|T_{U,i}|_{ub} = \sum_{k=F+1}^{N} \pi_k^i T(k) .$$ (9)

The upper boundedness of $|T_{U,i}|_{ub}$ can be easily justified using $T_U^s \leq T(|F(s)|)$:

$$T_{U,i} = \sum_{s \in U} \phi_s^i T_U^s = \sum_{k=F+1}^{N} \sum_{s \in U_k} \phi_s^i T_U^s \leq \sum_{k=F+1}^{N} \sum_{s \in U_k} \phi_s^i T(k) = \sum_{k=F+1}^{N} \pi_k^i T(k) .$$

Giving the relationships between Y and X_i', it is clear that $|T_{U,i}|_{ub}$ is the upper bound for $T_{U,i}$ implicitely used in [15].

Although the bounds $|T_{U,i}|_{ub}$ can be computed directly as the mean times to absorption of Y with initial distributions $P[Y(0) = u_k] = \pi_k^i$, this procedure requires $|S|$ solutions of Y (one for each state $i \in S$) and a more efficient approach when $|S| > 1$ is to compute $|T_{U,i}|_{ub}$ from $T(k)$, $k = F+1, \ldots, N$ using (9). $T(N)$ can be computed solving Y^N as $T(N) = \sum_{j=F+1}^{N} \tau(u_k, Y^N)$. Denoting by $\lambda(k)$ the output rate of u_k in Y, the remaining $T(k)$'s can be computed exploiting the following relations, which result from a conditional path analysis of Y:

$$T(k) = \frac{1}{\lambda(k)} + \frac{g(k)}{\lambda(k)} T(k-1) + \sum_i \frac{f_i(k)}{\lambda(k)} T(k+i) , \quad F+1 < k < N ,$$

$$T(N) = \frac{1}{g(N)} + T(N-1) ,$$

yielding:

$$T(N-1) = T(N) - \frac{1}{g(N)} ,$$

$$T(k) = \frac{1}{g(k+1)} [\lambda(k+1)T(k+1) - 1 - \sum_i f_i(k+1)T(k+1+i)] ,$$

$$k = N-2, \ldots, F+1 .$$

3.3 Computation of $|C_{U,i}|_{ub}$

The strategy to find a bound $|C_{U,i}|_{ub}$ potentially smaller than $|T_{U,i}|_{ub}$ is to exploit the fact that many of the states in U are operational and, thus, do not contribute to $C_{U,i}$. As we shall show, the strategy can be implemented using the concept of *failure distance*, which has been useful to speed up the simulation of the type of models considered in this paper [3]. The failure distance from an state x, $d(x)$, is defined as the minimum number of components which have to fail (in addition to $F(x)$) for the system to go down ($d(x) = 0$ for $x \in D$).

Let $U_{k,d}$ be the subset of U including the states with k failed components and failure distance d and let $\pi_{k,d}^i$ be the probability that X_i enters U through $U_{k,d}$. We have:

$$\pi_{k,d}^i = \sum_{s \in U_{k,d}} \phi_s^i . \tag{10}$$

Assume that upper bounds $C(k,d)$ to the mean down time in U conditioned to entry in U through any state $\in U_{k,d}$ are available. Then, an upper bound for $C_{U,i}$ can be computed as:

$$|C_{U,i}|_{ub} = \sum_{k,d} \pi_{k,d}^i C(k,d) . \tag{11}$$

Since $\pi_k^i = \sum_d \pi_{k,d}^i$, it is clear (9) that $C(k,d) \leq T(k)$ implies $|C_{U,i}|_{ub} \leq |T_{U,i}|_{ub}$. If, in addition, $C(k,d) < T(k)$ for some pair (k,d) with $\pi_{k,d}^i \neq 0$, $|C_{U,i}|_{ub} < |T_{U,i}|_{ub}$.

Our approach to obtain bounds $C(k,d) \leq T(k)$ includes two steps. In the first step, we obtain upper bounds to the mean down time in U conditioned to entry in U through U_k. Then, we let $C(k,d) = C(k)$ and improve iteratively $C(k,d)$. The bounds $C(k)$ are $\leq T(k)$ and, as a result, $C(k,d) \leq T(k)$. Thus, our bounds $|C_{U,i}|_{ub}$ are always $\leq |T_{U,i}|_{ub}$ and our upper bound $|UA|_{ub}'$ is never worse than $|UA|_{ub}$.

Let L be the minimum number of components which have to fail to take the system down ($L = d(o)$). With the reward rate structure:

$$r(u_j) = \begin{cases} 0 & \text{if } j < L \\ 1 & \text{if } j \geq L \end{cases} ,$$

the mean reward to absorption of Y^k provides a suitable bound $C(k)$. To justify this, let C_U^s be the mean down time in a stay in U since entry through state s. C_U^s is the mean down time of the transient CTMC Y_U^s. Using exact aggregation results for transient CTMC's, Lemma 1, and the fact that all states in U_k, $k < L$ are operational:

$$C_U^s = \sum_{j \in U \cap D} \tau(j, Y_U^s) \leq \sum_{k \geq L} \tau(U_k, Y_U^s) = \sum_{k \geq L} \tau(u_k, Y_U^{s\prime}) \leq \sum_{k \geq L} \tau(u_k, Y^k) = C(k) .$$

For $F + 1 \geq L$, $C(k) = T(k)$. Otherwise, $C(k) < T(k)$. $C(N)$ can be easily computed from the mean times to absorption vector of Y^N as $C(N) = \sum_{i=L}^{N} \tau(u_k, Y^N)$. The remaining $C(k)$'s can be computed using the following recursive equations (analogous to the equations giving $T(k)$, $k < N$), where $I(c)$ is the indicator function which returns 1 if c is true and 0 otherwise:

$$C(N-1) = C(N) - \frac{1}{g(N)} ,$$

$$C(k) = \frac{1}{g(k+1)}[\lambda(k+1)C(k+1) - I(k+1 \geq L) - \sum_i f_i(k+1)C(k+1+i)],$$
$$k = N-2, \ldots, F+1.$$

Let FC be the set of different cardinalities of the failure events of the model. Let $F(k,d,i,r)$, $i \in FC$, be upper bounds for the sum of failure rates involving i components from any state in U with k failed components and failure distance d to states with failure distance $\leq r$, let $w = \min\{i,d\}$, and let:

$$f_{i,j}(k,d) = F(k,d,i,d-j) - F(k,d,i,d-j-1), \quad 0 \leq j < w,$$

$$f_{i,w}(k,d) = F(k,d,i,d-w).$$

The iterative improvement procedure of $C(k,d)$ is based on the following result (proved in [4]), where in the expression for $C'(k,d)$ the terms $C(k,d)$ corresponding to unfeasible pairs (k,d) have to be set to 0. The feasible pairs (k,d) are given by $F+1 \leq k \leq N$, $\max\{0, L-k\} \leq d \leq \min\{L, N-k\}$.

Proposition 1 *Let $C(k,d)$ be upper bounds for C_U^s, $s \in U_{k,d}$ and assume that $C(k,d)$ is decreasing on d. Then, for any $s \in U_{k,d}$:*

$$C_U^s \leq C'(k,d) = \frac{I(d=0)}{g(k)} + \max\{C(k-1,d), C(k-1,d+1)\}$$
$$+ \sum_{i \in FC} \sum_{j=0}^{w} \frac{f_{i,j}(k,d)}{g(k)} C(k+i, d-j). \tag{12}$$

The iterative improvement procedure can be implemented using (12). At each step, $C'(k,d)$ is computed for each feasible (k,d) pair and accepted as new $C(k,d)$ if $C'(k,d) < C(k,d)$. The procedure can be finished when no bound $C(k,d)$ has been reduced significantly during a step. It is important to note that the correctness of the bounds $C'(k,d)$ requires that the available set of $C(k,d)$ bounds be decreasing on d. It is proved in [4] that this is satisfied if 1) the bounds $F(k,d,i,r)$ are decreasing on d, and 2) the bounds $C(k,d)$ are reviewed grouped by k. In our implementation the bounds are reviewed by increasing values of k and, for a given k, by increasing values of d. This ordering has been proved effective, in the sense that very few improvement steps (typically < 10) are required to reach stable values for the bounds.

It is possible to argue that the bounds $C(k,d)$ obtained at the end of the iterative improvement procedure for $d > 0$ are potentially much smaller than the original $C(k)$ if $\sum_{j=0}^{w} f_{i,j}(k,d) = F(k,d,i,d) \ll g(k)$. Consider $C'(F+1,d)$ with $d > 0$ and $C(k,d) = C(k)$. For such a case, the first two terms of $C'(F+1,d)$ are 0 ($C(k,d) = 0$ for non-feasible (k,d) pairs) and only the last term remains, but even considering that $C(k) > C(F + 1)$ for $k > F + 1$, the last term can be much smaller than $C(F + 1)$ if $\sum_{j=0}^{w} f_{i,j}(k, d) \ll g(k)$. Consider now $C'(F + 2, d)$ with $d > 0$. A similar discussion can be made except that the second term will not be null, but since this term corresponds to revised values $C'(F + 1, d)$ with $d > 0$, it is potentially much smaller than $C(F + 1)$, and thus than $C(F + 2)$. The argument can be iterated for increasing values of k.

Combining (7), (8), (9) and (10) $|T_{U,i}|_{ub}$ and $|C_{U,i}|_{ub}$ can be formulated as:

$$|T_{U,i}|_{ub} = \sum_{j \in G} \tau(j, Y_G^i)\alpha(j) \,,$$

$$|C_{U,i}|_{ub} = \sum_{j \in G} \tau(j, Y_G^i)\beta(j) \,,$$

with

$$\alpha(j) = \sum_{s \in U} \lambda_{j,s} T(|F(s)|) \,,$$

$$\beta(j) = \sum_{s \in U} \lambda_{j,s} C(|F(s)|, d(s)) \,.$$

Note that $\alpha(j)$, $\beta(j)$ are independent on i and the above formulations are used with advantage when $|S| > 1$.

3.4   Computation of failure distances and bounds $F(k, d, i, r)$

The computation of $|C_{U,i}|_{ub}$ requires the knowledge of the failure distances from the states in the frontier of U. The failure distance $d(x)$ from a state x can be computed from $F(x)$ if the minimal cuts of the structure function of the system [1] are known. Let MC be the set of all minimal cuts of the structure function of the system, using standard bag notation, we have:

$$d(x) = \min_{m \in MC} |m - F(x)| \,. \tag{13}$$

Although (13) can be used to compute all the required failure distances, most of the transitions from G to U will be of the failure type (all if G contains all states up to a given number of failed components K) and a more efficient procedure can be established introducing the notion of "after" minimal cuts associated with a given failure event e. Let $MC_e = \{m' \mid m' = m - e, m \in MC, m \cap e \neq \phi\}$ be the set of "after" minimal cuts associated to e, the failure distance from any state reached from x through a failure transition with failure event e can be computed as:

$$ad(x, e) = \min\{d(x), \min_{m \in MC_e} |m - F(x)|\} \,. \tag{14}$$

The cardinality of MC_e is in general much smaller than the cardinality of MC. Then, for each state x in the frontier of G we can compute its failure distance using (13), and use (14) to compute the failure distances for the states in U reached from x through failure transitions. If some state y in U is reached from x through a repair transition, then we can construct $F(y)$ and compute $d(y)$ using (13).

The tightness of the bounds $C(k,d)$ depends on the tightness of the bounds $F(k,d,i,r)$. In general, better bounds $F(k,d,i,r)$ require a more detailed analysis of the model and thus their computation requires more effort. The bounds $F(k,d,i,r)$ used here are relatively easy to compute and, as the examples in the next section will show, provide good results. The bounds are based on two structural properties of failure events. The *importance* $I(e)$ of a failure event e is defined as the minimum number of components which are left unfailed in any minimal cut affected by the failure event. The *activity* $A(e)$ of a failure event e is defined as the maximum number of components of the failure event in any minimal cut. From their definitions, $I(e)$ and $A(e)$ can be computed by:

$$I(e) = \min_{m \in MC, m \cap e \neq \phi} |m - e|,$$

$$A(e) = \max_{m \in MC} |m \cap e|.$$

Consider a state with k failed components and failure distance d and another state reached from it through a failure event e. The number of components left unfailed in any minimal cut m is $\geq |m - e| - k$, since at most k components not included in $m \cap e$ were failed before e. Then, $d' \geq I(e) - k$. Also, $d' \geq d - A(e)$, since at most $A(e)$ components in the same minimal cut will be failed by e. Imposing $d' \leq r$ results in:

$$I(e) - k \leq r,$$

$$d - A(e) \leq r.$$

Then, the failure rate from any state with k failed components and failure distance d due to failure events with i components leading to states with failure distance $\leq r$ is bounded above by:

$$F(k,d,i,r) = \sum_{e \in E_1, A(e) \geq d-r, I(e) \leq k+r} \lambda_{ub}(e).$$

It is easy to check that these bounds are decreasing on d, as required for the correctness of the iterative improvement procedure for $C(k,d)$.

4 Numerical Analysis

In this section our bounding method is compared with the method proposed in [15] using the large model described there and a variation of it to explore the impact of the redundancy level L on the relative tightness of the bounds given by the methods. We use the same state generation

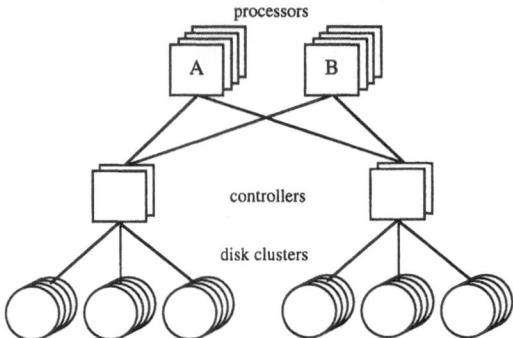

Figure 3: Fault-tolerant database system from [Mun89] (model 1).

strategy as in [15], i.e., G includes all the states with up to K failed components. The large model considered in [15] is the fault-tolerant database system shown in Figure 3. The system includes two processor types (A and B), two sets of dual-ported controllers with two controllers per set and six disk clusters with four disks. Each set of controllers controls three clusters. Each processor type has three spares. The system is operational if at least one processor of any type is unfailed, at least one controller in each set is unfailed and at least three disks in each cluster are unfailed. Thus, $L = 2$. A failure in the active processor A is propagated to the active processor B with probability 0.10. Processors and controllers of one set fail with rate 1/2000, controllers of the other set fail with rate 1/4000. Disks fail with different rates from one cluster to another. These rates are 1/6000, 1/8000, 1/10000, 1/12000, 1/14000, and 1/16000. Any component fails in one of two modes with equal probabilities. The repair rate is 1 for one mode and 0.5 for the other. Components are repaired by a single repairman who chooses components at random from the set of failed components. Unfailed components continue to fail when the system is down. This model has about 9×10^{10} states, clearly illustrating the "largeness" problem. A slight variation of this example is also considered. We call the original model from [15] model 1, and call model 2 its variation. Model 2 is obtained from model 1 by increasing the number of controllers in each set to 3 and the number of disks in each cluster to 5, without modifying any other aspect. For model 2, $L = 3$.

Tables 1 and 2 give the number of generated states, the steady-state unavailability bounds and bands under both methods, and the improvement measured as the band ratio. Our method always gives significantly smaller bands. Thus, for model 1, our method for $K = 2$ (231 states) gives bounds wich can be considered tight enough for most purposes, whereas the bounds given by the method proposed in [15] are quite loose. Using that method, 1,763 states ($K = 3$) should be generated to achieve bounds of acceptable quality. For model 2, our method with any K gives tigther bounds than the other method with $K + 1$. The improvement of our method decreases for larger values of K and is considerably larger for model 2. Both behaviours can be explained by the relative sparsness of down states in U.

Table 1: Comparison of the bounding methods for model 1.

$\lvert G \rvert$	K	F	Muntz et al.	proposed	improvement
231	2	0	3.2313×10^{-6}	3.2313×10^{-6}	23.7
			8.9886×10^{-6}	3.4746×10^{-6}	
			5.7573×10^{-6}	2.4322×10^{-7}	
1,763	3	0	3.3167×10^{-6}	3.3167×10^{-6}	14.2
			3.4182×10^{-6}	3.3239×10^{-6}	
			1.0155×10^{-7}	7.1676×10^{-9}	
10,464	4	0	3.3192×10^{-6}	3.3192×10^{-6}	9.44
			3.3208×10^{-6}	3.3194×10^{-6}	
			1.5547×10^{-9}	1.6469×10^{-10}	

Table 2: Comparison of the bounding methods for model 2.

$\lvert G \rvert$	K	F	Muntz et al.	proposed	improvement
231	2	0	0	0	522
			8.5262×10^{-6}	$1.63.4 \times 10^{-8}$	
			8.5262×10^{-6}	1.6324×10^{-8}	
1,771	3	0	4.5418×10^{-9}	4.5418×10^{-9}	202
			1.6621×10^{-7}	5.3420×10^{-9}	
			1.6167×10^{-7}	8.0016×10^{-10}	
10,616	4	0	4.7214×10^{-9}	4.7214×10^{-9}	95.6
			7.4986×10^{-9}	4.7504×10^{-9}	
			2.7773×10^{-9}	2.9058×10^{-11}	
52,916	5	0	4.7277×10^{-9}	4.7277×10^{-9}	55.2
			4.7736×10^{-9}	4.7286×10^{-9}	
			4.5912×10^{-11}	8.3150×10^{-13}	

It has been observed that the tightness of the bounds derived in [15] increases with F. This is not typically the case with ours. Table 3 gives $\lvert S \rvert$ and the lower and upper bounds obtained with both methods for model 1, $K = 3$ and all possible values for F. The lower bound (identical for both methods) does not experiment variations at the level of the 6th significant digit. The upper bound given by the method proposed in [15] experiments some improvement when F increases. Our upper bound experiments a slight improvement from $F = 0$ to $F = 1$, but deteriorates considerably with further increase of F. This behavior can be explained as follows. Given the orders of magnitude difference between failure and repair transitions, the model reaches state o with high probability and in short time for any $i \in S$ and $T_{G,i}$ tends to depend vary little on the "return" state i. Then, the dependency of $|UA_i|'_{ub}$ (6) on i comes mainly through $C_{G,i}$ and $|C_{U,i}|_{ub}$. The latter is determined by the exit distribution $\pi^i_{k,d}$ (11). For larger F, S includes states with more failed components and smaller failure distances, the distributions $\pi^i_{k,d}$ are more shifted to high values of k and smaller values of d, and since $C(k,d)$ increases with higher k and smaller d, the corresponding $|C_{U,i}|_{ub}$ are larger. When $F \geq L$, S includes down states and the shift of $\pi^i_{k,d}$ for such states is specially significant, since all failure transitions from i give contributions to $\pi^i_{k,d}$ with $d = 0$. Also, the corresponding transient CTMC's Y^i_G include visits to down states with probability 1, yielding larger $C_{G,i}$ values. For $0 < F < L$ the small dependency of $T_{G,i}$ on i may outweight the other factors and yield a slightly tighter upper bound than for $F = 0$. Both behaviors are clearly supported by the results in Table 3 ($L = 2$ for model 1). The cost in time of the bounding method is very sensitive to F, since $|S|$ CTMC's Y^i_G have to be solved, and $F = 0$ should be the reasonably choice for our method.

Table 3: Impact of F on the bounds for model 1 and $K = 3$.

F	$|S|$	Muntz et al.	proposed
0	1	3.31670×10^{-6}	3.31670×10^{-6}
		3.41825×10^{-6}	3.32386×10^{-6}
1	20	3.31670×10^{-6}	3.31670×10^{-6}
		3.39292×10^{-6}	3.32373×10^{-6}
2	210	3.31670×10^{-6}	3.31670×10^{-6}
		3.39275×10^{-6}	3.34645×10^{-6}
3	1532	3.31670×10^{-6}	3.31670×10^{-6}
		3.39272×10^{-6}	3.36944×10^{-6}

Our bounding method is more complex both theoreticaly and computationaly. The last aspect requires some discussion. The only storage and time overheads of our method which can be significant are related to the computation of the failure distances: generation and storage of minimal cuts and computation of failure distances using (13), (14). Efficient algorithms (see [13] for a review) exist which will find all minimal cuts very fast even when their number is of the order of several thousands. Thus generation "per se" does not seem to be an important problem. Since a minimal cut requires less storage than a state and the information associated to it, the requirement of storing the minimal cuts can only be significant when the number of minimal cuts is substantially larger than the number of states of the model. Regarding the cost in time associated to the computation of failure distances, it represented a 5% overhead for the examples used here which have 9 minimal cuts. When the number of minimal cuts is large the method described here for failure distances computation can be time consumming. However, the techniques proposed in [3] can be used to reduce drastically the number of minimal cuts which have to be "touched" to compute the failure and "after" failure distances from a particular state. Using these techniques, storage and time overheads will only be significant when a number of minimal cuts in the order of several thousands has to be managed. We also note that knowing *all* minimal cuts is not a requirement of the method. We can simply consider the minimal cuts with up to a given number M of components and assume that the system is down for all combinations of more than M failed components to obtain a looser upper bound (but never worse than the bound obtained using [15]). Thus, a tradeoff can be made between tightness of the bounds and overhead caused by the management of the minimal cuts.

5 Conclusions

In this paper we have proposed a method to bound the steady-state unavailability of repairable fault-tolerant systems using CTMC's which, with the same number of generated states, can give significantly smaller (and *never* worse) bands than a method previously proposed [15]. Using the failure distance concept we have obtained an upper bound exploiting the fact that, typically, the system is operational a large portion of the time the model is out of the generated state space. The quality of our upper bound depends on the tightness of the failure rate bounds $F(k, d, i, r)$. The bounds $F(k, d, i, r)$ we have used here are relatively simple and we plan to consider in the future the use of more precise $F(k, d, i, r)$ bounds. We are also interested in studying the behavior of our bounding method and how it compares with the method proposed in [15] in combination with state exploration techniques recently proposed [17].

Appendix

Sketch of the proof of Theorem 1

Let C_i (T_i), $i \in S$ be the expected reward (time) in X between entry in i and the next entry in S from U. Using results from semi-regenerative process theory (Theorem 6.12 of [Cin75, Chapter 10]) and using the fact that X is irreducible and finite, it is easy to show that:

$$R = \frac{\sum_{i \in S} \psi_i C_i}{\sum_{i \in S} \psi_i T_i},$$

where ψ_i, $i \in S$ is any invariant measure of the embedded discrete-time Markov chain Π of X. Being Π finite and irreducible, there exists an invariant measure for Π satisfying $\psi_i > 0$, $\sum_{i \in S} \psi_i = 1$. Using this, it can be shown by induction on $|S|$ that:

$$\min_{i \in S}\{C_i/T_i\} \le R \le \max_{i \in S}\{C_i/T_i\}.$$

Being X irreducible, S is reached in X from i with probability 1. Then, i is recurrent in X_i. Assuming $X_i(0) = i$, it is easy to check that X_i is recurrent aperiodic. C_i and T_i are, respectively, the expected reward and time between recurrences. Then, by regenerative theory, $R_i = C_i/T_i$ and the result follows. $\bigcirc$

References

[1] R. E. Barlow and F. Proshan, *Statistical Theory of Reliability and Life Testing. Probability Models*, McArdle Press, Silver Spring, 1981.

[2] M. A. Boyd, M. Veeraraghavan, J. B. Dugan, and K. S. Trivedi, "An approach to solving large reliability models," in *Proc. AIAA Components in Aerospace Conference*, pp. 245–258, 1988.

[3] J. A. Carrasco, "Failure distance-based simulation of repairable fault-tolerant systems," in *Proc. 5th Int. Conf. on Modelling Techniques and Tools for Computer Performance Evaluation*, pp. 351–365, Elsevier, 1992.

[4] J. A. Carrasco, "Tight steady-state availability bounds using the failure distance concept," Research report, September 1993, submitted for publication.

[5] E. Çinlar, *Introduction to Stochastic Processes*, Prentice-Hall, Inc., New Jersey, 1975.

[6] M. Dal Cin, "Availability analysis of a fault-tolerant computer system," *IEEE Trans. on Reliability*, vol. R-29, no. 3, pp. 265–268, August 1980.

[7] P. J. Courtois and P. Semal, "Bounds for the positive eigenvectors of nonnegative matrices and for their approximations," *Journal of the ACM*, vol. 31, no. 4, pp. 804–825, October 1984.

[8] P. J. Courtois and P. Semal, "Computable bounds for conditional steady-state probabilities in large Markov chains and queueing models," *IEEE J. Selected Areas in Communications*, vol. SAC-4, no. 6, pp. 926–937, September 1986.

[9] P. J. Courtois and P. Semal, "Bounds for transient characteristics of large or infinite Markov chains," in W. J. Stewart, editor, *Proc. 1st Int. Conf. on Numerical Solution of Markov Chains*, pp. 413–434, Marcel Dekker, 1991.

[10] A. Goyal, P. Shahabuddin, P. Heidelberger, V. F. Nicola, and P. W. Glynn, "A unified framework for simulating Markovian models of highly dependable systems," *IEEE Trans. on Computers*, vol. 41, no. 1, pp. 36–51, January 1992.

[11] A. M. Johnson Jr. and M. Malek, "Survey of software tools for evaluating reliability, availability and serviceability," *ACM Computing Surveys*, vol. 20, pp. 227–271, 1988.

[12] J. G. Kemeny and J. L. Snell, *Finite Markov Chains*, Springer-Verlag, New York, 2nd edition, 1978.

[13] W. S. Lee, D. L. Grosh, F. A. Tillman and C. H. Lie, "Fault Tree Analysis, Methods and Applications—A Review," *IEEE Trans. on Reliability*, vol. R-34, no. 3, pp. 194–203, August 1985.

[14] J. C. S. Lui and R. R. Muntz, "Evaluating bounds on steady-state availability of repairable systems from Markov models," in W. J. Stewart, editor, *Proc. 1st Int. Conf. on Numerical Solution of Markov Chains*, pp. 435–454, Marcel Dekker, 1991.

[15] R. R. Muntz, E. de Souza e Silva, and A. Goyal, "Bounding availability of repairable computer systems," *IEEE Trans. on Computers*, vol. 38, no. 12, pp. 1714–1723, December 1989.

[16] R. A. Sahner and K. S. Trivedi, "Reliability modeling using SHARPE," *IEEE Trans. on Reliability*, vol. R-36, no. 2, pp. 186–193, June 1987.

[17] E. de Souza e Silva and P. M. Ochoa, "State space exploration in Markov models," *Performance Evaluation Review*, vol. 20, no. 1, pp. 152–166, June 1992.

Author Index

Abouelnaga, A.	185		Leu, M.	69
Abraham, J.A.	327		Levitt, K.N.	373
Anderson, S.	15		Littlewood, B.	219
Apostolakis, G.	161		Lunt, T.	211
Azadmanesh, M.H.	251		Lynch, N.	399
Ball, D.	185		Lyu, M.R.	459
Babaoğlu, Ö.	271		McLean, J.	223
Boucher, P.K.	49		Meadows, C.	227
Bruns, G.	15		Meadows, C.	383
Butler, R.W.	31		Millen, J.K.	93
Carrasco, J.A.	479		Millen, J.K.	229
Cheung, S.	373		Miller, S.P.	33
Clark, R.K.	49		Miremadi, G.	435
Coenen, J.	309		Morley, M.J.	37
Dacier, M.	215		Nair, V.S.S.	349
Deswarte, Y.	379		Neumann, P.G.	387
Dugan, J.B.	459		Ramos-Thuel, S.	411
Echtle, K.	69		Randell, B.	389
Garrett, C.	161		Raynal, M.	271
Gligor, V.D.	109		Rowell, R.	349
Gligor, V.D.	139		Saias, I.	399
Goemans, M.	399		Schepers, H.	309
Gong, L.	139		Schneider, F.B.	43
Greenberg, I.B.	49		Shankar, N.	41
Guarro, S.	161		Strosnider, J.K.	411
Hooman, J.	291		Stubblebine, S.G.	109
Hugue, M.M.	233		Suri, N.	233
Jensen, E.D.	49		Torin, J.	435
Kailar, R.	109		Turski, W.M.	3
Kailar, R.	139		Walter, C.J.	233
Kanawati, G.A.	327		Wells, D.M.	49
Kanawati, N.A.	327		Wilken, K.D.	393
Kieckhafer, R.M	251		Yau, M.	161
			Zhou, P.	291

Dependable Computing
and Fault-Tolerant Systems

Edited by A. Avižienis, H. Kopetz, J. C. Laprie

Vol. 8:

C. E. Landwehr, B. Randell, L. Simoncini (eds.)

Dependable Computing for Critical Applications 3

1993. 91 figures. XII, 383 pages.
Cloth DM 198,–, öS 1386,–. ISBN 3-211-82481-2

Prices are subject to change without notice

This book contains the papers presented and discussed at the 3rd IFIP Working Conference on Dependable Computing for Critical Applications. Based on feedback at that meeting, these papers were then revised and updated prior to inclusion in this volume. The topics addressed span the spectrum of dependable computing, from design methods for distributed, fault-tolerant systems to formal and experimental validation techniques. The unique focus of this forum on critical applications is what distinguishes many of these papers from those found elsewhere. This book is of interest to individuals involved in the development of computing systems where dependability attributes such as reliability, safety, and security are a major concern.

Springer-Verlag Wien New York

Sachsenplatz 4–6, P.O.Box 89, A-1201 Wien · 175 Fifth Avenue, New York, NY 10010, USA
Heidelberger Platz 3, D-14197 Berlin · 3-13, Hongo 3-chome, Bunkyo-ku, Tokyo 113, Japan

Dependable Computing and Fault-Tolerant Systems

Vol. 7: H. Kopetz, Y. Kakuda (eds.)
Responsive Computer Systems
1993. 96 figures. XII, 377 pages.
Cloth DM 150,–, öS 1050,–. ISBN 3-211-82458-8

Vol. 6: J. F. Meyer, R. D. Schlichting (eds.)
Dependable Computing for Critical Applications 2
1992. 114 figures. XIV, 439 pages.
Cloth DM 172,–, öS 1204,–. ISBN 3-211-82330-1

Vol. 5: J. C. Laprie (ed.)
Dependability: Basic Concepts and Terminology
In English, French, German, Italian, and Japanese
1992. 3 figures. XII, 265 pages.
Cloth DM 128,–, öS 896,–. ISBN 3-211-82296-8

Vol. 4: A. Avižienis, J. C. Laprie (eds.)
Dependable Computing for Critical Applications
1991. 88 figures. XIII, 431 pages.
Cloth DM 162,–, öS 1134,–. ISBN 3-211-82249-6

Vol. 2: U. Voges (ed.)
Software Diversity in Computerized Control Systems
1988. 41 figures. VII, 216 pages.
Cloth DM 75,–, öS 530,–. ISBN 3-211-82014-0

Prices are subject to change without notice

Springer-Verlag Wien New York

Sachsenplatz 4–6, P.O.Box 89, A-1201 Wien · 175 Fifth Avenue, New York, NY 10010, USA
Heidelberger Platz 3, D-14197 Berlin · 3-13, Hongo 3-chome, Bunkyo-ku, Tokyo 113, Japan